Impact of E-Business Technologies on Public and Private Organizations:

Industry Comparisons and Perspectives

Ozlem Bak
University of Brighton, UK

Nola Stair
University of Greenwich, UK

BUSINESS SCIENCE Reference

BUSINESS SCIENCE REFERENCE
Hershey · New York

Senior Editorial Director:	Kristin Klinger
Director of Book Publications:	Julia Mosemann
Editorial Director:	Lindsay Johnston
Acquisitions Editor:	Erika Carter
Development Editor:	Myla Harty
Production Coordinator:	Jamie Snavely
Typesetters:	Michael Brehm, Milan Vracarich, Jr., & Deanna Zombro
Cover Design:	Nick Newcomer

Published in the United States of America by
Business Science Reference (an imprint of IGI Global)
701 E. Chocolate Avenue
Hershey PA 17033
Tel: 717-533-8845
Fax: 717-533-8661
E-mail: cust@igi-global.com
Web site: http://www.igi-global.com/reference

Library of Congress Cataloging-in-Publication Data

Impact of e-business technologies on public and private organizations :
industry comparisons and perspectives / Ozlem Bak and Nola Stair, editors.
p. cm.
Includes bibliographical references and index.
ISBN 978-1-60960-501-8 (hbk.) -- ISBN 978-1-60960-502-5 (ebook) 1.
Information technology. 2. Electronic commerce. 3. Educational technology.
I. Bak, Ozlem, 1973- II. Stair, Nola, 1963-
T58.5.I45 2011
338'.064--dc22

2010042259

British Cataloguing in Publication Data
A Cataloguing in Publication record for this book is available from the British Library.

Editorial Advisory Board

Table of Contents

Detailed Table of Contents

This chapter underlines the e-landscape and its impact on the agri-food industry and asserts the dynamics of sustainable production, the presence of integrated food supply systems, globalised markets, key players in the agri-food industry, and food security.

This chapter highlights the importance of environmental factors and decisions made by organizations when implementing e-business applications and analysis based on real-life data its impact on the environment.

In this chapter, Dr. Sharon Nachtigal explores, in a more overarching manner, the concept of e-business, its development and implications on organizations, its barriers, and the general principles on how organizations overcome such barriers. The chapter first discusses e-business and its impact on the modern business and social environments involving practitioners ranging from executives to developers.

Security issues affect the large-scale deployment of Web services and e-business technologies. In the rapidly evolving Web environment, how can this be addressed effectively? This chapter focuses on initiatives for mitigating security risks in Web services and e-business technologies.

This chapter explores e-business security software landscapes and organizational tools, describing the need for more e-business functionality integrated to the organizational processes. Dr. Nachtigal challenges the traditional notion of relying on software rather than organizational decision-making.

Preface

BACKGROUND TO E-BUSINESS APPLICATIONS AND THEIR IMPACT ON PUBLIC AND PRIVATE ORGANIZATIONS

Defining the impact of e-business applications in a multi-disciplinary field with cross industry contexts has its own challenges. In this book, our attempts to compile a book started from the diversity on the use of e-business applications. In our own research on the use of technology in higher education and the automotive industry, we have seen that the impact of e-business applications may vary between industries and sometimes in organizations within the same industry in which it has taken place. In our literature review about the impact of e-business technologies on organizations, we have observed that there were accounts of e-business applications, useful cases indicating why e-business application implementations fail or succeed, but the absence of cases compiled under one roof prompted us to write this book. As editors, we have not sought to predefine the contextual setting or the area of impact, nor provide a certain definition or area of interest; the only boundaries we set forth were the assessment of e-business applications and the impact on organizations.

The purpose of this book initially was twofold: to assess the impact of e-business, and to provide two distinctive angles between private and public organizations. However, having looked at our contributors coming from 10 different countries with a wide array of e-business applications and implications on diverse organizations, we decided to group them into industries, and within them provided a public and private division. We believe that the changes reflected in the edited book provided a more richer picture and a chest of resources with concepts which may have already been published in various journals, but many of the perspectives are new, which makes this book both suitable for practitioners and academics at the same time. Twenty-four authors have contributed to this book, which are operating as either practitioners or academics in the fields, who are involved with various aspects of the impacts of e-business applications and strategies.

E-Business Applications and Impact Studies

E-business applications are defined with a wide array of tools, and a variety of definitions, in this book, similar to Pant (2003) we consider e-business applications (based on various trading partners) to include the following classifications: business-to-consumers (B2C), business-to-business (B2B), consumer-to-business (C2B), consumer-to-consumer (C2C) and intra-business (organization unit to organization unit) Intranet, Extranet and other contexts. The impact is not new as previous literature reviews in the 1990s have centered around the debate on the impact of IT[1] (Vollman, 1996), where others have focused on efforts to define the impact of e-business applications, implementation projects, and still in the academic research agenda, there are debates on the impact of e-business. The "Electronic Markets" journal in 2002 dedicated a special issue solely to the debate on "B2B E-Commerce Revisited: Revolution or Evolution?" In an ongoing effort to translate the impact of e-business applications on particular sectors, "On the Horizon" (2008) included a special issue on the "Library and the Internet: introduction to the special issue". The themes within these special issues ranged from the content to the applications used and its challenges. Another platform that we have witnessed the debate was on the political agenda, wherein Laratta (2010) discussed the use of e-business applications in political agenda, and to what level it impacted the content and context of non-profit organizations (NPOs) which provides services for social improvement, that have utilized electronic techniques for advocacy purposes, ranging from application to the impact of e-business applications and its content. These discussions and ongoing debate is not exclusive to the above mentioned authors and journals; however, conclusions from these sources show a variance and underline the different angles to e-business applications impact. Similarly, the diversity of our cases in this book indicates also diversity in the contextual settings ranging from the impact of e-learning platforms on education, to health record database systems, to the agri-food industry, which enables the reader to gain a richer picture of the impact of e-business technologies from the user interface to security. Hence the structure for the edited book has been divided into three main areas:

1. E-business Applications and its use in Higher Education
2. E-business Impact on Public and Private Organizations
3. E-Business Impact Definitions and Conceptual Discussions

Although we have categorized the chapters in three distinctive sections, each section in itself contains synergies that can be cross-linked across the book, which shows the synergies that can be gained through having a broader picture of implementation stories to conceptual discussions.

E-Business Applications and Its Use In Higher Education

The first section involves a wide array of e-business applications, and although discussed within the higher education setting, it is interesting to see that there are similarities as well as differences in the impact of e-business applications and how far the descriptive accounts of implementation cases differ within the same region. E-business no longer just applies to conducting business online. Higher education has recognized the role that technology can play in re-shaping the entire teaching and learning process.

The first chapter introduced by Debbie Holley and Dr. Martin Oliver, investigates the UK's promise to make equality a priority in respect to student access to electronic curriculum resources (i.e. e-resources). Their investigation at a London-based Business School uncovered key points that need to be considered when designing curricula for a diverse group of students, in which a "three stage model" emerges. The model introduces the three stages of high, medium, and low risk in terms of students' retention of learning materials when dealing with their own expectations, use of space and control. The debate of the research introduces the reader to ways of designing more inclusive business schools curricula for students who face different challenges according to their personal, social, or economic situation. The development of e-resources is a central e-business strategy to add value at this institution.

From this discussion of the inclusiveness aspect of electronic curriculum development in UK, the second chapter introduces us to the context of Higher Education at Spanish universities, which has been researched by Drs. Ana Jimenez-Zarco, María Pilar Martínez-Ruiz, Virginia Barba-Sánchez and Alicia Izquierdo-Yusta. Their research indicated that the e-business applications in two distinctive universities have been used in a different manner, showing that the contextual setting, even though present in the same industry, can show different pictures. Their chapter investigates the impact of e-technologies based on the standard model for generic use of ICT in higher education versus advanced or radical models, and it proposes as an intensive use of ICT to assess the impact on teaching-learning process.

In their chapter, Dr. Petros Ieronmachou and Nola Stair investigate the effective use of e-learning technologies in Web-based course delivery and the inter-related impact on academic and administrative processes at a London-based Business School in the United Kingdom. The resulting case study identifies the importance of holistic implementation strategies and uses a LOGIC evaluation model to identify key inputs and outputs for future research. While much has been written about Web-based delivery of courses in countries such as the UK and United States of America (USA), this chapter contributes to the growing need to plan and manage the institutional challenges of e-business strategies.

Following with a chapter assessing the color impact on e-learning platforms to encourage memorization of content and trust, Drs. Pelet and Pappandolou investigate several e-learning platforms and assess the role color and display have on trust and memorization, especially whether or not the impact of color in an e-learning platform increased the likelihood of readability and suitability. In addition, Pelet and Papandolou challenged the role of color on the online interface in e-learning platform and the notion of whether color can enhance student trust. This study has been evaluated under the framework of Knowledge Management and content management research agenda.

E-Business Impact On Public and Private Organizations

From Higher Education, the next section moves to the diverse industries starting with the banking industry. Chapter six, introduced by Dr. Rachel Johns, explores the use of self-service technologies in-depth with a small sample of Australian business bank customers. The exploratory nature of this research bases its premises on assessing the underlying impact on the banking organization in case of self-service technologies enabled by e-applications. The research gives a self-service technology introduction and investigates the relationship to relationship marketing and service dominant logic with two dimensional reasons for introductions (cost saving, relationship, convenience and switching cost) and reasons for use (co-creation and collaboration, value, resource focus, relationship requirements).

The next chapter is introduced by Dr. Raechel Johns, who investigated the impact of e-business technologies in the realm of an Australia e-banking initiatives and the impact on their business customers. The exploratory research explores how utilising an e-banking context impacts relationships between the bank and its business customer, and challenges the notion the use of self service e-technologies have on relationships within a business to business context. This chapter evaluates three main areas of e-banking in B2B (Business-to-Business): (1) relationship marketing in the context of self-service technologies, (2) face-to-face contact versus online contact and its impact on B2B relationship, and (3) the use of technology and its impact on trust and relationship commitment in B2B.

Dr. Jiaqin Yang, in his chapter, investigates the realm of the banking industry, in which he explores e-banking within the context of development issues, focusing on small to medium sized banks in remote rural areas, and the impact of cultural, economic, and political factors on e-banking systems in different nations. The comparison is based on results of an exploratory study that has taken place in the US and China. This chapter underlines the challenges of small and community banks located in rural areas and their attempt to catch up their counterparts in larger cities in terms of the application of e-banking. In this study, Dr. Jiaqin Yang focuses on emerging issues and challenges, such as Internet accessibility, culture, and tradition playing a vital role in the development of e-banking industry among different nations.

Chapter 7 investigates microfinance and the impact of information and communication technologies (ICTs) on social inclusion. Having examined the literature and secondary sources, Dr. Ana Marr and Dr. Lin Yan create a framework for ICT and its impact on microfinance in developing countries. The framework highlights how the capability generation is interlinked with social inclusion, microfinance, and the use of ICT. They categorize this framework in an effort to contribute to the existing body of work, as "people's capability building at the heart of the impact chain" for microfinance in developing countries.

Gary Stair, in his chapter, explores a pharmaceutical company that successfully implements an Enterprise Risk Management (ERM) program to assess the inherent risks, in order to provide guidance to executive management and governance entities on critical aspects of organizational governance, business ethics, internal control, fraud, and financial reporting. This chapter also discusses an approach to build an ERM implementation plan within a company by outlining the pharmaceutical industry and its key informants (sales forces, middle-management, senior leadership). In assessing the ERM, the author also investigates also the potential use of e-media and virtual sales representatives in the context of pharmaceutical industry.

In the following chapter, Georg Juelke discusses the staffing industry and underlines the need of business intelligence to optimize process, reduce costs and develop new services. Juelke identifies the root causes that hinder the staffing industry's ability to develop analytics, reflecting upon the market conditions to the design of IT systems, and discusses the need to generate a common nomenclature to classify job categories to enable data integration in a global world, where the staff mobility currently is rather dispersed. This chapter argues the need to deploy common job classifications across IT staffing systems with real-life examples.

Dr. Brian Gugerty and Dr. Michael J. Maranda introduce the reader to the impact of e-technologies on healthcare from the perspective of practitioners, while exploring the application of Information Technology with the introduction of electronic health records (EHR) and their impact to healthcare in the context of the United States. Their research investigated four main areas of implications, namely: surveillance allowing systematic monitoring, implications for reporting, verification of implications of patients' data, and response to the potential challenges stemming from the EHRs. Finally, but not lastly, they explore the use of personal patient records on social, legal, technical/ administrative, digital indifference and consumer indifference factors and suggest possible solutions for improvement from the patients' perspective.

The next chapter in this section, by Dr. Chad Lin, introduces us to e-business in healthcare, which can potentially improve the operations of health service providers and the quality of the delivery of the internal process flows. Dr. Chad Lin explores the benefits and assesses the challenges of the e-business applications in healthcare. The findings underline that in order to realize the e-business benefits, four main building blocks are needed: (1) IT investment evaluation methodology (IEM) which enables the selection and monitoring of e-business investments, (2) an IT benefits realization methodology (BRM) which evaluates the process needs and the project life cycle, (3) IT Maturity which provides crucial information about the organization's ability to effectively utilize IT, and (4) company policies, business objectives, and/or strategies that would provide a sound basis for building and improving healthcare organizations' level of organizational IT maturity.

Chapter 12, by Dr. Kim Bryceson, underlines the e-landscape of the agri-food industry and asserts in the food for thought section that there are five key drivers: (1) the dynamics of sustainable production, (2) the presence of integrated food supply systems, (3) globalised markets, (4) key players in the agri-food industry and (5) food security. Although these key drivers are all important in the e-landscape of agri-food business, Dr. Kim Bryceson also assesses that e-readiness and knowledge management are the other roadblocks to establishment of a global e-landscape for agrifood business.

E-Business Impact Definitions and Conceptual Discussions

Chapter 13, by Dr. Lan Yi, investigates Information and Communication Technologies (ICT) in organizations and the impact to the carbon emission. In his research agenda, Dr. Yi introduces the Internet's lifecycle starting from manufacturing of Internet equipment/hardware to its usage and final disposal. This chapter highlights the importance of the environmental factors and the decisions made by organizations when implementing e-business applications.

In the next chapter, Dr. Sharon Nachtigal explores in a more overarching manner the concept of e-business, its development, implications on organizations, and its barriers She also discusses the general principles on how organizations overcome such barriers. The chapter first discusses e-business and its impact on the modern business and social environments involving practitioners (i.e. executives, developers and users). This is followed by descriptions of changes in the business world, based on the premises that integration of information systems (IS) into the business environment is crucial, involving examples of variety of e-business models the e-business characteristics.

Next, the chapter by Dr. David Anyiwo and Dr. Sumana Sharma introduces the readers to a large scale deployment of Web services and e-business technologies in general, and discusses the decision making processes of e-solutions impacted by security standards, business continuity planning, and cyber legislation. Their research indicates that eight measures needs to be embedded, which are: employee training, introduction of innovative approaches, incorporation of user feedback, implementation of "the triad" (confidentiality, integrity and authentication), high quality data encryption techniques, maintenance, and implementation of a security culture.

Following, in Chapter 16, Dr. Sharon Nachtigal explores e-business security tools including software landscapes and organizational tools, describing the need for more e-business functionality integrated to the organizational processes. She challenges the traditional notion of relying on software rather than organizational decision making. In this chapter the discussion revolves around academic research papers and cases from real-life e-business security flows. Furthermore, this chapter discusses the major e-business security challenges, evaluates the current e-business security requirements for organizations, challenges this notion driven towards a process-based security, and presents alternative approaches to e-business security.

The final chapter is introduced by Dr. Wilhelmina Djoleto with four cases in the USA where the application of e-business etiquette is analyzed as part of decision-making procedure and assesses the success of organizations and institutions as e-societal members with the use of e-business tools. Her research addresses that evaluations of e-etiquette must be conducted to inform decision makers on the enhancement, which will impact the application of e-business solutions. The cases include two private and public institutions.

FUTURE DEVELOPMENTS

Trying to forecast what future developments hold for e-business applications is, like any area, a risky business; rather than forecasting, based on the contributing authors work, we have seen that there are learning opportunities from e-business applications and their impact on diverse public and private organizations. We tried to combine our learning into three distinctive areas;

- Divergence of contextual settings: although in the same industry, the higher education and e-business technologies applied within the education sector indicate a divergence. In which, the applications have been used as standalone tools to gather information and data, to an interactive learning platform. Although there are curriculum development platforms such as those based on UK Higher Education, a further research agenda might be to explore to what extent these networks or EU-wide teaching platforms have an impact in creating a coherent basis for the use of e-technologies in higher education.
- Need for integration of platforms in a regional global context: the research on agri-food business and health providers indicates that the e-business tools have a rather limited impact if used as standalone system, and not integrated, neither regionally, nor globally. A further research agenda might be necessary for the evolution of such networks and their benefits to each industry.
- Increasing the inclusiveness of service providers: the chapter on banking, microfinance and e-etiquette indicates that the societal aspects of e-business applications allow a more inclusive distribution of information and resources capability. It would be interesting to see whether the models generated in this chapters could be replicated in other areas in testing the inclusiveness of the research agenda.

The above mentioned further developmental areas are just of some of the issues, which were interlinked in the edited book, but are not limited to other subject areas. This was also one of are main goal to create a cross-disciplinary book with academic and practitioner contributors, to provide case studies on e-business applications impact. Our hope is that this edited book can partially open the doors for further research areas and generate new inquiries.

Ozlem Bak
University of Brighton, UK

Nola Stair
University of Greenwich, UK

REFERENCES

Dai, Q., & Kauffman, R. J. (Eds.). (2002). B2B e-commerce revisited: Revolution or evolution? [Special Issue]. *Electronic Markets*, *12*(2). doi:10.1080/10196780252844508

Laratta, R. (2010). Political advocacy or electronic advocacy? The use of electronic techniques by Japanese NPOs for political advocacy purposes. *Journal of Global Responsibility*, *1*(1), 85–97. doi:10.1108/20412561011039726

Pant, S., Sethi, R., & Bhandari, M. (2003). Making sense of the e-supply chain landscape: An implementation framework. *International Journal of Information Management*, *23*(3), 201–221.

Sax, B. (Ed.). (2008). *The library and the internet: Introduction to the special issue. 16*(1)

Vollmann, T. E. (1996). *The transformation imperative*. Boston: Harvard Business School Press.

ENDNOTE

1 The transformation study is a result of "Manufacturing 2000," which is a collaborative project between leading multinational companies (including automotive companies such as Volkswagen/Audi) and the International Institute of Management Development (IMD) in Switzerland. The results have been published in "The Transformation Imperative" 1996 by Harvard Business Press.

Acknowledgment

In addition to the main contributors and our editorial board members, we would like to thank Myla Harty from IGI Global, for her encouraging words and patience in the preparation of this book.

Ozlem Bak
University of Brighton, UK

Nola Stair
University of Greenwich, UK

Chapter 1
Diversity and Design:
An Emergent Model of Matching Curricula Design to Student Need

Debbie Holley
London Metropolitan Business School, UK

Martin Oliver
London Knowledge Lab, UK

ABSTRACT

Higher Education Institutions have worked to make equality of access to electronic curriculum resources the 'status quo'. However, there is evidence that simply providing e-learning – no matter how well intentioned – is insufficient to address the problems that students are experiencing. A three stage model has been developed through analysis of students' learning experiences at an inner-city, post-1992 University to illustrate how students have to negotiate their engagement with Higher Education. The model provides a way of mapping aspects of course design to different portraits of students, enabling students to be considered as high, medium and low risk in terms of retention. The value of this model for design and analysis of courses is located within the debate of how inclusive business schools curricula are for a diverse student body.

INTRODUCTION

This chapter contributes to current debates about curriculum offerings by Business Schools in the UK. This represents a significant and growing area within UK Higher Education. From the latest data available, in the academic year 2007/08, almost 1 in 7 of all students were studying business and management; a significant proportion of which studied at undergraduate level. During the past thirteen years undergraduate study of this subject has risen by 66 percent (Association of Business Schools 2008:19). In the light of this growth, the Governmental view of higher education as primarily about feeding the needs of industry can be seen as highly contentious. In particular, there can be tensions between this economic imperative and

DOI: 10.4018/978-1-60960-501-8.ch001

social agendas such as widening participation. It is this tension that sets the context for the chapter.

Higher Education Institutions have made it a matter of principle that all students should have equality of access to e-learning. However, there is evidence that simply providing e-learning – no matter how well intentioned – is insufficient to address the problems that students are experiencing. Indeed, it has been suggested that blanket provision may cause as many problems as it solves. Daniel (1996) in his book 'Knowledge, Media and Mega-Universities' calls for a dialogue about the use of the new technologies with the aims of high volume, low cost, and high quality. The industrialistion of technology for learning follows many principles of technological implementation for business in that it focuses on economies of scale, best suited to a 'one-size-fits-all' operation. For example, post the National Committee of Inquiry into Higher Education (1997) and its recommendations for resource based learning, universities in the UK and elsewhere have invested heavily and systematically in Virtual Learning Environments (VLEs). A VLE can be defined as, "a collection of integrated tools enabling the management of online learning, providing a delivery mechanism, student tracking, assessment and access to resources" (Jisc Infonet). Thus this tool matches a managerial agenda with the needs of providing a robust system for content delivery to students.

A three stage model has been developed (Holley & Oliver 2009) through cross case analysis of students' learning experiences in a Business School to explain some of the tensions between the delivery of a standardised blended learning offering, for example via a VLE, and the frustrations and difficulties some students experience in their efforts to participate fully in the blended curriculum. This model illustrates how, even when issues of access are 'solved', students have to create new and innovative ways to negotiate their engagement with Higher Education. This negotiation involves their individual expectations of:

- Their ability to control technology
- Their forthcoming educational experience
- Managing their 'learning space', including the times and settings when they study

This allows aspects of course design to be mapped against different portraits of students, resulting in an assessment of risks in terms of retention. The value of using this model when designing and analysing courses will be argued.

CONTEXT

The UK's attempts to develop alternatives to traditional classroom teaching can best be understood within wider international concerns about meeting the needs of a "knowledge economy" (Hodge, 2002).

"Education and skills are the key to a competitive and efficient workforce. Top quality management and strong leadership is a crucial factor in this. Business and Management continues to be one of the most popular University subjects at both undergraduate and postgraduate level." Rammell (2009)

Within this context, political intervention in higher education is, from governmental perspectives, justified in terms of enabling the UK to compete within the international trading environment. For example, White & Davis (2002) portray technology as breaking down international barriers to education. Computer–mediated learning environments are argued to make possible whole new ways of learning. They create global learning communities that can bring together academics and professional practitioners. Rudestam & Schoenholtz-Read (2002:18) foresee a future where the distinctions between classroom based and internet based education is completely eroded, and teaching methods become tailored to the needs of the subject and students. However, with pedagogic choice arguably becoming a matter of strategy, rather than tactics, the choice of teaching techniques is becoming what Noble (2001:3) calls the increasing commodification of education: an educational experience that has been disintegrated and distilled into "discrete, reified, and ultimately saleable things or packages of things". ICT has moved from being associated with peripheral innovations and developments to affecting all aspects of learning and teaching.

Disempowering strategies such as those outlined in 'Harnessing Technology' represent for Conole, White & Oliver (2007:12) "knee-jerk policy which does not take account of evidence emerging from research" but which still have a huge impact on students. Noble (2001) argues that whilst e-learning is akin to training, which is purely for the benefit of others and where any assertion of the self would become a subversive activity, 'education' involves the integration of knowledge with the self – where knowledge is defined by and helps to define the self. He stresses that whilst typically the push for e- learning is predicated upon a belief in cost cutting, staff reduction and so forth, education relies on the quality of interpersonal relationships offered – and that to date educational research has demonstrated that good education requires a labour-intensive, personal relationship between students and quality academics.

UK University management, under the auspices of a heavy audit culture and a diminishing unit of resource, has pushed academics towards the use of online resources (Holley & Oliver, 2000) which are believed to offer economies of scale. Virtual learning examples offer a useful illustration of this. The Universities and Colleges Information Systems Association Survey (UCISA 2008) identified Moodle, an open source VLE, as the most commonly used platform in pre -1992 (the ex-polytechnics) institutions, with 56% reporting this as their main platform. Of the commercial platforms, Blackboard remains the market leader and is widely used – a rise to 50% from its 43% score in 2005, with 61% of post-92 institutions reporting its use. WebCT features strongly amongst Scottish institutions (71%), and a merger between WebCT and Blackboard make this the dominant commercial VLE. Only 3 institutions surveyed reports no VLE in use. Students are coming to expect materials to be made available online, and the growth in VLE use reflects this. Moving aspects of teaching and learning online is believed to help meet sector wide challenge of creating a climate in which students develop as independent learners (Higher Education Academy 2010).

However, tutors vary in their approaches to pedagogic design in the classroom, and it is no different when designing for online learning. These new environments are not always designed with the student in mind, and their responses can be restricted to reacting to ready-prepared online materials. As Smith & Oliver (2002) suggest, the active, engaged aspects of learning can easily be ignored, reducing the idea of what it takes to be a student to a matter of selecting which of a limited range of learning modules to consume.

Diversity in the Classroom

Considerations of diversity can be limited to individual organisational policies of 'diversity' understood in terms of age, race, disability, ethnicity, economic status, family/marital status, nationality, religious belief, sexual orientation, spent convictions, political affiliation, and gender reassignment. However, there is also a far wider discourse, drawing upon emergent research from the USA which claims to recognise broader dimensions of inequality than those typically considered in organisational equal opportunities policies (Shaw; Brain; Bridger; Foreman and Reid 2007:32/33). The report concludes that the evidence for a link between student diversity and positive teaching and learning outcomes is still limited and remains under-researched (Shaw et al 2007:3). Thus it is timely to explore issues of widening participation and student diversity in terms of the curriculum, including the business curriculum.

Historically, much of the focus on work-based education and skills development within UK education was undertaken by Polytechnics which, under the governorship of the Councils for National Academic Awards (the CNAA), awarded more 'vocational' degrees and Higher National Diplomas to students who, in the main, were non-traditional (by today's definition). However there was a perceived image that these qualifications were of a lower quality than those degrees awarded by the traditional university sector. This perception often meant that students who graduated from the Polytechnics were discriminated against in the graduate job market (Evans 2004).

In an attempt to remedy this situation the government of the day decided to remove this 'binary' approach to higher education. In 1992, all polytechnics and a few higher education colleges were granted charters to become universities and award their own degrees. At the same time, all universities were encouraged to grow significantly in order to widen access, with the majority of this increase occurring in the new university sector. The pre-1992 universities, however, did not grow anywhere near as rapidly as the post-1992 universities. Moreover, if anything, their exclusivity grew (Evans 2004).

Overall, however, the numbers of students entering higher education grew rapidly during this period: the proportion of school leavers entering higher education doubled. However, the numbers of non-traditional students did not really change (Forsyth & Furlong, 2000:1). The increase in numbers came primarily from the middle socio-economic backgrounds, which were traditionally associated with universities anyway. Forsyth & Furlong argue that, despite a decade of educational policy change, 80% of HE students in the mid-1990s still came from the most affluent geographic areas. This situation was highlighted by the National Committee of Inquiry into Higher Education (1997) which laid the foundation for the current widening participation policy. It also set the government target for entry into higher education by 2010 to be 50% of all 18–30 year olds. An increasing number (and percentage) of these students were to be non-traditional in profile. The Higher Education Funding Council for England (HEFCE) publication M8/97 (June 1997) defined non-traditional students as having at least one of the following characteristics:

- Being from an ethnic minority group;
- Having a long-term disability;
- Possessing non-standard qualifications on entry to higher education;
- Being aged over 25 years on entry to university;
- Coming from lower socio-economic groups of origin.

Haselgrove (1994:171) outlines some of the consequences of this shift from an elite to a mass system of participation. The impact of these changes for students "will depend on what resources they bring to negotiating and controlling their experience of the system." As noted, older universities have grown less than post-1992 Universities and their student profile has typically changed less, although they have been more successful in recruiting from overseas markets. Other higher education institutions have diversified their profile to include greater proportions of older people, women, ethnic minority, home students and those studying part time.

In summary, the widening participation agenda in the UK has had a significant impact on many HE institutions. Perhaps the most important aspect of this is the diversity that has arisen within the student population. One driver for this has been the government's concern about the UK's capacity to compete in a globalised market where technological advances and the information revolution are seen as increasing the demand for a skilled workforce (Blunkett, 2000). In post-1992 universities, normally characterised by a focus on teaching and vocational relevance rather than research, this agenda is an important force for change that affects both traditional and online teaching. One consequence of this is that students enter Higher Education with increasingly diverse biographies, expectations and constraints (Holley & Oliver, 2009). Curriculum planning has to be rethought (Barnett & Coate, 2005), and this is something that academics can find challenging. For this reason, research has been undertaken to create a model that can support academics in working through these issues.

Personalisation of Learning

"Policy makers and educationalists have come to realise that a 'one-size-fits-all' approach to education is inappropriate and won't meet either individual or societal needs" (Conole 2010:2).

Conole (2010:3) goes on to outline the emergence of the concept of personalised learning, explaining that this is partly a response to the challenges of living and working in modern society; in part a way of capitalizing on opportunities for learning using new technologies; and in part a recognition that current educational provision is too narrow and too restrictive for many who wish to participate. As we think about how to design more effective online learning environments, one thing is clear: we need to treat students as individuals rather than as homogenous groups. Rather than maintaining a fixed view of what all students want or what all students need, we need to be flexible and create environments that enable greater choice for individual students (Twigg, 2005:5). For this to happen, curricula and materials need to be designed in new and innovative ways, to enable students to learn 'any time, any place', and giving the learner a degree of ownership of the learning process (Holley; Greaves; Bradley & Cook, 2010:286).

The barriers to these kinds of developments are reflected in Hopkins (2006:18) who comments on the limits of the traditional educational context; namely physical space, the lack of innovative use of technology, the constraints arising from the rigid nature of the curriculum, the problems associated with teaching large classes, and the conservative nature of many educational institutions. A recent BecTA report (2008:27) outlined the experiences of young adults in the classroom. While 55% of them enjoyed groupwork and a further 35% enjoyed 'working with a friend', their most common experience was of copying from the board (52%) or listening to the teacher talk for a long time (33%). This is a long way away from the Learner Entitlement Framework BecTA (2008:33) envisage: 'Personalized learning which reflects learners' interests, preferred approaches, abilities and choices, and a tailored access to materials and content'.

Yet academics are not well placed to respond to these problems. Curriculum design is not frequently discussed in Higher Education (Barnett & Coate, 2005), and academics may receive little or no support for this aspect of their work (Wisdom, 2006). As a result there have been several initiatives which have tried to create resources that can be used to support the process of curriculum planning (e.g. Oliver & Conole, 2000; San Diego; Laurillard; Boyle; Bradley; Ljubojevic; Neumann; & Pearce, 2008). However, while there is a growing interest within these in designing curricula that meet the needs of learners, these resources do not directly address issues of access, participation and retention. As a result, they run the risk of perpetuating rather than redressing the problems that excluded students encounter. For this reason, an approach to curriculum planning and analysis was developed that focused on these issues, based on studies of student exclusion and the struggles that individuals faced when trying to participate in Higher Education (Holley et al, 2006). In particular, the three stage model which is described below, focuses on the potential issues that specific students are likely to face in relation to a course that has already been designed, rather than (as with the others tools of this kind) focusing on the planning stage, before the needs of students participating in the course can be known.

A MODEL FOR CURRICULUM PLANNING

The model for curriculum planning draws upon a study of student engagement with their online learning in two modules located within a post-1992 Business School (that is, an ex-polytechnic with a vocational tradition). The first study is based on a compulsory 'Business Skills' Module studied by all first year students entering the Business School (around 800 students). The second study consists of a specialist third year 'International Purchasing' module, this was selected because most International Business degree students take this, and it thus offers a very diverse range of students (around 70 students).

The case studies from which the model was derived involved a series of interviews exploring students' areas of concern. A very brief 'pen portrait' of each student is now provided. (All names are pseudonyms, for further information see Holley, 2008). Charles was a professional student who usually attended University on a day-release basis, and at the time of the research, negotiated to study the module on a completely 'virtual' basis as his recent promotion involved significant overseas travel. Juanita is an overseas student who attended regularly and took advantage of the blended learning as well as face-to-face provision. Kwame, an overseas student, attended University to obtain a Higher Diploma in Purchasing and Logistics, and had significant problems and challenges with accessing the curriculum. Marco is characterised by his 'juggling' of his life, and giving up sleep to manage his combined roles of working and studying. A refugee from Somalia, Nyela starts her narration by talking about how hard she has found the move to the UK both in terms of social and educational integration. Her story is one of difficulties – with making friends, settling into a different education system, with not having the familiar extended family support network around her and sharing very close space with her siblings and her mother in a small flat. Joanne is a single parent who started University six years ago, and had to give it up when she became pregnant. During her time at home, she feels she matured and started to learn to manage her time (Holley & Oliver, 2009).

An analysis of student biographies (Holley & Oliver, 2009) revealed how students develop distinctive, idiosyncratic study patterns, in order to fit what they experienced as an homogeneous curriculum offering to their individual circumstances. This involved, for example, studying after-hours at work and negotiating access to a shared computer at home. The analysis of these accounts led to the development of a model that could account for these variations in terms of what they expected and what they experienced.

What is significant in the development of the model was the 'gap' revealed between students and staff. Students very seldom negotiate with staff in order to find their own spaces for learning – indeed, this process is rarely explicit, is much more subtle than direct questioning or challenge, and involves a combination of factors including:

- Their expectations about education mapped against home spaces;
- Their expectations about education mapped against their control of technology;
- Their expectations about control of technology mapped against home space;
- These ideas, and their relationships, will be outlined in the sections that follow.

Expectations about Education

Elton (1988) suggests that student learning falls into three main patterns of study strategy: personal meaning (which, he argues, university teachers favour), reproducing (which university teachers deplore) and achieving (which university teachers tolerate). This argument is developed further in Elton (2001) where it is claimed that in the current teaching culture of universities it is only the 'best students' who will learn in ways that are valued by academics. This attitude is captured in Trout (2000), who comments, "There is a lack of students still wanting to pursue challenging courses, and their numbers are shrinking relative to the number of those who are 'disengaged' from academic pursuits and values" (Trout, 2000:1). He highlights some of the 'problems' with students:

Here is a typical complaint: 'Last semester, many of my students drifted in late, slumped into chairs, made excuses to leave early and surrounded my desk when papers were due.... [complaining that] "Nothing interests me." ... When called upon, the students often slump in their chairs, shrug their shoulders and mumble, "I dunno." Often, I feel as if I am invading their space simply by asking them a question. When they do interact with me, they sometimes seem resentful and angry. Other teachers complain of unprecedented numbers of students who rarely attend class, do not read the texts, and show scant interest in mastering required material.'

As policy initiatives bring increasing numbers of non-traditional students into a changing higher education sector, Haggis (2004:182) suggests that there is the "need to look again at what is understood by learning (and, indeed non-learning) in this context." Haggis suggests that the implications of the current populist models of deep and surface learning (Entwhistle, 2000) are less than helpful, as they reveal a rather elitist set of assumptions about student purpose and motivation, and thus reflects a rather one-dimensional perspective, which implicitly follows institutional agendas.

Haggis offers a different view of the learner, in which a person who is experiencing difficulty in engaging with the unexplained norms and values in higher education. "May be exhausted from part-time work or parenting, distracted by family or financial problems, or lacking the fundamental confidence, self-esteem or health to engage in the ways that are assumed to be both desirable and possible" (Haggis, 2004:98). Thus, the experiences of these different kinds of learners, now pushed through a mass education system by a plethora of government initiatives; do not meet with the expectations of academics driven by disciplinary concerns and a culture that values research over teaching. New and different ways of teaching and learning are needed.

"[If] the system is to grow into a genuinely accessible form of education for 50% of 18 year olds, in addition to the widest possible range of adults learning throughout their lives, it is going to have to find new ways of conceptualising its core values and activities" (Haggis, 2004:102).

The literature indicates that previous educational experience is important, but for widening participation students, it is their expectations about the forthcoming educational experience that determines their reaction to the blended learning experience they eventually have. The students in the case studies (Holley & Oliver, 2009) were thus mapped according to their articulation of surprise, confusion or difficulty with the module studied, in order to explore how these difficulties were experienced and reacted to. What emerged was the importance of a 'match' between the expectations of the tutor and the student, in terms of what it meant to participate in education, and in particular, what a student would need to do to learn successfully online. This match/mismatch was a powerful explanation of the frustration and confusion that some widening participation students experienced.

CONTROL OF TECHNOLOGY

Sharples (2005:2) considers that "the most successful learning comes when the learner is in control of the activity, able to test ideas by performing experiments, to ask questions, collaborate with other people, seek out new knowledge and plan new actions." Thus, the teacher has no ontologically privileged position, but, as Sharples suggests, is "simply another participant in the conversation of learning. This does not fit easily with traditional classroom schooling". This approach can also be seen in the work of Garrison & Anderson (2003:7), who state that blended learning is "about doing things differently, not about entrenching deficient face-to-face approaches such as lecturing by using e-learning to access more incomprehensible information. Nor that e-learning is about having students accessing the same deficient approaches through a different medium". Arguably, the focus of classroom practice and the conventional power relations of the classroom need to change to embrace the possibilities of students controlling their own learning.

The lecturer is, at present, doing what Collis & Van der Wende (2002) call "stretching the mould" to encapsulate the dominant use of Information and Communication Technology (ICT) in higher education. Three main areas emerged from their study of the ICT experience of teaching institutions across Europe, the USA and Australia. The first area encompasses the pace of change that the authors categorise as slow rather than radical and based on a business approach. The second main area is that ICT in teaching and learning has become part of the blend of learning, but the "core medium" of instruction, most valued by institutions, remains the lecture. The third area is that instructors are gradually doing more, but with no reward. Thus instructors are "stretching the mould" by incorporating more IT into their daily practices; but they are not radically changing the ways in which they teach.

True power and participation for e-learners can be achieved by careful design of ICTs, suggest Cook and Light (2006). In this context, we still need to understand what is required to design digital media that can assist the motivations of 'real people' in a way that empowers them (Cook *et al,* 2006). This clearly demonstrates the need to develop both new tools and new conceptions of pedagogic space, and the three stage model outlined below is designed to assist in this process.

The students in our case studies encountered various challenges with controlling their technology; Kwame, for example had physical difficulty because his technical college in Ghana had very few computers, and students learned to program using 'cut out' cardboard keypads, thus he found the actual 'spring' in the keyboards difficult to master, leading to intense frustration in his efforts to take part in an online discussion. Nyela had to manage access at home with two siblings and one computer, yet felt safest working in her home space; the struggles of these students can be contrasted by the ease of access of our part time professional student, Charles, whose employment status allows a comfortable and well equipped home space with his employer meeting all of his costs of study.

Space

Our understanding of Higher Education is shaped by the ways we think about time, space and our ability to build relationships with learners. Technology has an important role to play in this, making it easier to bridge discontinuities in time and space. This bridging, Ehrmann (1999) suggests, has a fundamental impact on our organisation of knowledge for research and teaching. He identifies three revolutions that have enlarged and reorganised higher learning. The first was from the oral dialogue of Socrates' day toward educational forms that included reading and writing. This developed with the reading-writing revolution where key technologies include pen, paper and, later, printing presses. The second was the geographical change in the early Middle Ages, from the independent scholar teaching the independent learner to scholars and students grouping together in universities. In the 20th century this campus revolution featured lecture halls, chalkboards, laboratories, dormitories and transport that could bring scholars from their home. The third is characterised by silicon chips, and a digital world of communications and data storage. The digital age envisaged here is not a virtual world such as Second Life, where students and teachers would meet only virtually, but a dissolving of the physical and time barriers that constrain more traditional teaching in educational establishments today.

Current research into the digital divide advocates that technologies are not value-neutral, and can only be understood when placed "within a wider political context of an unequal and changing pattern of power relationships" (Loader, 1998:7). That the digital divide exists is acknowledged by the Organisation for Economic Co-operation and Development (OECD) who refer to the divide in terms of the gaps in access to information and communication technology, which, "to some extent is simply a deepening of existing forms of exclusion" (OECD, 2002:11). This is seen to be of growing importance in a society that has been positioned as a "knowledge economy", and is something seen as having implications across the educational system (Peters & May, 2004). Thus, arguably, the divide should be of concern to government policymakers, as without policy interventions, Information and Communication Technology (ICT) could intensify societal divisions, and increasingly marginalize those who are, "unemployed, poor, housebound, disabled, less educated, members of ethnic and cultural minorities - and in many countries, women" (OECD, 2002). This politicises access to technologies, making Lefebvres' analysis useful, notably his ideas on where capitalism has "now laid its imprint upon the total occupation of all pre-existing spaces and upon the production of new space" (Lefebvre, 1994:326). These ideas of 'space' not being value free and, indeed, having a value as a commodity impacts upon our students. Having fast internet connections, up-to-date equipment and access at a time and place of the student's choice start to delineate those who 'have' and those who 'have not' in terms of access.

For the students in the case studies, 'space' was not value free. For example Joanna was only able to use University space due to the demands of a family at home; Juanita found the social aspects of learning difficult at University and used social networking sites to compensate for this; while Marco used his employer's well equipped office (with permission) after long shifts as a bar manager, and forfeited sleep in the process.

It is in this context that students have the least support from Universities, since the standard policy seems to be confined to some initial Information Technology training around the VLE. The assumption is made that enabling access to well (or poorly) equipped computer laboratories fulfils institutional obligations and responsibilities to equality of access for all students.

An Outline of the Model

Student cases, such as those analysed by Holley & Oliver (2009), have proved valuable in conceptualizing the complex set of factors influencing student expectations of their higher education experiences. Students bring their own life experiences to academic study, and those better prepared by work and previous positive educational history will be advantaged, more easily 'matching' the learning demanded by a module that incorporates elements of blended learning. The converse is also the case: those poorly equipped in terms of previous life skills and educational experiences are most likely to experience difficulties with the course because its demands and their expectations are mismatched.

This idea of match and mismatch formed the basis of a classification of particular students in terms of risk management. Three areas were identified where difference between demand and expectation could put students at risk of non-completion. Since the interaction between these areas was also important in understanding students' experiences, their interaction was mapped as a series of grids, in which individual cells were flagged as being low, medium or high in terms of risk of failure and withdrawal (Holley & Oliver, 2010).

The models presented below are generalisations of this mapping that have been derived from this for use by other educators. Thus, students in the 'low risk of failure' squares, would have a propensity to show characteristics of a good match between their expectation of education and the courses they select, especially if any have a blended learning component. Similarly they would typically have a high degree of control over their home environment, or the confidence to negotiate successfully with those they share the home with. They would also show a high degree of control over their use of technology, and this could be expressed as a preference towards using technology for private study in isolation, or for using technology to replicate some of the social networks usually found in the classroom.

The students placed in the high-risk squares would struggle to use technology effectively and be more likely to claim that their expectations of the course and the course offering were not met. Thus the tutor could, hypothetically, start to develop a framework where students are supported to enable them to achieve the optimum learning for a particular module, or even to locate the course on the grid and then adapt its design to the different needs of the student body.

Stage One: Mapping Educational Expectations and Use of Space

The axis for Figure 1 is defined by how the students like to use their home space and the online study tasks they undertake from this location. 'Private' indicates a preference for utilisation of home space for private study; 'social' indicates a preference for replicating more social aspects of learning. The students located in the middle of the axis could be classified as being at greater risk of having problems with study, as they lack an awareness of how their personal space could be colonised for learning.

Figure 1. Expectation and space

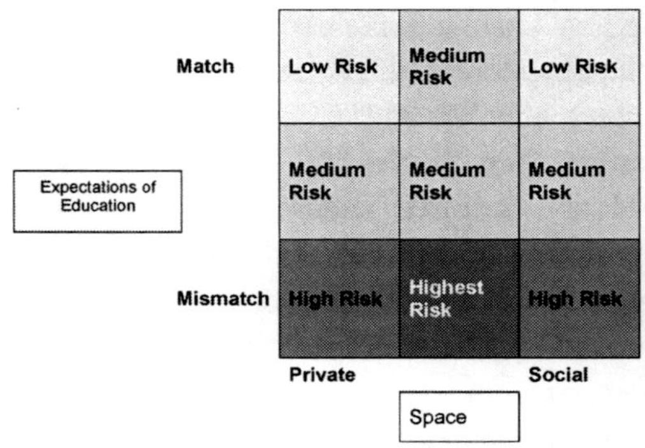

Students that are not confident in their use of technology for social or private study could thus be identified. The further up the middle column they are placed, the higher their chances of success, because they can anticipate and plan for the experiences that will challenge their peers. Students falling within the bottom-middle square would therefore be at the highest risk of having difficulties with their course/module as they are unable to colonise their home space for successful study, and are mismatched in terms of educational expectation.

Stage Two: Mapping Educational Expectations and Control of Technology

The axis for Figure 2 is defined by how the students are matched/mismatched by their expectations of their course, and as well as this, their ability to control the technologies they will be expected to work with.

Students placed at diagonal opposites on the grid would represent the most and least likely to

Figure 2. Expectation and control

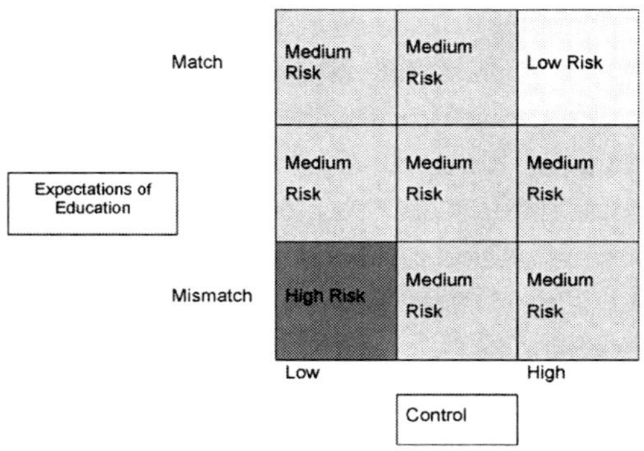

Figure 3. Control and space

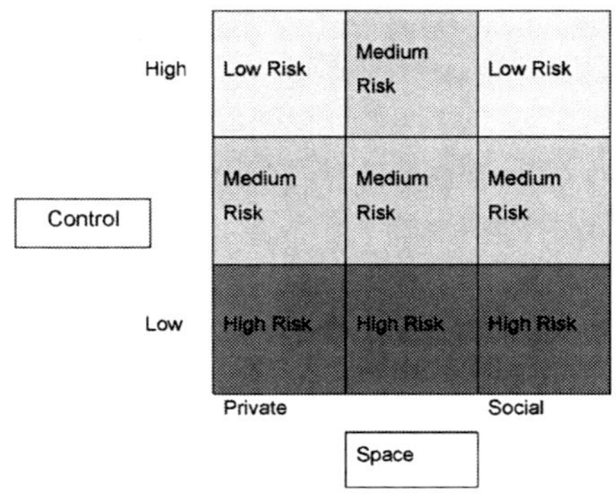

pass the course. Those with less confidence in independent learning outside the classroom and a mismatch with their expectations of university life would be placed in the bottom left cell, being most likely to experience problems that lead to their withdrawal or failure. Those with confidence and a match of experience of education would be represented in the top right cell.

The implications of this would be that some thought might have to be given towards meeting the learning needs of students who might be classified as being in the bottom left corner. This could involve offering pre-course IT support, orientation sessions, communication with students who have already undertaken the module and so on. Such measures could improve their confidence in their ability to control the technologies they will need to use in their studies, and also help change their expectations before the course starts.

Stage Three: Mapping Control of Technology and Use of Space

The axis for Figure 3 is defined by how the students like to use their home space and the online study tasks they undertake from this location, and how confident they are in control of technology. For example, this could be in terms of accessing interactive materials a tutor has placed within a VLE, or using VLE tools for discussion or online assessment.

Students located in the two cells labeled 'low risk' at the top left and right would be most likely to have the necessary control of technology for their studies, and be confident in their expressed preference for private or social online learning. The implications of this, as in the previous figure, would be some consideration as to how the curriculum could be developed to meet the different needs of students with very different abilities in terms of control over their technology. Again, this might involve support for technology use. The three 'high risk' cells along the bottom of the grid indicate that a student placed here could have misunderstandings about the nature of their course, and upon embarking upon his/her studies would find the technology beyond their capabilities. In common with Figure 1, these students could be classified as being at greater risk of having problems with study, as they lack an awareness of how their personal space could be colonised for learning.

More challenging and less familiar, however, might be the need to support students in thinking through and managing their patterns of study, not least in terms of their ability to create times and spaces in which this study can take place. Such support rarely happens proactively, although it might take place through student counseling once problems have been encountered. Due to the sensitivities of advising students about the use of home spaces, it might be that such support might be offered through some kind of mentoring or peer support, rather than by a member of institutional staff.

Using the Model to Revise the Curriculum

The model described above is an analytical tool, which may be generally useful for framing thinking about curriculum design, or of use in working with specific cases. However, it is also possible to develop the model further, by using it as a resource for designing curricula.

To apply the concepts discussed above to a course intended for a new set of students, the group of students would need to be located on the axes by ordinal data type so as to identify groups who were at risk or in need of particular forms of support. It is, however, likely that any tutor would be either designing a course without knowing their specific students, or else would have interviewed a large group of students but not yet know enough about any individuals to undertake any kind of credible mapping.

However, it would be possible to supplement student interviews or enrollment data with specific questions relating to the axes, allowing at least some of the students to be plotted along them. This could be useful for clustering groups of students in terms of the support and guidance that they might need. Applying a simple three-category framework to each axis (as illustrated in the Figures 1-3) results in nine possible options for each figure allowing a simplistic but nevertheless useful identification of students who might be at risk of failure on the course.

Practically, this could be achieved simply by using a blank template based on the figures, with prospective students clustered within the appropriate cells on grid. The advantage of the simple grid over a finer-grained classification is that high student numbers can be catered for, and broad categories applied, rather than very specific categories drawn from a set of rich case analyses. Richer analysis would be valuable, but is unlikely to be practical; the interviews required for detailed cases required many hours each for analysis (Holley & Oliver, 2009), and this amount of time is unlikely to be available to tutors at the point of admission. The grid therefore represents a compromise between simplicity and explanatory power.

The 'blank template' model has been used in this way to inform the curriculum redesign of the certificate level (first year) 'Business Skills' module, a core module for all first year business school students. On the basis of the redesign the course was offered as series of interactive workshops, instead of the traditional large lecture/seminar format. This enabled tutors to get to know students better and support them one-to-one and in small groups. To achieve this curriculum change, the student cohort was analysed in terms of intake and strategies for engaging students, influenced by the work of Johnson et al (2009); and the report of the Higher Education Academy/Equality Challenge Unit Final Report (2008). In this initial pilot it was not possible to 'match' the individual student preferences through all three stages of the model. However, early evaluation of the cohort indicates significant improvements to the curriculum in terms of student engagement and progression.

Expectations about Education Mapped against Home Spaces

Students were encouraged to attend class regularly and all materials were available via Blackboard, the University Virtual Learning Environment to enable home study where class attendance was not possible. Tracking tools were used to track cohort, and in some cases, individual, performance and attendance.

Expectations about Education Mapped against their Control of Technology

Currently the University does not offer pre-session IT support, but this is planned for the future as part of the welcome week schedule. In class with the students, tutors demonstrated Blackboard materials each week to familiarise students with the format and tools available. Students were encouraged to reflect upon their experiences of developing a critical awareness of their discipline by creating and sharing a wiki site with their tutor.

Expectations about Control of Technology Mapped against Home Space

Students were offered information about the business school peer support system, where students could access trained peer mentors and discuss any difficulties with their studies. The module assessment changed from a presentation, formal report and a piece of reflective writing to a presentation and three 'patchwork' texts, integrated through a short reflective commentary. To start to encourage students to 'blend' their own learning for themselves, one 'patchwork' briefing was around selecting and evaluating a Web 2.0 technology that they would find of use in their studies in future.

CONCLUSION

This chapter has identified problems faced by Universities as they seek to adapt their teaching for mass education and an increasingly diverse student body. It has illustrated the possibilities of using an inclusive, student centered model to enhance curriculum design and accessibility, particularly in relation to the development of an electronic curriculum. What our model gives is a way of anticipating problems for particular students, and supporting and supplementing the existing curriculum. It does this by identifying 'at risk' students, drawn from case study research on issues of concern to students. There are three significant findings from the research that can inform practice. The first is the opportunities offered by the model to the institution; namely exploring the balance between the student centred approach and the necessities of economies of scale required by an organisation funded by the public purse. The second is the importance of what can be understood as cultural capital (Bourdieu 1983) focusing on students who may find an adapted and scaffolded curriculum enables them to settle quickly into what may be a 'new, cold climate' (Sinfield *et al* 2004). The final outcome is a framework of what is possible in terms of developing inclusive practice; this can be explored and further developed across both public and private organisations seeking to embrace diversity.

FUTURE TRENDS

The new Horizons Report (2010:3) is the source of information for future trends in technologies, and these are the current key themes featured:

- People expect to be able to work, learn, and study whenever and wherever they want to.
- The technologies we use are increasingly cloud-based, and our notions of IT support are decentralized.
- The work of students is increasingly seen as collaborative by nature, and there is more crosscampus collaboration between departments.
- New scholarly forms of authoring, publishing, and researching continue to emerge but appropriate metrics for evaluating them increasingly and far too often lag behind.
- Digital media literacy continues its rise in importance as a key skill in every discipline and profession.

- Institutions increasingly focus more narrowly on key goals, as a result of shrinking budgets in the present economic climate.

In the short term (up to 12 months), mobile computing and open software are key technologies to watch, and on the far-term horizon, set at four to five years away for widespread adoption, but clearly already in use in some quarters, are gesture-based computing and visual data analysis (New Horizons:6/7).

For Business students, the merging of where and when they chose to study, as well as the technologies (much more sharing) personalised (individual to them) and cloud based will be important for study, but also, crucially for access to graduate level employability as they complete their studies.

REFERENCES

Association of Business Schools. (2009/10). *Pillars of the sustainable economy: sustainability and current market.* Retrieved on May 11, 2010, from www.the-abs.org.uk/files/ABS%20Pillars_Annual%20Report%2009.pdf

Barnett, R., & Coate, K. (2005). *Engaging the curriculum in Higher Education. Berkshire.* McGraw-Hill.

Bec, T. A. *(2008).* Harnessing technology: Next generation learning. *Retrieved onJanuary18, 2010, from*http://publications.becta.org.uk/display.cfm?resID=37348&page=1835

Bennett, R. (2002). Lecturers' attitudes to new teaching methods. *International Journal of Management Education, 2*(1), 42–58. doi:10.3794/ijme.21.36

Blunkett, D. (2000). *Secretary of State for education speech given at the University of Greenwich.*

Bourdieu, P. (1983). *The forms of capital* (pp. 183–198).

Collis, B., & Van der Wende, M. (2002). *Models of technology and change in higher education: An international comparative survey on the current & future use of ICT in higher education.* University of Twente, Enschede: Centre for Higher Education Policy Studies. Accessed on October 26, 2005, from http://www.utwente.nl/cheps/The%20CHEPS%20Team/Staff/Chairs/wende.doc/

Conole, G. (2010). Personalisation through technology enhanced learning. In O'Donoghue, J. (Ed.), *Technology-supported environments for personalized learning method and case studies* (pp. 1–14). Hershey, PA: IGI Global.

Conole, G., White, S., & Oliver, M. (2007). The impact of e-learning on organisational roles and structures. In Conole, G., & Oliver, M. (Eds.), *Contemporary perspectives in e-learning research: Themes, methods and impact on practice. Part of the Open and Distance Learning Series.* London: RoutledgeFalmer.

Cook, J., Holley, D., Smith, C., Bradley, C., & Haynes, R. (2006). *A blended m-learning design for supporting teamwork in formal and informal setting*s. Mlearn Conference, Banff, Alberta, Canada.

Cook, J., & Light, A. (2006). New patterns of power and participation? Designing ICT for informal and community learning. *E-learning, 4*(1), 51–61. doi:10.2304/elea.2006.3.1.51

Daniel, J. S. (1996). *Mega-universities and knowledge media: Technology strategies for higher education.* London: Kogan Page.

Ehrmann, S. C. (1999). Access and/or quality? Defining choices in the third revolution. *The Educom Review, 34*(5). Retrieved on October 25, 2005, from http://www.educause.edu/ir/library/html/erm/erm99/erm9956.html.

Elton, L. (1988). Student motivation and achievement. *Studies in Higher Education*, *13*(2), 215–221. doi:10.1080/03075078812331377886

Elton, L. (2001). Research and teaching: Conditions for a positive link. *Teaching in Higher Education*, *6*(1), 43–56. doi:10.1080/13562510020029590

Entwhistle, N. (2000). Approaches to studying and levels of understanding: The influences of teaching and assessment. In J. Forest (Ed.), *Higher education: Handbook of theory and research.* (p. 15). New York: Agathon Press.

Evans, M. (2004). *Killing thinking: The death of the universities*. London: Continuum.

Forsyth, A. & Furlong, A. (2000). *Socioeconomic disadvantage and access to higher education.*

Garrison, D.R. & Anderson, T. (2003). *E-Learning in the 21st Century*. London: Routledge Falmer.

Haggis, T. (2003). Constructing images of ourselves? A critical investigation into approaches to learning research in higher education. *British Educational Research Journal*, *29*(1), 89–104. doi:10.1080/0141192032000057401

Haselgrove, S. (1994). *The student experience*. Buckingham, UK: SRHE & Open University Press.

Higher Education Academy. (2010). *Subject centre spotlight: ESCalate*. Retrieved on May 10, 2010, from http://www.heacademy.ac.uk/resources/detail/ourwork/ipp/Issue3_Escalate

Higher Education Academy/ Equality Challenge Unit. (2008). Ethnicity, gender and degree attainment project. Retrieved from http://www.heacademy.ac.uk/assets/York/documents/ourwork/research/Ethnicity_Gender_Degree_Attainment_report_Jan08.pdf

Hodge, M. (2002). *Minister for lifelong learning*. Paper presented at Education Conference.

Holley, D. (2008). Using biographic narrative to explore students' experiences of online learning. In Frame, P., & Burnett, J. (Eds.), *Using auto/biography in learning and teaching* (pp. 53–58).

Holley, D., Greaves, L., Bradley, C., & Cook, J. (2010). You can take out of it what you want: How learning objects within blended learning designs encourage personalised learning. In O'Donoghue, J. (Ed.), *Technology-supported environments for personalized learning method and case studies* (pp. 285–304). Hershey, PA: IGI Global.

Holley, D., & Oliver, M. (2000). Pedagogy and new power relationships. *International Journal of Management Education*, *1*(1), 11–21. doi:10.3794/ijme.11.b

Holley, D., & Oliver, M. *(2009). A private revolution–how technology is enabling students to take their work home.* Journal of Enhancing Learning in the Social Sciences, 1*(*3). *Retrieved onApril30, 2009, from*http://www.eliss.org.uk/

Holley, D., & Oliver, M. (2010). Student engagement and blended learning: portraits of risk. [Elsevier Publishing.]. *Computers & Education*, *54*, 693–700. doi:10.1016/j.compedu.2009.08.035

Holley, D., Sinfield, S., & Burns, T. (2006). It was horrid, very very horrid: A student perspective on coming to an inner-city university in the UK. *Social Responsibility Journal*, *2*(1), 36–42. doi:10.1108/eb045820

Hopkins, D. (2006) *Choice and voice in personalised learning*. (pp 1-22). retrieved January 18, 2010, from http://titania.sourceoecd.org/vl=2327409/cl=25/nw=1/rpsv/cw/vhosts/oecdthemes/99980029/v2006n3/contp1-1.htm

Jisc Infonet. (2010). *Effective use of VLEs*. Retrieved on January 27, 2010, from http://www.jiscinfonet.ac.uk/InfoKits/effective-use-of-VLEs

Johnson, L., Levine, A., Smith, R., & Stone, S. (2010). *The 2010 horizon report*. Austin: The New Media Consortium.

Johnson, L., Pokorny, H., Djebbour, Y., & Curran, J. (2009). *Look who's here: Engagement and attendance initiatives*. HEA Teaching and Research Development Grant. Retrieved from http://www.heacademy.ac.uk/business/projects/detail/trdg/2008-09/0809trdg_Linda_Johnson_LMU

Lefebvre, H. (1991). *The production of space (trans D.Nicholson-Smith)*. Oxford: Blackwell.

Loader, B. (Ed.). (1998). *Cyberspace divide: Equality, agency, and policy in the information society*. London, New York: Routledge. doi:10.4324/9780203169537

National Committee of Inquiry into Higher Education. (2010). *Chair information*. Retrieved on May 10, 2010, from https://bei.leeds.ac.uk/Partners/NCIHE/

Noble, D. (2001). *Digital diploma mills: The automation of higher education*. New York: Monthly Review Press.

OECD. (2000). *Schooling for tomorrow: Learning to bridge the digital divide. Centre for Educational Research and Innovation. The National centre on Adult Literacy*. NCAL.

Oliver, M., & Conole, G. (2000). Assessing and enhancing quality using toolkits. *Quality Assurance in Education*, *8*(1), 32–37. doi:10.1108/09684880010312677

Peters, M., & May, T. (2004). Universities, regional policy and the knowledge economy. *Policy Futures in Education*, *2*(4), 263–277. doi:10.2304/pfie.2004.2.2.4

Rammell, W. (2009). *MP, Minister of State (lifelong learning, further and higher education)*. Retrieved on July 23, 2009, from http://www.theabs.org.uk/files/PILLARS-AR

Rudestam, K., & Schoenholtz, J. (Eds.). (2002). *Handbook of online learning*. London: Sage Publications.

San Diego, J., Laurillard, D., Boyle, T., Bradley, C., Ljubojevic, D., & Neumann, T. (2008). Towards a user-oriented analytical approach to learning design. *ALT-J: Research in Learning Technology*, *16*(1), 15–29.

Sharples, M. (2005). *Learning as conversation: Transforming education in the mobile age*. Paper presented at the 'Seeing, Understanding, Learning in the Mobile Age' Conference, Budapest, Hungary.

Shaw, J., Brain, K., Bridger, K., Foreman, J., & Reid, I. (2007). *Embedding widening participation and promoting student diversity*. Retrieved on May 10, 2010, from http://www.heacademy.ac.uk/assets/York/documents/resources/publications/exchange/embedding_wp_business_case_approach_july07.pdf

Sinfield, S., Burns, T., & Holley, D. (2004). Outsiders looking in or insiders looking out? Widening participation in a post-1992 university. In Satterthwaite, J., Atkinson, A., & Martin, W. (Eds.), *The disciplining of education: New languages of power and resistance* (pp. 137–152). Trentham Books.

Smith, H., & Oliver, M. (2002). University teachers' attitudes to the impact of innovations in ICT on their practice." In C. Rust (Ed), *Proceedings of the 9th International Improving Student Learning Symposium*, (pp. 237-246). Oxford: Oxford Centre for Staff and Learning Development.

Trout, P. (2000). Flunking the test: The dismal record of student evaluations. *Academe: Bulletin of the AAUP, 86*(4). Retrieved on November 2, 20015 from http://www.aaup.org/AAUP/pubsres/academe/2000/JA/Feat/trou.htm

Twigg, C. A. (2001). *Innovations in online learning: Moving beyond no significant difference. National Center for Academic Transformation (NCAT)*. Monograph.

Universities and Colleges Information Systems Association Survey. (2008). Retrieved from http://www.ucisa.ac.uk/publications/~/media/groups/tlig/vle_surveys/TEL%20survey%202008%20pdf.ashx

White, S., & Davis, H. (2002). Harnessing Information Technology for learning. In Ketteridge, S. M., & Fry, H. (Eds.), *The effective academic: A handbook for enhanced acdemic practice*. London: Kogan Page.

Wisdom, J. (2006). Developing higher education teachers to teach creatively. In Jackson, N., Oliver, M., Shaw, M., & Wisdom, M. (Eds.), *Developing creativity in higher education* (pp. 183–196). London: Routledge.

ADDITIONAL READING

Abercrombie, M. (1993). The Human nature of learning (ed). Buckingham, SRHE and Open University Press.

Akerlind, G. S., & Trevitt, C. (1999). Enhancing Self-Directed Learning through Educational Technology: When students resist the change. *Innovations in Education and Teaching International*, *36*, 2. doi:10.1080/1355800990360202

Anderson, T., & Elloumi, F. (Eds.). (2004). Theory and Practice of Online Learning available electronically from http://cde.athabascau.ca/online_book/. Athabasca, University of Athabasca.

Beetham, H., McGill, L., & Littlejohn, A. (2009) Thriving in the 21st century: Learning Literacies for the Digital Age (LLiDA project). Available online: http://www.caledonianacademy.net/spaces/LLiDA/uploads/Main/LLiDAreportJune09.pdf

Bird, J. (1996). *Black Students and Higher Education Rhetorics and Realities*. Buckingham: SRHE and Open University Press.

Bradwell, P (2009) Edgeless University: why higher education must embrace technology Demos available from: Czerniewicz, L. and Brown, C. (2005) Access to ICTs for teaching and learning: From single artefact to inter-related resources. International Journal of Education and Development using ICT, 1 (2).

Creanor, L., Trinder, K., Gowan, D., & Howells, C. (2006) The Learner Experience of e-Learning Final Project Report. Available online: www.jisc.ac.uk/uploaded_documents/LEX%2520Final%2520Report_August06.pdf

Dreyfus, H. (2001). *On the Internet*. London, New York: Routledge.

Elton, L. (1988). Student motivation and achievement. *Studies in Higher Education*, *13*(2). doi:10.1080/03075078812331377886

Freire, P. (2000). *Pedagogy of the Oppressed*. New York: Continuum International Publishing Group.

Garrison, D. R. and T. Anderson (2003). E-Learning in the 21st Century. London, RoutledgeFalmer.

Hoggett, M. (1995). *The decline of professional autonomy. The politics of Quality in the Public Sector. I. L. Kirkpatrick, M.* London: Sage.

Jilik, B. (2006). Spaces, Places and Future Learning. Spaces, Places and Future Learning, Rich Mix London, Futurelab conference report available electronically from http://www/futurelab.org.uk/events/listing/spaces_places_and_future_learning/ accessed 29/07/08.

Johnson, L., Levine, A., Smith, R., & Stone, S. (2010). *The 2010 Horizon Report*. Austin, Texas: The New Media Consortium.

Knight, P., & Yorke, M. (2003). *Assessment, Learning and Employability*. Open University Press, McGraw Hill Education Maidenhead.

Laurillard, D. (2002). *Rethinking University Teaching A conversational framework for the use of learning technologies* (2nd ed.). Routledge-Falmer. doi:10.4324/9780203304846

Leathwood, C. and P. O'Connell (2003). "' It's a struggle': the construction of the 'new student' in Higher Education." Journal of Education Policy pp 597-615 in, 18(6)(2).

Lockard, J., & Pegrum, M. (2006). *Brave New Classrooms*. Peter Lang.

Noble, D. (2001). *Digital Diploma Mills The automation of Higher Education*. New York: Monthly Review Press.

O'Donoghue, J. (Ed.). (2010). *Technology-Supported Environments for Personalized Learning Method and Case Studies*. Hershey, USA: Information Science Reference (an imprint of IGI Global).

Pachler, N., Bachmair, B., Cook, J. and Kress, G. 2009 *Mobile Learning:*

Quinn, J., & Thomas, L. (2005). *From life crisis to lifelong earning:rethinking working class 'dropout' from higher education*. Joseph Rowntree Foundation.

Rowland, S. (2000). *The Enquiring University Teacher*. Buckingham: SRHE and Open University Press.

Sharples, M., Arnedillo Sánchez I., et al. (2007). Mobile Learning: Small Devices, Big Issues. Technology Enhanced Learning: Principles and Products.

Structure – Agency – Practices. New York: Springer.

Thorpe, M. (2001) From independent learning to collaborative learning: new communities of practice in open, distance and distributed learning. In M.Lea and K. Nicoll (eds), Distributed learning: social and culturalapproaches to practice, 131-151. Routledge, London, UK.

Wengraf, T. (2001). *Qualitative Research Interviewing*. London: Sage Publications.

Wonga, D., Shephard, K. L., & Phillips, P. (2008, March). The cathedral and the bazaar of e-repository development: encouraging community engagement with moving pictures and sound ALT-J. *Research in Learning Technology*, *16*(1), 31–40.

Yorke, M. (2004). Retention, persistence and success in on-campus higher education, and their enhancement in open and distance learning. *Open Learning*, *19*(1). doi:10.1080/0268051042000177827

KEY TERMS AND DEFINITIONS

Curriculum: A course of study within a disciplinary area, often understood in terms of plans, designs or specifications.

Diversity: Increasing variation in the characteristics of learners in a way that celebrates rather than negatively positions the variations in perspective and background that this may lead to; see also, widening participation.

Massification: The move from an elite system of education to one in which participation is more common or even universal; in UK Higher Education, this was interpreted in terms of a 50% participation rate amongst 18-30 year olds.

Personalisation: Within education, the idea that the educational process and resources should be adapted to individual learners, often relying on common-sense decisions about how learners work or simplistic models of learner 'types' based on psychometric tests.

VLE: Virtual Learning Environment, a web-based, database-driven and password protected environment in which educational resources and process are made available to groups of learners.

Web 2.0: A phrase used to describe internet-based applications that emphasise social connectivity, user-generated content and sharing.

Widening Participation: Increasing the representation of groups within the student population, typically understood in terms of categories such as age, gender, ethnicity, social class and/or prior qualifications.

Chapter 2
ICT Use in Universities: An Educational Model for Digital Natives

Ana Jimenez-Zarco
Universitat Oberta de Catalunya, Spain

María Pilar Martínez-Ruiz
University of Castilla-La Mancha, Spain

Virginia Barba-Sánchez
University of Castilla-La Mancha, Spain

Alicia Izquierdo-Yusta
Universidad de Burgos, Spain

ABSTRACT

This chapter analyzes the adequacy of two university education models according to their Information Communications Technology (ICT) usage: One model is based on the use of ICT as a cornerstone and a cohesive element of the teaching-learning process, whereas the other considers ICT a mere teaching support tool. The preliminary results obtained from a review of case studies reveal that ICT use is not only a means for improving the teaching–learning process but also an emerging prerequisite, demanded by digital natives in their choice of a university. Furthermore, it represents an additional valued outcome of the teaching–learning process.

1. INTRODUCTION

In the current socio-economic and technological environment, with persistent questions about the competitiveness of productive systems, higher education attains special relevance as a strategic and adaptive option. European universities, including those in Spain, are therefore undergoing a process of deep renovation, especially under the influence of the European Space for Higher Education (ESHE).

DOI: 10.4018/978-1-60960-501-8.ch002

In the case of students born in the digital age, and who require continuous training throughout their lives, universities must change their educational systems, aiming for a new model in which: (a) the student becomes the main focus of teaching and learning, (2) the teacher acts as a guide for students, facilitating the use of learning resources and tools, and (3) there is an emphasis on competencies and skills that students should acquire, and which are demanded by the market.

Salinas (2002) shows how the use of ICT by teachers and students facilitates teaching and learning. Nevertheless, when faced by similar situations, each educational centre may show different levels of ICT use. In such a case, as Roberts et al. (2000) point out, this affects the educational model proposed and consequently the role of the student and teacher, as well as the technological applications used.

Given the above situation, this paper examines the suitability of two models of university education, which are defined in terms of their use of ICT. First, the so-called *standard model* of education is a university in which various ICTs (such as computers, etc.) are merely additional tools (like a blackboard, books, etc) which back up the teaching-learning process. The second model, known as *advanced or radical*, proposes an intensive use of ICT, given that the internet is the environment in which it develops the teaching-learning process. In this sense, the ultimate goal of this chapter is to show the great value of the intensive use of ICT in the learning processes, especially when students and teachers show a high degree of knowledge and active participation towards ICT. Preliminary results suggest that the use of ICT provides a means of improving the teaching-learning process. Furthermore, the "digital natives" who attend universities often acquire the use of ICT before they attend college, because they value the use of ICT as additional skills obtained through the process of teaching and learning.

2. BACKGROUND

Never has education been so much in the spotlight, nor has it previously captured quite so much attention or spurred so much controversy. A growing conviction asserts that education must serve as an engine of social development, increasing levels of welfare and prompting new kinds of economic growth.

Even if everyone agrees about the importance of high-quality education for all citizens though, opinions about how to deliver it vary widely. Education and training take place in multiple venues and persist in a lifelong process, marked by a vast number of different types of stakeholders. The traditional roles of the family and the school are becoming more blurred, even as a host of other stakeholders, whether intentionally or not, enter the arena and play educational roles in the contexts of leisure activities, the media, the corporate world, and other public spheres.

For these more and less conventional stakeholders, information and communication technologies (ICT) offer networks with diverse purposes and sizes, creating, disseminating, reshaping, and exchanging information freely. The phenomenon of the Internet has brought a new dimension to higher education—a dimension that remains difficult to grasp in its entirety (Cebrian De la Serna and Vain, 2008).

In fact, the Internet has changed the role played by universities in the educational process. Thus, today, the traditional educational centres have gone from being the hub of the educational system to becoming one of the nodes of the network of networks where the "user-student" moves in a more flexible way through cyberspace. Moreover, the elimination of time and space barriers, which the Internet provides, brings about the emergence of new educational organizations which are constituted as partnerships or networks of institutions and educational systems, characterized by modularity and interconnection. However, this situation requires educational institutions to be more flexible in their procedures and administrative structures in order to accommodate alternative training systems which are more responsive to the needs which this new situation presents (Bautista, 2006).

For Higher Education Institutions to respond effectively to the challenge of the intensive use of

ICT they should review their existing educational models and promote innovative experiences in teaching and learning. In this sense, the change in the instructor's teaching methods and the use of different types of educational resources (video and audio materials, chat, wiki, etc.) will be essential to this process of educational innovation (Salinas, 2002).

On the other hand, these projects require a change at the institutional level, in such a way that these are considered as institutional and global strategies, in which the entire organization should be involved (Bautista, 2006). In our universities we can find many experiences of 'virtual learning', 'virtual classrooms', etc. These projects can be interesting but on many occasions they constitute projects which are isolated from the general dynamics of the institution itself and respond to particular initiatives and in many cases there can be a difficulty in their generalization, since they are not taken on by the organization as global projects (Pavon and Casanova, 2007). This particular type of initiative reveals the rigidity of university structures when it comes to integrating into their daily operations the use of ICT in the teaching-learning processes. Therefore, in order to develop an efficient and effective ICT integration, the active participation and motivation of teachers is required, but a strong institutional commitment is also needed. In this sense, the university culture promotes the production, research (often to the detriment of teaching) and the processes of innovation in this field. However, these are processes that make our universities more innovative institutions and better adapted to the current, new realities.

3. A NEW EDUCATIONAL ENVIRONMENT

Duderstand, (1997) notes that in today's knowledge society there are four important issues which directly affect the role that universities play in society:

1. The importance of knowledge as a key factor in determining security, prosperity and quality of life.
2. The global nature of our society.
3. How technology, computers, telecommunications and multimedia make the exchange of information easy and rapid.
4. The degree to which informal collaboration (especially through networks) between individuals and institutions are replacing more formal social structures such as corporations, universities and governments.

However, the response of universities to these challenges cannot be standard. Each university must respond by considering its own specific features, the situation in which it is, the student profile at which it is directed and its available resources. In this context, it is worth noting that Spanish universities are undergoing a profound process of transformation, following the introduction of ESHE. Moreover, and as far as the student profile is concerned, a new generation of students is entering the educational system. Known as "digital natives" or the "net generation", these young people are said to have been immersed in technology all their lives, imbuing them with sophisticated technical skills and learning preferences for which traditional education is unprepared (Bennet *et al.*, 2008). Finally, as regards resources, it should be noted that Spanish universities have begun to incorporate ICT, to a greater or lesser extent intensively, in its normal activity and, consequently, the role of the teacher has changed, becoming a students' guide.

3.1. The European Space of Higer Education (ESHE)

It can be said that ESHE is one of the most important educational reform movements in the history of European institutions of higher education. Created in 1999, following the signing of the Bologna Agreement, ESHE seeks to increase the comparability and compatibility of the European systems of higher education. The building of ESHE is based on agreements and commitments on objectives to be achieved among the signatory countries.

The ESHE is seen as an open space where there are no obstacles to the mobility of students, graduates, teachers and administrative staff, and is based on the recognition of degrees and other higher education qualifications, transparency and European cooperation in order to guarantee quality. Thus, this process involves:

- The desire to develop a common system of degree programs which will ensure recognition and compatibility with the studies carried out in other countries.
- A thorough rethinking of the structure, content, degrees and materials which make up the university system.
- Revision of testing systems and methods.
- Transparent public accountability of systems and degree programs.
- Recognition of both academic and professional qualifications in order to respond better to the reality of a unified, European labour market.
- Enhanced competitiveness in training and research.
- Increased mobility for students and teachers.
- Improved competitiveness of university graduates.
- Quality accreditation.

3.2. The Digital Native: Characteristics and Needs as Students

The international academic community faces the need to determine who twenty-first century students are and what they need. The nature of students has changed, but not as a voluntary act or because of a specific new need. Rather, the development of ICT, and young people's role in expanding their acceptance, has shifted the core values of education, from the pursuit of access to the knowledge that universities possess to a recognition of the collaborative, co-productive dissemination of knowledge (Bennet *et al.*, 2008).

Modern, connected, mobile students are accustomed to experimentation, online socialisation, and immediacy. Referred to as "digital natives" (Prensky, 2001) or "Net generation" (Tapscott, 1998). This generation born roughly between 1980 and 1994 has grown up in digital world surrounded by and using computers, videogames, digital music players, video cams, cell phones, and all the other toys and tools of the digital age.

The Centre for Educational Research and Innovation (CERI) of the Organisation for Economic Cooperation and Development (OECD) provides a profile of the digital native, which assumes that they approach learning differently. (Bennet *et al.,* 2008; Collins, 2008) As students, they prefer images to text, would rather have a multimedia than a static presentation, need to change their activities frequently and cannot sit through long classes, and need to master the learning process through multi-tasking instead of in a linear fashion, such that they can jump across and among subjects (Prensky, 2001), The programme's empirical research also debunks some myths about these digital natives, such as the implication that their technological dependence has altered their cognitive development. Bates (2000), or Moran and Myringer (1999), among others, stress that the use of ICTs has a positive impact on their learning; generally speaking, students who use ICTs include the highest achievers, regardless of their socio-economics status. Other experts (Bruns *et al.*, 2007) instead designate new technologies handy educational tools that can be counterproductive for learning. The effects depend on the use of ICT; the overwhelming use of PowerPoint presentations, for example, means that students might just copy the words they see on the screen instead of synthesizing and assessing knowledge from various sources.

If digital natives are the modern-day students, which educational model is most appropriate? New technologies continue to spread rapidly; people interact through chats and text messaging, posting and responding to videos on YouTube, and commenting on their status in social networks such as Facebook and Tuenti. Institutions and educators therefore need to study how and to what extent these and similar tools might be included in the day-to-day activities of primary and secondary schools; at universities, such applications are already being demanded by restless digital natives.

If ICT integration into university education is necessary, the next question involves the appropriate extent. Should ICT complement traditional education model? Should it lead to a new totally virtual model? Demand for virtual offerings from universities is increasing, and many digital natives regard the availability of virtual courses as a criterion of choosing where to pursue their higher education. These young adults use new technologies intensively; they do not consider whether using the Internet is good or bad, because it is so much a part of their lives.

Nowadays, intensive ICT use by educational institutions is quite common; in addition to the uses that the current generation of students makes of ICT, other entities value it as a means to develop the learning process. For example, the need to continue learning throughout a person's career and life has introduced a new type of nontraditional student to the university--older to 30 years of age and often forced to reinitiate studies they might have abandoned years ago. For these students, technological ignorance can add further difficulties to the learning process. However, the possibility of developing specialized coursework that acknowledges their (often complex) personal and job circumstances may encourage these students to choose virtual classes.

But virtual learning must go beyond posting courses online; the idea is to change the learning process. For example, one new method replaces the professor with the student as the core of the educational process, as the European Higher Education Area requires. The reform of universities initiated through ESHE aims to provide students with certain competencies that will support their professional life. Therefore, universities must alter their curriculum and teaching material. The structure of curricula according to the competencies required in the professional world attempts to provide students with not only theoretical knowledge, as in traditional academics, but also the abilities, skills, procedures, and techniques that they will need to acquire and keep a job.

3.3. ICT Use at the University

Society demands that new university graduates should have a set of skills which involves the effective use of information and communication technologies. It is believed that professionals should be able to use ICT, to use knowledge, to know how to find, process, interpret and rule that the information is not relevant.

This means that universities have begun to use ICT in different areas such as:

- Innovative teaching with the incorporation of ICT in teaching and learning.
- Modification of university structures, as a beginning to considering these technologies in the organizational and management bodies of the universities.
- Innovative experiences of all kinds, relating to the operation of the communicative possibilities of ICT in university teaching.

The level of ICT use in universities varies greatly, as well as the way that students and teachers approach it. Roberts et al (2000) identify four different models, depending on the degree of incorporation of ICT in the normal activity of the institution:

1. **The initiation model:** this is characterized by providing notes and other material in any web format. This model does not provide opportunities for either interaction or dialogue, nor extra resources.
2. **The standard model:** which attempts to use the advantages provided by technology to allow a certain degree of communication and interaction between students and teachers, as well as providing other electronic resources (in the form of links, electronic copies of all printed course materials, class slides, notes from classes, workshops, tasks and solutions, guidelines for activities, an electronic discussion list for the course, etc.).
3. **The evolved model:** which improves the standard model by introducing additional elements to increase the teaching (student tracking, electronic management, etc.), and learning environments (CD-ROM distribution, pre-recorded classes on audio, animation, etc.).
4. **The advanced or radical model:** In this model, the concept of classes is ignored and only makes reference to a "virtual space", where students, teachers and educational resources interact. Here, students are divided into groups and learn from interacting with each other and using a vast amount of existing web resources with the teacher acting as a guide, adviser, facilitator, etc.

In the next section, we analyse two of these models: the standard and the radical ones which are implanted in the Spanish universities of Burgos, Castilla-La Mancha and Oberta de Catalunya.

4. THE UNIVERSITY OF BURGOS AND UNIVERSITY OF CASTILLA-LA MANCHA: ICT AS A COMPLEMENT TO TRADITIONAL UNIVERSITY EDUCATION

4.1. The Universities of Burgos and Castilla-La Mancha

The Universities of Castilla-La Mancha and Burgos are two obvious examples of public university policy developed in Spain during the last 30 years. Both universities are young, but despite having a very similar profile between them, there are also some differences relatives to ther degree of ICT use.

Located in the town of Burgos, The University of Burgos was born in 1994. It currently has almost 10,000 students spread among its 30 bachelor's degrees, 20 doctoral programs, and several other officers and Masters graduate courses. The University has five centres: the Science Faculties, the Faculty of Economics and Business Studies, the Faculty of Humanities and Education, the Faculty of Law, and the Higher Polytechnic School, as well as at three associated schools, the School of Nursing, the School of Labour Relations and the School of Tourism.

On the other hand, The University of Castilla-La Mancha was created 1982 through the union of different university centres into a single institution. Currently, it has almosts 27,500 students, and offers offer with a total of up to 49 graduate degrees and more than 60 postgraduate, Masters and PhD programmes. One special trait of the University of Castilla-La Mancha is a multicampus structure (the university is located at four Campuses, everyone located in one different city). All of them have a great autonomy of management, and their interrelations and communications are made in real time by videoconferencing and a series of internal voice and data communications

4.2. ICT as a Complement to Traditional University Education

The traditional university education model relies on the physical presence of students and teachers in classrooms and on and the role of the teacher as source and transmitter of knowledge through the development of master class. As note Horno and Ibarretxe (2008), in this model, ICT plays two secundary roles: (1) to facilitate the class and (2) to guarantee students access to certain resources or information related to the subject (e.g., Intranet, Web sites, library catalogs).

This usage is common to many universities, including the University of Burgos and the University of Castilla-La Mancha. Computers, video, and audio equipment facilitate class presentations, such that instead of using a chalkboard, instructors show dynamic PowerPoint presentations. The classes seem more enjoyable, though students may interact less with the class contents. In addition, instructors make frequent use of the Internet in class to find theoretical or practical points or even develop their seminars. For example, to advertising, videos on www.youtube.com give the teacher easy access to various advertising campaigns that market the same brand in different countries.

In addition, the instructor might use ICT to ensure certain resources or information are available to students. Many universities maintain intranets (i.e., a "virtual campus") that enable students to access class notes or complementary material. Furthermore, it is increasingly common that students can submit their homework and check their grades through such systems.

In some cases, ICT use has intensified so strongly that some content is available only online. Some theoretical content gets communicated in the class, but practical work and team projects take place online. At the University of Burgos and University of Castilla-La Mancha, Moodle software supports Web sites on which students and teachers interact through features such as dedicated chatrooms and wikis.

Moreover, ICT use facilitates management processes, performed by both students and teachers. For example, on the UNIVERSITAS XXI Web site, faculty from the University of Burgos can manage their teaching and research activities. This software application allows the university research staff manages all information relating to their curriculum, and also fills and submits all necessary documentation to apply for research projects. On the other hand, with the PICASSO data warehouse system, the academic authorities at the University of Burgos can simplify student registration, plan study designs, and so forth.

A low use of ICT limits and ignores many potential applications -as we can see in the next section- often because teachers and/or students are not aware of them. A growing number of universities are establishing mixed models, offering many credits through online courses. However, this model still places too much importance on physical class meetings, which is not compatible with one of the main ESHE objectives, namely, the continuous learning process. This discrepancy is especially detrimental to those students who cannot attend classes (e.g., because they work, live elsewhere).

5. A NEW MODEL BASED ON COLLABORATIVE LEARNING ENVIRONMENTS: THE UOC CASE

As Garrison and Anderson (2005), and Gan and Zhu (2007) show, an effective practice for online learning requires a pedagogy that supports the development of learning communities. This perspective establishes two requirements: (1) the need to adapt technology to improving interactive processes online and thereby improving the quality of the learning processe and (2) the use of tools which enable communication, collaboration and the production of knowledge, which are fundamental to improving the learning process.

The evolution of virtual learning environments (the e-learning platforms) has made it possible for students and teachers to communicate and work collaboratively. Although important, this evolution remains insufficient because virtual learning environments must keep evolving into systems that can enable the management of higher education, given the magnitude of changes that involve new educational systems (Ehlers and Carneiro, 2008).

5.1. Implications for Universities

A model of advanced ICT use involves major changes in the university. These changes affect the role of the teacher and the student, as well as the teaching methods followed, together with the actual management of the institution.

Changes in the Teacher's Role

The changes that occur in the institution, among which we can highlight the impact of ICT, inevitably leads to bringing about a change in the role of the teachers and the role they play in the teaching-learning system. It is considered that the role of the teacher changes the transmission of knowledge to the students and thus become a mediator in the construction of knowledge as such (Pérez, 2002). This is a new vision of education in which students are at the centre and focuses attention on the role which the teacher plays.

The teachers act first as a person and then as a content expert. They promote personal growth in students and facilitate learning rather than the transmission of information. Thus, the university and the teacher cease to be sources of all knowledge and the teachers should move on to act as a guide for students, encouraging them to use resources and tools which they need to explore and develop new knowledge and skills, and beginning to act as a manager of the plethora of learning resources and emphasizing their role as counsellor (Salinas, 2002).

Changes in the Student's Role

There is no doubt that students who are in contact with ICT benefit in several ways and move into this new vision of user training. Obviously, this system requires educational activities related to the use, selection and organization of information, so that the student is trained as a mature citizen of the information society. The support and guidance they receive in each situation, and the differing availability of technology, are crucial elements in the exploitation of ICT for training in this new situation. Therefore, flexibility is needed to move from being a visiting scholar to being at a distance, and vice versa, with the flexibility to use a variety of materials independently (Perez, 2002).

Changes in Methodological Methods

Mason (1998) notes that the use of ICT in education opens up new opportunities for a better education. Thus, the students are supported by online environments whose strategies are common practices in classroom instruction, but are now simply adapted and rediscovered in its virtual format.

On the one hand, decisions related to designing education are defined by aspects related to the type of institution (distance, type of certification offered, etc.), to the design of teaching itself (teaching methods, strategies, etc), to aspects of the student-user of the system, and to learning (motivation, specific training needs, resources and equipment available). On the other hand, decisions regarding the technology itself involve the selection of the communication system via the computer or communication tools that are best suited to supporting the teaching-learning process. These decisions are based on the knowledge of technological advances regarding the possibilities of technology for content distribution, access to information, interaction between teachers and students, course management, the ability to control the users during course development, etc.

Changes in the Institution

This new situation means that universities have to be involved in innovative ICT-supported teaching. Influenced, among other factors, by the huge impact of the information age, universities loose the monopoly of production and transmission of knowledge, which simultaneously generates opportunities for markets and competencies new to the sector, and by a widespread requirement that students receive the skills necessary for lifelong learning (Salinas, 2002).

5.2. The Open University of Catalunya Case

In the last two decades, demand for university and post-graduate education by people between the ages of 30 and 50 years keeps increasing, yet traditional university models, cannot keep pace with this demand. In particular, many of these students work during the day, may live far from the school, and have family responsibilities. To face this growing need, the Government of Catalonia (Spain) created the Open University of Catalunya (UOC) in 1995. As an institution created in the midst of the digital society, with the purpose of supporting lifelong education, UOC is endowed with a flexible structure and supports its educational model with the intensive use of ITC. On its virtual campus, students and professors interact to teach, learn, and research. Officially, the UOC states the following objectives:

- The impulse, the improvement, and the innovation in teaching and learning in life through non-present means.
- The formation as well as the development of scientific research in all research areas in which the university is active, especially in the study of the knowledge society.
- The knowledge diffusion, the technology transfer, and innovation in the fields of formation and culture associated with the intensive use of ICT.

The success of UOC came from its pioneering use of a new university concept, with the student as the central element (Duart and Sangrá, 2000). In this sense, both its virtual presence and its centralisation of the student align UOC with the use of the Internet, including its strong communicative and information access potential. The organizational and learning model was designed specifically to work with this formative approach. The UOC model also recognises the central position of faculty, which play a key role as facilitators of the learning process. Even if the instructor does not deliver the material in person, he or she develops the evaluation activities and creates the virtual classrooms.

Furthermore, due to its objective of meeting personal and professional students' needs, according to the evolution of technology, the requirements of the business environment, and the demands of society in general, the UOC education model rests on five main principles: dynamism and flexibility, to adapt to students' needs and realities; cooperation among different agents in the teaching–learning process, to enable knowledge generation; interaction among all agents involved, including students, teachers, staff, and so on, to generate community; and personalized student attention.

A key element in the learning process is the virtual campus, where all the elements combine and interrelate. The life of the university community takes place in the virtual campus, a community made up of students, teachers, researchers, collaborators, and administrators. The student has access to all virtual classrooms through the virtual campus.

The classrooms themselves are learning spaces that host information, tools, spaces, and functionalities to enable formative action and organize the set of agents and elements. Each class corresponds to a student group, though the same subject can have more than one classroom, because all classrooms are managed by a teaching plan.

Cooperative work is the primary strategy to ensure the continuous generation of knowledge, learning, and motivation among long-distance students. The group concept makes it possible to engage in discussion and debate, share opinions and viewpoints about the subject, and achieve an element of social relationships. However, the volume of communication interchanges was initially so high that UOC created different spaces for various purposes, such as notice boards, discussion groups, and lists of classmates to facilitate students' communication with advisors, classmates, and teachers. To address some of the limitations of existing communication channels, while also considering the specific nature of its education model, the UOC also chose to introduce support systems that favour more efficient management and facilitate collaborative knowledge building. These actions consist of two main types:

a. New tools such as universal "notice boards" and "forums" in all the classrooms pertaining to the same subject, as well as chatboards, blogs, LANGblog and wikis.
b. Aspects related to forum usability that, together with improved navigability, provides more efficient information management, especially by incorporating search, filter, or perpetual updates via RSS feeds.

The virtual classroom also offers access to other spaces, such as teaching materials (usually in pdf and txt formats), statistical software (e.g., Minitab), and/or library materials (databases of journals and books). Moreover, students can consult continuous evaluation practices, the final grade earned in each subject, and their academic records and registration.

Beyond virtual classrooms, the UOC depends on technology overall, so maintaining a solid and modern technological infrastructure has always been a strategic goal. This effort includes hardware and software assets that support the various services UOC offers to conduct teaching, research, and internal administration. The hardware includes air conditioning and current stabilizers in data centres, sensors, cameras, large-scale computers that act as applications servers, network elements (e.g., routers, firewalls), personal computers, printers, telephones, and so on. The software ranges from operating systems (i.e., computer programs designed to perform basic functions essential for the administration of the equipment) to systems software (i.e., general purpose applications needed for the specific computer applications of the services to work, such as databases, application servers, and office automation tools).

Two applications are particularly essential to the current operations of UOC: GAT and TREN. The telematic academic management (GAT) system supports the management of academic and administrative processes related to students, teaching staff (including counsellors and tutors), and administrative users. It works in various languages and multiple currencies. The essential aim is to manage, in an agile way, the administrative tasks carried out by UOC and ensure they can be conducted virtually through the Web.

In turn, GAT consists of several modules:

- **Plans and Academic Regulations.** It defines any plan of study for any cycle (e.g., which subjects form the curricula, the order they should be taken, the requirements necessary to matriculate).
- **Academic Record.** Both students and counsellors can use this resource to locate academic information easily. It also includes all administrative requests made by the student.
- **Secretary's Office and Administrative Transactions.** This resource manages and resolves any administrative request made by a student (e.g., requests for certificates).

The TREN system provides basic services in support of other applications, with the following functions:

- Control of access to applications, whether market-based or in-house and regardless of the platform on which they may have been developed, according to a single user name per person.
- Scalability through load distribution among servers, and re-direction of processes to the application server from the front-end servers.
- Deliveries by mail from management applications to the virtual campus.
- Centralised maintenance management, including necessary information for obtaining access to a specific application. These maintenance efforts vary, ranging from a URL address to look up to a combination of users/password to connect to a database. Its centralisation allows for the relocation of applications, depending on need and without too much inconvenience.

6. RESULTS

According to Webometrics Ranking of World Universities, which ranks 15,000 institutions from all over the world, UOC is one of the top, attaining a rank of 672. In addition, UOC data pertaining to the 2007–2008 academic year offer further evidence that its education and management model has been a success.

In particular, UOC has more than 54,000 students enrolled in its 1,167 courses (including masters', postgraduate, and extension programmes), across 1,167 virtual classrooms, with 200 faculty, and 2,346 teaching collaborators. During the previous year, 4,400 students graduated, and more than 1,475,000 articles, studies, and teaching materials were downloaded from the virtual campus. Moreover, UOC has created an Internet-based community of more than 100,000 students, graduates, professors, researchers, and collaborators in more than 50 countries; signed more than 600 agreements with other universities, institutions, and companies around the world; and developed a network of more than 125 firms that constitute UOC associates. Finally, from a technological perspective, UOC sends approximately 42,000 final grades to mobile phones each semester; 10,100 readers subscribe to its digital publications; its YouTube channel receives more than 65,000 channel views monthly; and even its Flickr photo galleries prompts approximately 30,000 each month.

Yet to measure these results against the achievements of other, more traditional universities, it is important to acknowledge various dimensions of student life: academic, professional, and personal. In this respect, the results achieved by the Open University of Catalonia are high and satisfactory, whether its comparison to other Spanish universities or based on student evaluations.

The results in this section come from an analysis of various information sources, including the 2003 report of the Agency for Quality Assurance in the Catalan University System Agency (AQU) and the 2004 report of the Spanish Conference of University Rectors (CRUE). These publications provide academic, productive, and financial information about Spanish public universities. Therefore, the results reflect the scores on a set of indicators that reflect the relative academic position of UOC compared with other universities. Furthermore, student satisfaction questionnaires from UOC, including a 2005 report about the degrees and diplomas awarded and a report with a sample of 313 graduates from the field of economic studies (12.3% of total graduates in this knowledge field) suggest means to evaluate different indicators regarding the impact of learning in a virtual university, on both professional and personal levels.

The global evaluation of UOC is satisfactory. The time students need to complete their studies is similar in both traditional universities and this virtual university. Other academic indicators, such as the abandonment rate of students, number of graduates, and academic achievement, also are similar across both forms.

The professional achievement ratings for UOC also are satisfactory. A high percentage of graduates indicate that their professional situation improved substantially after they began their studies at the virtual university. Moreover, the occupation rates of UOC graduates is quite high; 60% of students either found their first job or improved their labour position after graduating.

Finally, UOC graduates indicate the positive impact of the virtual university on their personal lives. They claim high degrees of personal achievements (e.g., 67% of graduates believe they improved personally because of their studies). Furthermore, regarding key competencies related to their particular knowledge area, they say they significantly developed their intellectual ability and other general competencies.

7. CONCLUSION AND FUTURE RESEARCH DIRECTIONS

An education model based on a intensive use of ICT provides students more prominence, participation, and autonomy than do traditional models. This provision, perhaps in contrast with a priori expectations, is highly valued by students, especially by those whose profile does not fit the "professional student" model. For someone with job and family responsibilities, a traditional education model offers few real benefits.

Independent of the education model employed, computers clearly offer an extremely powerful tool, but the teachers must know how to use them, and the teaching model has to be adapted to students' needs. Instead of continuing to invest resources in making ICT available to students, universities should focus on addressing other needs if they want to promote ICT usage in support of educational purposes. Beyond the indiscriminate increase in the number of computers, the question should be what types of technology would be most helpful in the classroom and how it should be applied.

Because the Internet is rare in school curricula, digital natives entering universities find themselves in another world, not their familiar world of the Web. These developments imply that in the future, higher education must find a way to blend teacher-led classes and virtual learning by becoming a venue of interaction.

The high benefits universities receive from the use of ICT makes their study constitutes a major focus of interest of academics and professionals. Among the likely topics are under study to analyze the way that traditional universities can combine an advanced use of ICT with a classical education model, and based on the presentiality. As regards the virtual universities, the main challenge lies in adapting new tools online, based on web 2.0 in order to make the teaching-learning system more effective and efficient.

REFERENCES

AQU. (2003). *Procés d'avaluació de la Qualitat del Sistema Universitari Catalan.* Barcelona: Agencia per a la Qualitat del Sistema Universitari en Catalunya.

Bates, A. (2000). *Managing technological change. Strategies for college and university leaders*. San Francisco: Jossey-Bass Inc.

Bautista, G. (2006). *Didáctica universitaria en entornos virtuales de enseñanza-aprendizaje*. Madrid: Narcea Ediciones.

Bennet, S., Maton, K., & Kervin, L. (2008). The digital natives debate: A critical review of the evidence. *British Journal of Educational Technology*, *39*(5), 775–786. doi:10.1111/j.1467-8535.2007.00793.x

Bruns, A., Cobcroft, R., Smith, J., & Towers, S. (2007). *Mobile learning technologies and the move towards 'user-led education'*. Paper presented at the Mobile Media 2007 conference, Sydney.

Cebrian de la Serna, M., & Vain, P. D. (2008). Una mirada acerca del rol docente universitario, desde las prácticas de la enseñanza en entornos no presenciales. *Pixel-Bit: Revista de Medios y Educación*, *32*, 117–129.

Collins, L. (2008). *Generation Y in cyberspace: How do digital natives manage social stigma on MySpace*. Paper presented at the Australian Sociology Association conference.

CRUE. (2004). *Información académica, productiva y financiera de las Universidades Españolas*. Madrid: Confederación de Rectores de Universidades Españolas.

Duart, J. M., & Sangrà, A. (2000). *Aprender en la virtualidad*. Barcelona: Gedisa.

Ehlers, U. D., & Carneiro, P. (2008). Entornos de aprendizaje personales. *Revista E-learning Papers*, *9*, 2.

Gan, Y., & Zhu, Z. (2007). A learning framework for knowledge building and collective wisdom advancement in virtual learning communities. *Journal of Educational Technology & Society*, *10*(1), 206–226.

Garrison, D. R., & Anderson, T. (2005). *El e-learning en el siglo XXI: Investigación y práctica*. Barcelona: Octaedro.

Horno, M. C., & Ibarretxe, I. (2008). *Metodologías educativas y uso de las TIC en el área de lingüística*. Paper presented at the II Jornadas de Innovación docente, Tecnologías de la información y de la comunicación e investigación educativa. Universidad de Zaragoza. Retrieved June 3, 2010, from http://ice.unizar.es/uzinnova/jornadas/pdf/27.pdf

Moran, L., & Myringer, B. (1999). Flexible learning and university change. In Harry, K. (Ed.), *Higher education through open and distance learning* (pp. 57–72). Londres: Routledge.

Pavon, F., & Casanova, J. (2007). Experiencias docentes apoyadas en aulas virtuales. *Revista Iberoamericana de Educación a Distancia*, *10*(2), 149–163.

Pérez, A. (2002). Nuevas estrategias didácticas en entornos digitales para la enseñanza superior. In: J. Salinas & A. Batista (Eds.), *Didáctica y tecnología educativa para una universidad en un mundo digital*. Universidad de Panamá: Imprenta Universitaria.

Prègent, R. (1990). *La preparation d'un cours*. Montreal: Editions de l'Ecole Polytechnique de Montréal.

Roberts, T., Romm, C. & Jones, D. (2000). *Current practice in web-based delivery of IT courses*. APWeb2000.

Salinas, J. (2002). Modelos flexibles como respuesta de las universidades a la sociedad de la información. *Acción Pedagógica*, *11*(1), 4–13.

Tapscott, D. (1998). *Growing up digital: The rise of the Net generation*. New York: McGraw-Hill.

Webometrics. (2009). *Ranking of world universities*. Retrieved from http://www.webometrics.info/top6000.asp

ADDITIONAL READING

Area, M. (2001). Las redes de ordenadores en la enseñanza universitaria: hacia los campus virtuales. In A. García-Valcarcel, *Didáctica universitaria*. Madrid: La Muralla.

Duart, J. M., & Kiselyova, E. (2003). The Open University of Catalonia: A Local-Global University Model. In Varis, T., Usumi, T., & Klemm, W. (Eds.), *Global Peace Through The Global University System*. Tampere: GUS.

Duart, J. M., & Lupiáñez, F. (2005). E-strategias en la introducción y uso de las TIC en la universidad. *Revista de Universidad y Sociedad del Conocimiento (RUSC)*, 2. Retrieved June 7, 2010, from www.uoc.edu/rusc/dt/esp/duart0405.pdf

Fischer, F. (2002). Fostering collaborative knowledge construction with visualization tools. *Learning and Instruction*, *12*, 213–232. doi:10.1016/S0959-4752(01)00005-6

Goffman, E. (1981). *Forms of talk*. London: Basil Blackwell.

Kirschner, P. (2003). Validating a representational notation for collaborative in virtual communities. *Journal of Computer Assisted Learning*, *19*(4), 472–487.

Kobbe, L., et al. (2005). *Framework for the specification of collaboration scripts*. Kaleidoscope Network of Excellence deliverable. Retrieved May 25, 2010, from http://www.iwm-kmrc.de/cossicle/

Lipponen, L. (2004). From collaborative technology to collaborative use of technology: designing learning oriented infrastructures. *Educational Media International*, 111–116. doi:10.1080/09523980410001678566

Majó, J., & Marquès, P. (2001). *La revolución en la era Internet*. Barcelona: CissPraxis.

Prenksy, M. (2001). Digital natives, digital immigrants. *Horizon*, *9*(5), 1–6. doi:10.1108/10748120110424816

Salmon, G. (2000). *E-moderating. The key to teaching and learning online*. London: Kogan Page.

Sancho, T., & Màsia, R. (2007). A virtual mathematics learning environment for engineering students. interactive. *Educational Multimedia*, *14*, 1–18.

Xin, C., & Glass, G. (2005). Enhancing Online Discussion through Web Annotation. In G. Richards (Ed.). *Proceedings of World Conference on E-Learning in Corporate, Government, Healthcare, and Higher Education* (pp. 3212-3217). Chesapeake, VA: AACE.

KEY TERMS AND DEFINITIONS

Asyncrony: The lack of coincidence in space and time of agents that in the teaching–learning process.

Chat: A written, instant communication via the Internet between two or more people.

LangBlog: Audioblog for practising orals skills, with posts and replies on audio and video

Learning Community: Group of people who share common values and beliefs, are actively engaged in learning together from one another, and use common tools in the same environment.

MOODLE: A course management system (CMS), also known as a learning management system (LMS) or virtual learning environment (VLE). It is a free Web application that educators can use to create effective online learning sites.

Picasso: (Burgos University Data Warehouse) A tool to extract information about the university by aggregating important data from the information systems of the university and external sources.

Rich Site Summary (RSS): A format for delivering regularly changing Web content. Many news-related sites, Weblogs, and online publishers syndicate their content as an RSS feed to whomever wants it. It allows people to stay easily informed by automatically retrieving the latest content from chosen sites, such that people no longer need to visit each site individually.

Teaching Plan: A document written by faculty members to provide students with the organization of the teaching–learning process from the first day of class. This basic guide makes it possible to access learning objectives, content, learning materials, activities, evaluation system, bibliography, and so on. It provides a reference document for the student to foster his or her learning process.

Wiki: A Web site in which the Web pages can be edited by multiple volunteers. Users can create, modify, or erase the same shared text, and the system keeps track of the changes made over time, which facilitates the recovery of the previous content.

Chapter 3
E-Technologies in Higher Education Provision:
Planning, Implementation and Management

Petros Ieromonachou
University of Greenwich Business School, UK

Nola Stair
University of Greenwich Business School, UK

ABSTRACT

Technology is always advancing; furthermore, the economy and demographics, both at global and local levels dictate changes that businesses need to consider for future planning. Like all businesses, most academic institutions need to follow trends and offer new and up-to-date learning initiatives. This chapter discusses the assessment and evaluation of Web-based postgraduate and undergraduate programmes, at the Business School of a London-based university. The study reviews relevant literature, investigates student and staff views on the use of technology in learning, and reflects on the planning and management of two Supported Open Learning (SOL) programmes. It also offers suggestions for future programmes and courses designed for use with e-learning technologies.

INTRODUCTION

As the world around us changes so does the teaching profession. The use of electronic gateways to access up-to-date information and news, to effectively communicate and even socialise in virtual networks is increasing with every new product and application that reaches the market. Staff abilities and student learning preferences are never constant. Academic institutions and educators, in order to survive in the academic business, need to continue to facilitate effective learning. While most – if not all – learning institutions are focused on face-to-face instruction, e-learning initiatives in academia have been increasing. Course curriculum often incorporates online media resources that students can access through university or personal equipment, allowing learning to take

DOI: 10.4018/978-1-60960-501-8.ch003

place at a distance or generally in different ways than traditional teaching environments.

This London based university has entered the digital era of knowledge dissemination, for a number of years now, but mainly targeted existing programmes of study. Design for learning not only becomes a way to re-engage teachers in the debate about effective learning, but a means to reconsider the boundaries between knowledge and power, the relationship between the teacher and the learner, who controls the technologies and the curricula and learning (Walker & Ryan, 2008). Part of the wider development towards the facilitation of e-learning through new technologies was the creation of Supported Open Learning (SOL) versions for both postgraduate and undergraduate programmes, at the Business School. There cannot be a universal or standardised approach to e-learning provision because of the diversity of subjects being taught, as well as the multitude of available platforms and equipment and the local contextual factors. There was, therefore, a need to plan, develop, implement and manage these programmes with the School's specific context in mind. Such processes often require a balance of pedagogic, technical and administrative concerns. Even when both technology and pedagogy are well supported, there may be structural problems that inhibit engagement with e-learning (Oliver & Dempster, 2003). The teaching technique is also another issue that relies manifestly on the individual educators' backgrounds, training and style as well as the ability to engage with the changing technologies. Wider factors affecting the decisions for development of e-learning programmes and the capability of the institution to provide support are also considered.

The use of technology aimed to provide alternative solutions to the delivery of learning material, standardisation of assignment submission processes and effective management of large courses, in attending mode settings. The direction for the use of technology was recently focused towards the development of Distance Learning (DL) programmes and use of Virtual Learning Environments (VLEs), and Course Management Systems (CMS). The development of e-learning does, however, have significant current and future resource implications in terms of managing VLEs and their integration with other Information Technology (IT) environments. The Business School, as identified in its strategic plan, working with administrative central departments and offices identified such implications as:

1. Delivery of a growing number of blended and online courses that are offered within the local geographic boundaries and across borders in varying academic calendars;
2. Implementation of new plagiarism procedures with anti-plagiarism software and supported integration into the Student Information System;
3. Coordination of VLE electronic coursework submission process and coursework marking with seamless integration into the Student Information System;
4. Improvement of business processes that support the School's work with students:
   a. Timetabling of full-time, part-time, flexible and short courses (with appropriate levels of technology functionality).
   b. Batch course creation processing that facilitates automatic course rollovers with instructional content available immediately – instead of any manual copying.
   c. Batch load of students into VLE courses that contain unique identifiers for different tutorials.
5. Incorporating electronic portfolios into Personal and Professional Development courses (Year 1, 2 and 3) and new employability courses.

This willingness to work collaboratively to identify the e-business challenges that would constrain the planning, implementation and management of using new technologies reflects a transition from 'top-down' to 'bottom-up' communication. Otherwise, the expected technology system processes will be fragmented and used in isolation to each other, rather than effectively integrated to enhance the delivery and management of teaching and learning. Technical problems are often not the hardest problems to solve. Rather, the related managerial and organizational processes are much harder to understand, coordinate, and integrate within the demands of developing e-learning environments.

Part of the wider development towards the facilitation of e-learning through new technologies was the creation of Supported Open Learning (SOL) versions for both postgraduate and undergraduate programmes, at this London based university. The term Supported Open Learning (SOL) is used to describe the constructivist, activity-based model, for the design and development of e-learning materials. The programmes offered were the MBA SOL and the third and final year of a BA (Hons) degree in business studies for direct entrants to the programme. Both programmes were offered as part-time modes of study. In the case of the MBA this was due to the programme requirements and expectations. For the undergraduate programme this was necessitated due to government regulations not allowing the delivery of a full-time degree using distance learning in only one academic year. The programmes did not run virtually in their entirety. Face-to-face study days were scheduled which required compulsory attendance. This was put in place to provide students with training for the use of systems and the university procedures as well as briefing them on effectively accessing the teaching material, understanding new study methods and using online discussion boards for communicating between peers and educators.

This chapter aims to present staff and student experience involved in the SOL programmes, focusing on the use of the e-learning technologies; problems encountered and solutions provided. The authors played an important role in terms of designing and running the SOL programmes. Information on many of the aspects concerning the tutors' and students' experiences were observed first hand. A focus group was used to document other tutors' views and a standardised evaluation questionnaire was used to capture students' opinions. An evaluation model is provided as a visual representation of links between resources, activities, outputs, audiences, and short/intermediate/long-term outcomes associated with the SOL programmes and the results of the research are grouped under these headings.

Background The topic of information and communication technology (ICT) and its consequent use in education is very large and has been investigated in depth over the last decades. Higher Education (HE) institutions realise the need to create more flexible learning environments for students through the use of ICT. This section concentrates on the use of ICT in achieving learning outcomes, an important step to understand before moving onto the planning and implementation of the e-learning programmes.

ICT Strategies

Information and Communication Technology consists of various audiovisual and technological tools. As noted in the introduction of this report, students have different learning styles. Some students learn best by putting their thoughts into words while others through modelling them, some learn acoustically and some visually. The use of ICT in this respect can alert the students on the process of learning and encourage them to learn independently. ICT is also an important aid to learning and efficient course delivery, e.g. VLEs which provide tools to manage access to online resources, track student progress, deliver assessments collaborate asynchronously or synchronously, and provide feedback.

The Open University, primarily a provider of distance learning, was a pioneer in developing

Virtual Learning Environments (VLEs). Nowadays, learners and teachers – not just in distance education – are experiencing these methods. As a result, "*more teachers and learners face questions about how to orientate themselves to VLEs*" (Petit and Mason, 2007, p.148). As reported by Unwin (2007) this can have "*multiple consequences: teachers do not have time to learn from experience and have difficulty keeping up to date and they may feel reluctant to invest time and effort with the latest technologies*". The opinions expressed above are not isolated but at the same time there needs to be recognition on the significant impact that ICT has had on pedagogy in the last two decades in particular. Modern VLEs have many different features that allow students to choose from a large range of study related options. Apart from accessing course handbooks and PowerPoint lecture notes, it is possible to submit assignments, view deadlines and grades, receive specific course related messages and access online tutorial groups where they can write their activity messages and read other students'.

Electronic discussion groups are a good way of making use of ICT in full time education. WebCT for example may promote learning when used as an online forum for students to exchange comments and good practice tips in order to facilitate better learning. When students doing teamwork were asked to post questions on-line, answers could be provided for all to see, reaching more than one student at a time and minimising repetition that often occurs in classroom settings. Part of the repetition often concerns some vital processes in student learning: Assessing and giving feedback.

Assessing

According to Wakeford (2007, p44) "*effective assessment will reflect ... some combination of an individual's abilities, achievement, skills and potential ... [and] ... will permit predictions about future behaviour*". Previously, Butcher *et al* (2006) suggested that the type of assessment influences styles of learning.

Assessment on the SOL programmes took place by on-line group-work, and individual tasks students needed to complete and reflect upon. Templates were provided in the course handbook as an example of what the students could include but only act as guidelines. The use of the templates, not only an e-learning attribute, relates to *validity*, in order to "*assess what it is that you really want to measure*" (Race, 2007, p29). Tasks were designed by taking into consideration that the SOL students come from a variety of backgrounds and were designed to assess their knowledge of topics taught in the curriculum. Both for the final year in Business Studies as well as the MBA, students came not only from a number of different countries but also had been educated differently, were of varying age groups and in many cases had to balance studying with work and family life. In this way, the assessment tasks are *fair* because "students have equivalence of opportunities to succeed even if their experiences are not identical" (Race, 2007, p.29).

Giving Feedback

Feedback is necessary for students to adjust their learning strategies and for educators to accommodate the students learning needs. Feedback should encourage self-reflection, raise self-awareness and help students plan for future learning (Vickery & Lake, 2005). The importance of feedback as an 'accelerant' to learning has been repeatedly stressed in the scholarship of learning and teaching in higher education over the last two decades (Hounsell, 2004).

Due to the large number of students undertaking the SOL programmes, not particularly in the UK but in several international collaboration partners, meant that there are also a large number of tutors involved in delivering the course. Moderation meetings are required in order to disseminate the same information to all tutors but not easy to manage. Much of this work was taking place online – through short teleconference meetings, or through electronic communication which created the need for devising and explaining regulations to diverse audiences. In many cases different tutors had varying views of issues regarding either topics delivered, the way of assessment and of course marking. In "traditional" practice, when attending marking moderation meetings, usually led by the course leader, examples of assignments are provided along with the specific marking criteria, to act as points of reference for each specific section expected to be addressed by the students. The use of the marking criteria as a guideline resolved issues of *reliability* within large groups of assessors required to mark over two hundred reports.

MAIN FOCUS OF THE CHAPTER

E-Learning

The Business School, at this specific university, has an active commitment to e-learning and recognizes that one of the strategic responses to a number of the challenges being faced by the HE sector will be to find a solution to the issue of value in e-learning provision. New technologies, which introduce new methods of teaching, and new expected learning outcomes, need new methods of assessment and evaluation to be devised and tested. Since 2004, the Business School has been increasing use of e-learning, in both volume and sophistication, to the point where all courses (estimated to approximately 600) have a Virtual Learning Environment (VLE) presence that enhances the student experience and student learning. During each term, approximately 10,000 pages of student coursework are submitted electronically using a student information system generated barcode for tracking purposes. The next stage of enhancement has been developing capacity in e-learning pedagogy through the School's 'Teaching and Learning Champions' (TLC) and Learning Technologists to develop a research model of continuous professional development that will assist staff in engaging with innovative concepts for design and delivery of blended learning courses. The School's development of e-learning practice and pedagogy has well positioned itself to develop part-time, flexible and short course variants of its programmes and support a growing number of international partners.

Higher Education (HE) institutions are increasingly understanding of the need to create more flexible learning environments for students. Collis & Moonen (2001) created a checklist in identifying the flexibility dimensions of learning. They point that flexibility relates to:

1. Time: a. tempo / pace of studying and b. time of study during the day / semester
2. Content: learning material of the course
3. Entry Requirements
4. Instructional Approach and Resources: (a) social organisation of learning (face to face, group, individual), (b) learning resources; modality, origina (instructor, learners, library, www) and, (c) instructional organisation of learning (assignments, monitoring)
5. Logistics: (a) time and place, (b) method, technology and, (c location
6. Delivery Channel

However, when introducing flexibility dimensions for learning within HE institutions, the underlying conceptual processes of pedagogy requirements and ICT capabilities within VLEs must be carefully considered. For example, in the case of social organisation of learning in groups, the term "group" may have different connotations in the VLE's technical functionality. Similarly, while the creation of "group discussion boards" can effectively allow instructors to interact with easily identifiable groups, the "group submission tool" may not provide instructors with the capability to filter individual student submissions allocated to different groups. All student submissions are compiled in the VLE's 'assignment drop-box' and while submissions can be sorted alphabetically by last name, there is no way to effectively select particular groups for specific tutors who are sharing one VLE course shell. The selection must be done individually which is a time-intensive task when managing several groups of 25-30 students in a 200-student cohort.

If the course has large student numbers or is spread over several geographic regions, separate course shells can be replicated for the same course that is being offered simultaneously within the VLE to split large cohorts into smaller groups for easier management by multiple. Instead of updating each shell whenever new course materials are added or changed, or when announcements are posted, a complete understanding of the "course cross listing" process within the VLE must be discussed to consider if all instructional needs are met and if additional complexities are likely to occur. The process known as "cross listing" can link multiple courses together for tutors who are teaching multiple sections/tutorials of the same course. This creates the ideal situation where all students in both sections/tutorials can access the same course material and allows the tutor to view and communicate with all sections/tutorials from one place. In terms of time management, another feature of the VLE is the ability to create public and private group discussion boards. Instead of having separate course shells for similar courses, group discussion boards can be created so that each group can have separate areas of interaction and still access the same instructional materials, announcements, etc. However, if all group discussion boards are publically accessible to all students, online etiquette issues such as improper posting in the wrong group discussion board are challenges for consideration and resolution. If private discussion boards are used, the advantages and disadvantages for tutor monitoring must be considered when determining the most time effective and efficient method that can be employed.

Looking into one aspect of the flexibility dimensions of learning requires significant consultation and discussion between academic, administrative, and information technology departments. All of the various dimensions represented within highly flexible learning environments require a sustainable and comprehensive framework to support the planning, implementation and management for effective delivery. The authors developed a simple LOGIC model to assess and evaluate the planning and implementation issues relating to the design, development, and delivery of e-learning at their University, in response to issues relating to the staff and student experience in the SOL programmes. In Table 1, the components of the LOGIC model are presented, adapted from a study from the W.K. Kellogg Foundation (2004).

The LOGIC model is a useful tool that provides a visual representation of the connection between resources, activities, outputs, audiences, and short, intermediate, and long-term outcomes related to a specific situation. Once a situation has been described in terms of the LOGIC model, critical gaps, missing links and/or measures of perfor-

Table 1.

Inputs	Activities	Outputs	Outcomes (Intended Results)		
			Short-Term	Long-Term	Impact
What resources (human, financial, organizational, technological) are needed to create the program?	What are the processes, relationships, and actions needed that will bring about the intended changes or results?	What are the direct results of the program's implementation?	What are the specific changes in program participants' behaviour, knowledge, skills, status and level of functioning?	What are the specific changes in program participants' behaviour, knowledge, skills, status and level of functioning?	What are the intended or unintended improved conditions, increased capacity, and/or changes as a result of activities?

mance can be identified. The LOGIC model is revisited in the Solutions section of this chapter.

"Most of the value in a logic model is in the process of creating, validating, and modifying the model....The clarity of thinking that occurs from building the model is critical to the overall success of the program" (W.K. Kellogg Foundation, 1998, p. 43).

To help support the development of the LOGIC model, existing research about the institutional evaluation of managing the integrated functions of academic, administration, and information technology within e-Learning environments was constructed. A study by Sharpe et al (2006) conducted by the Oxford Centre for Staff and Learning Development at Oxford Brookes University in the United Kingdom, revealed that institutional evaluation of e-learning is quite problematic because measures of success are often not clearly defined and relevant data may not be available. Quite often available tracking tools within the Virtual Learning Environment (VLE) are used to provide evaluative data. However, these data typically only provide snapshots of online activity, such as the quantifiable aggregation of log-ins, document access, and discussion board postings. These results revealed the extremely limited capabilities of VLE quantitative tools to identify what aspects of learning were focused on and discerned. Armellini and Jones (2008) explored embedding good e-learning practice across another UK University through the ADELIE project. Through this project they researched the change that occurs as a result of the normalisation of e-learning practice at different levels: institutional, teaching practice and learner experiences; from the limited evidence available they still reported positive impacts on quality of assessment as well as learner attitudes and experience.

However, the limited insight into institutional evaluation suggests that new paradigms could promote richer studies into the characteristics of staff and student experiences. Petrova and Sinclair (2008) conducted a study that focused on improving course design based on stakeholder perspective. Students' perceived e-learning value and their behaviour when using technology were investigated through usage patterns of the online platform alongside a satisfaction survey. They proposed a five-stage framework for analysing the collected data. In a different chapter of the same book, Halperin (2008) focused more on the institutional and socio-organisational factors influencing the use of technology in higher education. The argument here was that the role of the *"prevailing mode of hybrid (or blended) e-learning [...] becomes all the more important, as the technological environment is meant to complement, rather than replace, the existing and long established learning system"* (ibid, p97).

Zawacki-Richter, Bäcker, and Vogt (2009, p.4) conducted an international Delphi study to develop a classification system consisting of three broad meta-levels for research areas in distance education. Their taxonomy includes the following levels:

- **Macro-level:** distance education systems and theories;
- **Meso-level:** management, organization and technology; and,
- **Micro-level:** teaching and learning in distance education.

Each of these meta-levels has been broken down in sub-categories after analysis of over 600 articles from five peer-reviewed distance education journals. Zawacki-Richter et al. (ibid) found that the majority of research focused on the micro level, and the published research on the meso-level was disappointing. They concluded that "*sound research and empirical investigation on the effectiveness of managerial interventions for education innovation*" is needed (ibid, p.28).

The Programmes

Two programmes were offered online – the MBA and Year 3 of the BA (Hons) degree in Business Studies, from both of which the end-of-year evaluation was used to capture student feedback. A focus group involving the relevant members of staff was used to capture comments that helped construct this case study. While the full-time MBA was delivered face-to-face over a one-year period, the part-time mode is offered entirely online over a period of three years. The SOL version of the final year in the BA (Hons) degree in Business Studies consisted of four core courses and was possible to complete in one year. In addition, one of these four courses was delivered online to students studying in South Africa and this cohort was selected for analysing student feedback as it represented a stronger distance learning experience.

Inputs regarding the resources needed to develop web-based programmes have been widely documented and published (Beetham & Sharpe 2007; JISC, 2004, Power, 2008). However, there are many challenges that HE administrators face in terms of institutionalising change management and resulting resistance (at all levels), as well as addressing the policy issues (Niemiec & Otte, 2009). These challenges can be identified as the activities needed that will support and link the inputs to the critical areas of outputs and outcomes. Once identified, these challenges can be transformed into strategic benefits in the areas of efficiency, enhancements and transformation within one's institutional context (HEFCE, 2009).

The design and development of effective and engaging learning materials for e-learning has a direct impact on student learning (Beetham & Sharpe, 2007). Therefore, it is necessary to address an adequate infrastructure that is reliable and accessible for all learners. After identifying the online pedagogical considerations in terms of audience and technical challenges, specific instructional strategies can be designed to fulfil course goals and objectives. Learning materials are transformed into interactive online learning activities: Student to Student, Student to Tutor, and Student to Content. These interactive learning experiences for students can range from weekly online engagement on asynchronous discussion boards concerning course topics to weekly submission of assignments. In addition, it is necessary to provide students studying online with a reasonable level of instructional assistance consisting of highly structured written instructional guidelines and navigation tips for a variety of online activities. This will decrease the amount of questions and frustration experience by students.

As a result, students can focus on understanding their newly acquired knowledge and problem solving skills through a variety of business simulations and case studies. Ongoing feedback from tutors concerning weekly activities and assignments is provided individually during virtual office hours, discussed online with small groups during asynchronous discussions, and/or posted as summaries in online discussion forums. Therefore, students have many opportunities to contact the lecturer/tutor personally, discuss issues (either asynchronously or synchronously) with their peers and/or lecturers/tutors, and to learn from their experiences from reflection in either online private journals or public blogs – which are available within the courseware application.

The organisation required to make delivery of the SOL programmes possible involved three distinct categories: the management at programme level, course level and pedagogical aspect – grouped under teaching and learning activities in Figure 1. There was a need for communication and cooperation between all these levels. The allocation and preparation of appropriate levels of materials and interaction for a flexible learning environment contributes to the overall success of student learning and the course development. The Business School has developed the following:

- A policy to develop blended and web-based (i.e., e-learning); and,

Figure 1. Total replies from the student evaluation

- An action working group, working as a sub-committee of the School's Teaching & Learning Committee to investigate and implement best practice in terms of e-learning within the School.

The Teaching and Learning Committee tasks include:

- Monitoring teaching and learning activity and support, including assessment;
- Encouraging research into good practice;
- Encouraging sharing knowledge and information across Departments and other groupings;
- Assessing developments across the HE sector for local implications; and,
- Establishing procedures to enable these objectives.

The Focus Group

At the end of delivery of the programmes, the evaluation study involved both of staff and students but reflecting on the experience of the two groups was done in different ways. Staff were invited to attend a structured focus group entitled: *'Designing, managing and delivering the MBA and BABS SOL programmes'*. The focus group was conducted in order to obtain information from the main staff involved in the planning, management and delivery of the two SOL programmes. Staff were assured that the evaluation was intended to be formative and qualitative. The goal was purely academic and not related to a management activity or formal review process, but the intention was to use the information to help improve our practice. The information collected was by design descriptive rather than numeric. The focus group was after words and impressions.

Consent and anonymity: All information collected was treated with confidentiality as to who provided it. It was assured that if responses were to be used in any written way, it would not be disclosed who actually participated in this focus group nor would there be any attributions for quotes. This was done in order to encourage staff to speak freely. This meeting was voice-recorded as it would not be practical to rely on memory or notes for the information that may be provided. Staff we asked if anyone minds tape-recording the meeting and nobody opposed it. Following are some of the answers recorded during the focus group with staff involved in the planning and delivery of the SOL programmes. The classification at the end of each answer corresponds to the LOGIC model.

Q1. What was the reason in developing the SOL programmes?

A. "*The reason was to meet the needs of students who cannot engage full-time. [...] And because most students need to work at the same time particularly mature students who work full-time want to engage in full time education find it difficult and sometime impossible to engage in a traditional manner. [...] This has offered a much easier model of delivering learning in a more flexible mode that needs their needs.*" (OUTPUT)

Q2. How is technology involved in the SOL mode?

A. *"The reason for using technology in the SOL is firstly in terms of access to the materials so that the students in terms of their flexible access can access the material in their own time, anytime, anywhere. We do not define when they engage with the material but we do define when they engage with learning. The other advantage is that it allows collaboration between students who are geographically dispersed and also time dispersed and they cannot access the material at the same time. [...] there is the ability in which in fact we use the technology to supplement the traditional f2f delivery where in the SOL mode it seems to be the other way round, at the most the facilitation is via technology and we supplement that with face to face sessions or occasional workshops."* (ACTIVITIES)

Q3. Have the course leaders noticed any impact coming out of the use of technology? Better grades or better understanding of the learning outcomes of the course?

A. *"Because of the discussions and because they have been engaging more each week they have a shared understanding, I don't know if it's higher or lower but they have a more of a shared understanding than the other students, they are also less than the f2f students and seem to be discussing everything online."* (OUTCOMES: SHORT TERM)

Q4. What barriers did you encounter in trying to deliver the SOL programmes and courses?

A. *"[technology] itself is not so friendly. Students do not find it user friendly and as a tutor I don't feel that I am given enough power to do what I want. That is the main barrier. There are other barriers of course, such as the way to communicate with students."* (OUTCOMES: SHORT TERM)

Q5. A follow up question would be, what changes would you like to see implemented with regards to how technology is allocated or structured? So if you are having similar thoughts about [technology] do we want to change anything?

A. *"It would be good to have photos attached to our online students, to be able to recognise them. Also if we could email them through WebCT it would have been easier."* (ACTIVITIES)

Q6. How would you evaluate your SOL students in terms of their use of technology?

A. *"I think that most of the students will engage in whichever technology the University determines for use with online learning. I think the challenge particularly for SOL students has not been the technology itself but the management of their own time and their own learning process. They need to approach it with the right attitude and be very self-disciplined."* (OUTCOMES: IMPACT)

Q7. What personal impact has the use of technology had on you? Are we all e-literate in a way?

A. *"I think it hasn't been the technology per se but the whole learning process. Having had to redesign the courses for delivery to students in a more flexible mode has forced be to rethink how to achieve the learning outcomes in other ways. So, the pedagogical construct. That's the impact it had on me, the fact that we use this technology made me to rethink our models of learning delivery."* (OUTCOMES: IMPACT)

Q8. Have you done that with your other courses that are not SOL?

A. "Yes"

A. *"I agree with that. After going through the facilitation guide provided which was very clear I started doing this in my other courses as well, to give better directions on how the class would be."* (OUTCOMES: LONG-TERM)

Q9. Overall, in this way do you think the students on one of the SOL courses would have been impacted by the use of technology? What about problems faced identified for traditional students? Has technology helped them be more motivated? More engaged?

A. *"Certainly they have been more motivated. But the fact that in order to engage in the topic at all they have to engage through technology in the learning process in the learning tasks – they have no option but to do that."* (OUTCOMES: IMPACT)

Q10. I want to go back to that point on impact and the use of technology. Is there something specifically about technology, a tool, meaning using email, or the discussion boards, or marking online that the university needs to make mandatory that all lectures, all courses should do?

A. *"By having all lectured recorded and stored in one place would be good – or keeping all questions from students and answers by tutors and students to specific questions. This would help students that were absent to find out about those questions asked and what was answered."*

A. *"From my personal perspective, the ability to provide feedback online in an efficient manner has been very useful, but also from students' perspectives in that they can access feedback, the timeliness of feedback and the quality of the feedback. That is something that I think we need to take forward and use in other courses. The other element that is quite near to my heart is course design. It's to do with revisiting the pedagogy of course design and taking into account the learning that many academics have had by going through this process of having to rethink courses for delivery in a different way."* (INPUTS & ACTIVITIES)

Q11. Last question on technology: Was there a problem with technology while teaching online that was not resolved? Or adding to that was there a misunderstanding in communication because of using online feedback, instead of f2f communication?

A. *"I faced difficulties in not being able to email the students, through the course site. The other point is that in large courses I could not select to see only my tutorial students."*

A. *Similar point in communicating with students. For example as I have six courses active at the same time, some f2f and some SOL it's difficult sometimes to know when students are calling upon you for some support or advise because you need to go into each individual course and spend time looking at the discussion boards to know what's happening so it puts more onus and more time-management responsibility back onto the academics to manage that rather than having say a notification that "you are needed on this course" or "somebody has posted something here for you".* (ACTIVITIES & OUTPUTS)

Student Experience

Anonymous student evaluation (conducted annually by the Business School using a standard evaluation sheet) was used to obtain sample feedback. Feedback for the BABS SOL programme came from a specific course, Thematic Independent Studies, and a cohort of an international partner institution; this resembled distance learning better rather than the version run in London, which was designed as a blended learning environment. The MBA SOL feedback, coming from a longer-running programme, covered the overall programme quality by combining answers for all courses in all three years. The BABS SOL feedback, which related to the authors' work, is the focus in this section. These were analysed in two ways.

Firstly, percentages of results were presented against the four multiple-choice questions asked – see Figure 1 and Table 2 for more details. This was not intended to serve as a detailed quantitative analysis but rather to indicate the level of satisfaction of students and any issues of concern raised. Secondly, the comments from students, corresponding to the fifth question of the evaluation form and providing the core of the qualitative feedback, were mapped against some key criteria. These were mapped against simple criteria such as: 'things students like'; 'things they don't like'; and, 'other things that don't fit in any of the above'.

Firstly, the replies of the TIS course cohort of the BABS SOL programme to the four multiple-choice questions are shown in Table 2.

The fifty-five comments were marked up into three main areas – the comments comprised the core of the qualitative feedback. These key criteria were mapped on to various aspects of what students actually said – they look quite rich in themselves. A sample of these replies is presented in the following 3 areas:

Area 1: Things students liked/said we did well

"Enjoyed the course and learnt a lot; thank you to all our tutors from overseas."

"This course was much better than some of the other Universities that I have seen! Everything was more organized."

"This was a great learning experience!! UoG is very good with the feedback. They are up to date, assist students were they can, and is excellent in time management."

"The supervisor online was very professional and helped us a lot."

"Overall this was a very convenient way of completing our course since I can access the portal from work; it helped me a lot that I didn't have to attend lectures so thank you!"

"As a foreign student, I found the overall process of completing and handing in this assignment to be satisfactory. The overseas lecturer was a great help."

Area 2: Things students did not like/we did poorly/ or could be changed

"It was my first time having an online lecturer and I must say I enjoyed the course but it was a bit hectic."

"The course was an enjoyable one and I would like to thank all my tutors that provided us with feedback and insight. The one thing I would change is the time taken to provide feedback."

"The course was well managed, but the audio classes were not very well presented."

"When it came to the online lectures they were very difficult to hear the lecture and also it would have been easy to have your own textbook to help you."

Area 3: Other things which don't fit in either of the above

"The lectors must be more in departments and must give all the right information from the start."

Table 2. Detailed results from the student evaluation (questions 1-4 out of 5)

Value of Course		Lectures		Tutorials		Learning Materials	
very good	21	very good	18	very good	9	very good	15
Good	25	good	21	good	28	good	25
satisfactory	9	satisfactory	13	satisfactory	18	satisfactory	14
Poor	0	poor	3	poor	0	poor	1

Table 3.

Inputs	Activities	Outputs	Outcomes (Intended Results)		
			Short-Term	**Long-Term**	**Impact**
• Business School: Lecturers/Tutors, Technologists, Teaching Champions, Project Manager • Curriculum Resources and Research Base: JISC Models, Constructivism, Active Learning • Collaborating Partners: Library, IT, Student Services, Registration, Funders, etc. • VLE (WebCT) • BANNER • Self-directed learners • Marketing Tools	Determine impact on: • course/ program coordinators • Convert F2F courses to online/blended for BA/BS & MBA programs • Compile FAQs about course content and navigation • Convert formative to summative assessments (i.e., portfolio) • VLE Creation of courses and batch upload of students • Coursework: Submission, 1st/2nd Marking & Moderation Determine impact on: • validation requirements • online registration • tracking	What are the types, levels and targets of services? • increased retention rates with flexible course delivery options • increased efficiency of providing feedback • Assess student experience performance with Subject Area (SAP) & Progression and Awards Board (PAB) • Investigate mark entry and electronic transfer from VLE to Student Information System	• improvement in student / tutor attitude toward technology use • Increased student engagement contributing to shared / deeper understanding of course content • Improved communication between academic, administrative, and information technology departments to understand key processes for effective delivery	• increased # of online and blended courses taught • link critical employability skills for students (in jobs that require Higher Education) • increased likelihood and interest of students to continue education • increased tutor awareness and knowledge of course delivery impact • enhanced publicity and public relations	• # of BA BS or MBA degrees obtained by graduates • increased student satisfaction & demographics (see NSS results) • improved time management and judgment of student's own learning process • improved understanding of pedagogical concepts for learning delivery, which also lead to improved F2F course delivery

"Not all comments and questions are answered, due to the time differences, between south Africa and the UK; it can become a bit frustrating when you have a crucial question that you need help with."

Solutions and Recommendations

"Half the difficulty with doing it better is knowing what the real problem is" (Ramsden 1992 cited Casey/Wilson).

Often, academics, administrators, and staff don't know what the impact on underlying problems. Higher Education (HE) institutions are complex environments that require a holistic understanding of the Information Technology (IT) systems and underlying academic, as well as administrative processes that are critical to delivering blended and online courses within local geographic boundaries and across borders in varying academic calendars. The use of the LOGIC model has provided a mechanism to identify the critical areas of performance that are needed to assess and measure the outputs/outcomes from e-learning environments. However, due to the complex environment, it will be necessary to begin developing separate LOGIC models for each of the Inputs and Activities to fully understand the interlinking relationships between the resulting Outputs and Outcomes. Additional focus groups will need to be conducted with Administrative staff to correlate and understand their desired outputs and outcomes. Table 3 represents a draft LOGIC model that identifies and summarizes critical outputs and outcomes resulting from the planning, implementation and management of the e-Learning environments.

The next stage will be to identify the critical gaps, missing links and/or measures of performance that will ensure effective outcomes (i.e., intended results).

FUTURE RESEARCH DIRECTIONS

As noted by Zawacki-Richter, Bäcker, and Vogt (2009), empirical investigation at the Meso-level (i.e., management, organization and technology) is critically needed in the areas of web-based and blended learning.

- How have emerging technologies for the administration, infrastructure, and support services enabled the effective implementation of e-learning within other Higher Education (HE) institutions?
- What are the possibilities of integrating emerging technologies which enable the transferability of data/discussions in e-Learning environments in different applications and/or formats?
- What are the future plans for integration technological processes between traditional student information systems & new e-systems, such as Customer Relationship Management (CRM)?

The authors have also found that not all automated processes are effective. Some processes have created more manual processes that hinder the future expansion of e-Learning environments. How effective are the automated processes? What still needs to be done manually anyway? How can course elements within multiple courses share common links that avoid manual updates in several places and/or across different IT systems? Missing links that results from the creation of VLE courses from the Student Information Systems and course roll-over from year to year still need refining to enable efficiency gains. For example, the web-based course that was delivered to students studying in South Africa consisted of 200 students in 11 different locations. Although the separate VLE courses can be cross-listed and easily updated with course content, other elements of the VLE course must be manually set up:

1. Manual creation of groups for each city locations in the country
2. Manual assignment of individual students to each group
3. Manual creation of 8 discussion board topics
4. Manual copying/pasting of topics for 11 centres

Therefore, what are the advantages and disadvantages of separate course shells vs separate discussion boards within one course shell? These are considerations that will enable effective management and organizational decision making.

CONCLUSION

Technology is always advancing; furthermore, the economy and demographics, both at global and local levels dictate changes that businesses need to consider for future planning. Like all businesses, most academic institutions need to follow trends and offer new and up to date learning initiatives. Data from around the world suggest considerable need for e-business and technical education (Fusilier and Durlabhji, 2009). National strategies for e-learning have now formally recognised the importance of technology in learning, teaching and assessment in all sectors of education in the UK, and in response many institutions have embedded the enhancement of learning and teaching through technology into their strategic missions (HEFCE, 2009a,b).

E-learning initiatives in academia have been increasing. This university, for years has been using technology to provide alternative solutions to the delivery of learning material, standardisation of assignment submission processes and effective management of large courses, in attending mode settings. It is difficult to use a standardised approach to e-learning provision because of the diversity of subjects being taught, as well as the multitude of available platforms and equipment. Attention must be given on students' needs and preferences as well as knowledge and capabilities of staff. The teaching technique is an issue that relies manifestly on the individual educators' backgrounds, training and style as well as the ability to engage with the changing technologies. It have been reinforced that technical problems are often not the hardest problems to solve. Rather, the related managerial and organizational processes are much harder to understand, coordinate, and integrate within the demands of developing e-learning environments.

This particular topic was selected because of its direct relevance to work conducted by the author in the Business School. This work originated from day to day administration & annual monitoring at programme level as leader of the third and final year direct entry route to BA (Hons) in Business Studies. Part of continuing development included redesigning the delivery and material to address students' changing needs. The project in fact became the basis for our international collaborative BA (Hons) Business Studies (BABS), programme provision, now supporting several hundred students in a number of overseas partners. At the same time, this expansion also spurred the conception of an additional project, the Distance Learning (DL) or Supported Open Learning (SOL) mode.

It was therefore important for the department and the School but in fact also became personally important as the engagement with this process opened a new chapter in a long line of research interests – that of e-learning. E-learning is not a new topic in academia and this project did not in fact make any groundbreaking addition to the vast amount of material that exists in the literature. However, the development and materialisation of these ideas comprised pioneering work for this Business School. Perhaps the way that these e-learning processes were planned and carried out in these first e-learning facilitation projects was not the same way that they will be offered in the future. However, this has undoubtedly been the setting of the corner stone on which future learning environments will be created and delivered by the Business School.

REFERENCES

Armellini, A., & Jones, S. (2008). Institutional embedding of e-learning: Findings from the Adelie Pathfinder Project. In Walker, S., Ryan, M., & Teed, R. (Eds.), *Designing for Learning – the e-learning@greenwich/conference*. University of Greenwich.

Beetham, H., & Sharpe, R. (Eds.). (2007). *Rethinking pedagogy for a digital age*. Routledge.

Brown, G., Bull, J., & Pendlebury, M. (1997). *Assessing student learning in higher education*. London: Routledge.

Brown, M., Fry, H., & Marshall, S. (2007). Reflective practice. In Fry, H., Ketteridge, S., & Marshall, S. (Eds.), *A handbook for teaching and learning in higher education* (2nd ed., pp. 215–225). Oxford: Routledge-Falmer.

Butcher, C., Davies, C., & Highton, M. (2006). *Designing learning–from module outline to effective teaching, key guides for effective teaching in HE*. New York: Routledge.

Collis, B., & Moonen, J. (2002). *Flexible learning in a digital world: Experiences and expectations*. London: Kogan Page.

Fry, H., Ketteridge, S., & Marshall, S. (2003). *A handbook for teaching and learning in higher education*. London: Kogan Page.

Fusilier, M., & Durlabhji, S. (2009). E-business education worldwide: On the right track? *The International Journal of Management Education*, *8*(2), 23–30. doi:10.3794/ijme.82.253

Gibbs, G. (1981). *Teaching students to learn: A student-centred approach*. Milton Keynes: The Open University Press.

Halperin, R. (2008). *The role of institutional factors in the formation of e-learning practices, advances in e-learning: Experiences and methodologies*. Hershey, PA: IGI Global.

HEFCE. (2009a). *Effective practice in a digital age, a guide to technology-enhanced learning and teaching*. JISC.

HEFCE. (2009b). *Enhancing learning and teaching through the use of technology: A revised approach to HEFCE's strategy for e-learning*. Retrieved from ww.hefce.ac.uk/pubs/hefce/2009/09_12

Hounsell, D. (2004). No comment? Feedback, learning and development. In Slowey, M., & Watson, D. (Eds.), *Higher education and the life course*. Buckingham, UK: Open University Press.

JISC. (2004). *Effective practice with e-learning*. Retrieved from www.jisc.ac.uk/practice

Leiva, M. (2007). *Teaching strategies for English language learners in mathematics: Implications for teaching all students*. Houghton Mifflin. Retrieved on December 2, 2007 from www.Beyond-the-Book.com

Niemiec & Otte. (2009). An administrator's guide to the whys and hows of blended learning. *Journal of Asynchronous Learning Networks*, *13*(1).

Oliver, M., & Dempster, J. A. (2003). Embedding e-learning practices. In Blackwell, R., & Blackmore, P. (Eds.), *Towards strategic staff development in Higher Education*. SRHE/OUP.

Petit, J., & Mason, R. (2007). Virtual space, real learning: An introduction to VLEs. In Fry, H., Ketteridge, S., & Marshall, S. (Eds.), *A handbook for teaching and learning in higher education* (2nd ed., pp. 148–161). Oxford: Routledge-Falmer.

Petrova, K., & Sinclair, R. (2008). *E-learning value and student experiences: A case study, advances in E-learning: Experiences and methodologies*. Hershey, PA: IGI Global.

Power, M. (2008). The emergence of a blended online learning environment. *Journal of Online Learning and Teaching*, *4*(4), 503–514.

Prosser, M., & Trigwell, K. (1999). *Understanding learning and teaching: The experience in Higher Education*. Maidenhead, UK: The Open University Press.

Race, P. (2007). *The lecturer's toolkit-a practical guide to assessment, learning and teaching* (3rd ed.). Oxford: Routledge.

Ramburuth, P., & Mladenovic, R. (2004). Exploring the relationship between students' orientations to learning. *Accounting Education*, *13*, 507–527. doi:10.1080/0963928042000306774

Sharpe, R., Benfield, G., Roberts, G., & Francis, R. (2006). *The undergraduate experience of blended e-learning: A review of UK literature and practice*. A report to the Higher Education Academy. Retrieved November 20, 2009, from http://www.heacademy.ac.uk/assets/York/documents/ourwork/teachingandresearch/Sharpe_Benfield_Roberts_Francis.pdf

University of Greenwich. (2007). *Business School Strategic Plan*.

Unwin, A. (2007). The professionalism of the higher education teacher: What's ICT got to do with it? *Teaching in Higher Education*, *12*(3), 295–308. doi:10.1080/13562510701278641

Vickery, A.W. & Lake, F.R. (2005). Teaching on the run: Giving feedback. *eMJA*, 183(5), 267-268.

Wakeford, R. (2007). Principles of student assessment. In Fry, H., Ketteridge, S., & Marshall, S. (Eds.), *A handbook for teaching and learning in higher education* (2nd ed.). Oxford: Routledge-Falmer.

Walker, S., & Ryan, M. (2008). A post conference reflection and report. In Walker, S., Ryan, M., & Teed, R. (Eds.), *Designing for learning – the e-learning@greenwich/conference*. University of Greenwich.

W.K. Kellogg Foundation. (2004). *Logic model development guide*. Battle Creek, MI: WK Kellogg Foundation.

Zawacki-Richter, O., Baecker, E., & Vogt, S. (2009). Review of distance education research: Analysis of research areas, methods, and authorship patterns. *The International Review of Research in Open and Distance Learning*, Retrieved on January 22, 2009, from http://www.irrodl.org/index.php/irrodl/article/view/741/1433

ADDITIONAL READING

Caruso, J. B. (2006). Measuring Student Experiences with Course Management Systems, Educause Center for Applied Research, Volume 2006, Issue 2. Retrieved April 7, 2010 from http://net.educause.edu/ir/library/pdf/ERB0619.pdf

Dearing, R. (1997). Higher Education in the Learning Society – Report of the National Committee of Inquiry into Higher Education.

Gibbs, G. (1988). *Learning by Doing. A Guide to Teaching and Learning Methods Further Education Unit*. Oxford: Oxford Polytechnic.

Griffiths, S. (2007). Teaching and learning in small groups. In *Fry, H., Ketteridge, S. and Marshall, S. (2007) A handbook for teaching and learning in higher education* (2nd ed., pp. 91–104). Oxford: Routledge-Falmer.

JISC. (2009). Committee of Inquiry into the Changing Learner Experience, *Higher Education in a Web 2.0 World*, available at: www.jisc.ac.uk/publications/ documents/heweb2.

McVey, M. (2008). Writing in an online environment: Student views of "inked" feedback. *International Journal of Teaching and Learning in Higher Education*, *20*(1), 39–50.

Parchoma, G. (2006). A proposed e-learning policy field for the academy. *International Journal of Teaching and Learning in Higher Education*, *18*(3), 230–240.

Steen, H. L. (2008). Effective e-learning design. *Journal of Online Learning and Teaching*, *4*(4), 526–532.

KEY TERMS AND DEFINITIONS

E-Learning: (i.e., electronic learning) uses information and communications technology to facilitate and support teaching and learning.

LOGIC Model: Tool that provides a visual representation of the connection between resources, activities, outputs, audiences, and short, intermediate, and long-term outcomes related to a specific situation.

Virtual Learning Environment (VLE): An online software environment that provides tools to update and distribute course materials, post announcements, host online discussions asynchronously/synchronously, provide assessments, and submit/grade assignments.

Chapter 4
Investigating the Effect of Color on Memorization and Trust in E-Learning:
The Case of KMCMS.net (Knowledge Management and Content Management System)

Jean-Eric Pelet
Department of Marketing and Information System, France

Panagiota Papadopoulou
University of Athens, Greece

ABSTRACT

The objective of this chapter is to study the effect of the color of e-learning platforms as an atmospheric variable of the interface on the learning process. It focuses on two important variables: memorization and trust. First, it aims to explore the effect of color on the memorization of the educational content available in an e-learning platform. It investigates if the colored appearance of an e-learning system reinforces readability, suitability, and above all the memorizing process leading to learning. Second, it investigates the effect of color on trust in an e-learning platform and in the e-learning content. The chapter offers an examination of the role of the interface and the color in e-learning and memorization and presents trust in e-learning based on an exploratory qualitative study on how trust is developed in a website, as a result of the colors of the online interface. A case of an e-learning platform has been investigated to understand how the colors of the interface can enhance student memorization of the content as well as trust.

DOI: 10.4018/978-1-60960-501-8.ch004

1. INTRODUCTION

E-learning enhances the speed of training people, drastically reducing costs for educational institutions and businesses facing an increasing need to train people rapidly. E-learning often relies on the goodwill of some teachers or practical approaches of companies specialized in e-learning. A growing body of literature has studied e-learning with several studies focusing on principles for e-learning design and implementation (Scott, Shurville, Maclean *et al.*, 2007; Ettinger, Holton and Blass, 2006; Siqueira, Braz & Melo, 2007). Based on 29 research case studies of organizations implementing e-learning projects, Ettinger et al. (2006) present key areas that should be taken into consideration. These studies point out that quality content and user-friendly technology needs to be carefully chosen, as they are even more important than for traditional learning. Among the various issues that are associated with e-learning, the interface is of primary importance. As mentioned by Weinreich *et al.* (2006), a breakdown of page characteristics shows that users often do not take the time to read the available text or consider all links. An e-learning platform cannot suffer from this type of problem, since learners have to trust a system where the information is neither hidden nor difficult to identify. The latest can be partly solved by a professional use of colors, (a) of text and hyperlinks (foreground color) and (b) of background.

The web interface is the portion of the website that is visible to the web user (Dailey, 2004). In parallel, drawing from Kotler's (1973) definition of brick and mortar atmospherics, web atmospherics can be defined as the conscious design of web environments to create positive effects in users (e.g., positive affect, positive cognitions, etc.) in order to increase favorable responses (e.g., site revisiting, browsing, etc.) (Dailey, 2004). An important user behavioral response to a stimulus such as a course presented on an e-learning platform to learners, can be the memorization of educational content. Drawing from previous studies in e-commerce (Pelet, 2008, 2010b), the atmospherics of the e-learning interface can arguably have an important effect on memorization. However, with regard to research dealing with an element of the interface appearance, such as color, we have little information about the role in the process of memorization of e-learning content. Limited studies refer to this topic, and empirical studies dealing with the effect of interface consistency or inconsistency on learning processes for e-learning remain still quite rare.

With the large amount of information presented on e-learning websites, memorization becomes an important factor for online learning since studies are facilitated when the learner can retain information from one page to the next. This implies that memorization of information in an e-learning website may have an impact on learners ability to digest and understand the subject matter and may allow consequently for success at their exams. This can potentially be facilitated by the website colors. However, the relationship between memorization and learning online has not been investigated. In addition, there is a lack of research regarding color and its effect on memorization and learning in e-learning websites. Hence, the education industry would probably find interesting results linking color, memorization and trust, since e-learning is close to e-commerce in terms of looking for and retaining information from a screen. Results from a previous research on the effects of color have proven how important this attribute of the interface was for the perception of time for example. Gorn *et al.*, (2004) focusing on the impact of the three color components on downloading time perception demonstrate that a lengthy waiting time influences the user's appraisal of the web site and can lessen his/her desire to recommend it to others. The same could be evocated in a learning context, since learners increasingly go for "quick to acquire embrace" content.

This chapter investigates the effect of the color component on the memorization of online educational content and trust. As Tricot and Plégat-Soutjis, (2003) point out the visual use of a device such as the computer screen is only effective if it is perfectly consistent with the entire device. The user behaves with what he recognizes to make sense with the e-learning platform. This is the principle of affordance proposed by Gibson (1977), which corresponds to the suggestive capacity of action for an object, a button or a form. Affordance theory states that the world is perceived not only in terms of object shapes and spatial relationships but also in terms of object possibilities for action - perception drives action (Gibson, 1977). This is the reason why measuring the content of the information stored by respondents seems interesting. It serves to make assumptions regarding the learners' perception of the color appearance of the used e-learning platform since what a learner memorizes depends on what he understands according to the e-learning platform interface and its variables such as colors.

Following Rhee et al.work (2005), who studied the question of the effect of consistency of interface on learning and the potential differences between groups of students skilled in computers versus novice students, our aim is also to investigate the effect of the colors of an e-learning interface on user trust. We posit that the colors of an e-learning environment can influence user perceptions regarding the trustworthiness of the e-learning platform, the educational content as well as the actors involved in the e-learning process. Our findings could provide both researchers and practitioners with new criteria for color contrast design issues in the era of e-learning.

As a first step, we review the factors aiding the memorization of the educational content discussed in the literature. We will then investigate the research issues related to the effects of the colors of websites that could improve our understanding in the field of e-learning. The chapter then presents the concept of trust by linking it to e-learning and how it can be influenced by the use of colors. The link between online trust and color is further examined with a qualitative study that is presented next. The study continues by providing an exploration of the effect of color on memorization and trust of e-learning users using a case study of an e-learning platform and brief conclusions end our chapter.

2. E-LEARNING INTERFACE

E-learning platforms are information systems are increasingly used in organizations in the business as well as the educational sector (Pelet, 2010a). Most of the users of such platforms are known as *lifelong learners* as the skills they acquire mean that they are continuously challenged and search for information (Reisman, 2003, 240). Some of these *lifelong learners* can also be gamersplayers. Literature suggests that 3-D virtual worlds such as Second-Life can be well suited for experiential learning environments (Jarmon & al., 2009). This looks like a new approach of e-learning.

Literature in the area of e-learning points out that the quality of educational software is significantly related to its interface quality (e.g. Buzhardt *et al.*, 2005; Cantoni *et al.*, 2004; Chu and Chan, 1998; Hinostroza & Mellar, 2001). Regardless of the technical type of the platform, desktop software or internet website, the interface remains equally problematic, dealing with the behaviour of the user, depending on its perception of the latest. The interface quality of educational software or websites, moreover, has a serious impact on the learning outcome of the student (Gauss & Urbas, 2003; Jonassen & Wang, 1993). Crowther *et al.* (2004) argue that the impact of a poor interface design in education is more serious than in business. It impairs a student's overall motivation, as well as their learning performance, and has serious moral and ethical implications. In essence, interactivity between student and interface has been considered as the most important aspect in several studies on how to improve quality of education through e-learning (Cantoni *et al.*, 2004; Chou, 2003;

Ellis & Blashki, 2004; Gauss & Urbas, 2003). According to interface consistency theory, which deals with the interaction between the user and the interface, increasing the consistency levels of interface results in a significant decrease of error rates in computer and web-based tasks (Ozok & Salvendy, 2004).

On an e-learning platform, the learner views various web pages, which use different features such as color, animation, sound, text, animations, photos, textures, and graphics in order to draw the learner attention. Among the range of visual factors that define the web interface, color can be deemed to be of high importance. The interface is made of components which affect the visual and auditory ability to stimulate the senses of the learner, in order to perceive its emotional, cognitive, psychological, physiological and behavioural functions through their changes. This is even verified referring to the psycho-environmental model from Mehrabian and Russel (1974), also called Stimuli-Organism-Response (SOR) model.

Before succeeding scoring on an e-learning platform, learners need to read and memorize. Students in an e-learning course have to analyze and synthesize the information they read on e-learning websites. But they also have to memorize where this information was, and how to reach it quickly for getting results they want efficiently. They also have to memorize it if they want to fill a questionnaire. This is why memorization is so important for e-learning objectives, as a part of the entire process of learning online.

Literature on the development of computer-assisted courses has provided for a long time inquiries on the use of text, color and graphics (Aspillaga, 1991, Livingston, 1991; Rubens & Krull, 1985; Soulier, 1988; Steinberg, 1991; Szabo & Poohkay, 1994). The most important difficulty in the development of such courses is to facilitate learning as much as possible in order to reduce the external cognitive load which is associated to the use of the tool and to the type of presentation in order to benefit to the intrinsic cognitive load related to acquiring content (Sweller, 1999). For this reason many techniques, combining for example graphics, images, videos and sounds, must be developed in this direction in order to optimize the storage capacity of the learner. It was shown that the learner memorized more content if both memory channels, audio and visual, were prompted at the same time (Mayer, 2001; Mayer & Gallini, 1990; Clark & Paivio, 1991; Ando & Ueno, 2008).

The visual channel refers to the reading and retention of content on an e-learning platform. Readability represents the reaction time required to find a target word when searching in a website (Hall & Hanna, 2004). Readability is rather informative in this stage with respect to basic processing, it does not address outcomes such as retention, which is based on the cognitive architecture. The term "cognitive architecture" refers to the manner in which cognitive structures are organized. The two most important of those aspects of human cognitive architecture relevant to visually based instructional design and around which there is broad agreement are the working memory and the long term memory (Sweller, 2002). While considerable work has been carried to the organization of human cognitive architecture (Sweller, 2002), far less effort has gone into investigating the memorization of the information presented on e-learning platforms. De Groot's (1965) work on chess (first published in 1946) demonstrated the critical importance of long-term memory to higher cognitive functioning. He showed that memory of board configurations taken from real games was critical to the performance of chess masters who were capable of visualising enormous numbers of board configuration. The skills depended on schemas held in long-term memory, thanks to the retention of information.

Retention is a very important factor for a learner facing a large number of information on the websites he/she must visit before answering to questions and pass courses. It is an important factor for e-learning applications, since the user's goal is usually to retain the information beyond the time the page is being read: in order to score at exams based on reading and retention of information permitting to answer to questionnaires, memorization appears as the main way to score on e-learning courses. Thus, measures of higher level processing, such as retention remains an important topic in examining the effects of text-background color combinations, for the success of e-learning websites.

3. EFFECT OF COLOR ON E-LEARNING

Studies have shown that 80% of the information processed by the brain of the Internet user comes from the sense of sight (Mattelart, 1996). This implies that the visual function of the learner is a key factor for e-learning. The visual function is linked to light, color, and various other dimensions. This visual function is important to consider since it has already been established that human beings appeared to be exceptionally sensitive to visual learning (Kobayashi, 1986) and that any use of visual content could enhance its acquisition. On the contrary, over-cognitive load associated with excessive use of graphical means do not seem to be productive. Indeed, research has shown that if learners are sensitive to fixed or animated images, this is even more true if these images facilitate navigation without overloading its access or its clarity (Wright et al, 2000). This is especially important for schools or educational institutions interested in embracing e-learning. Color and typeface are two important characteristics of visual stimuli that may affect visual performance (Shieh et al. 1997). Color can be an effective means to improve human-computer communication (Pastoor, 1990; Silverstein, 1987). Color is also commonly expected to provide an additional subjective benefit by making work seem more pleasant and trustworthy.

Although color is widely researched topic (Divard & Urien, 2001), to this day there is a lack of studies focusing on color in the online context. Color in websites has been studied within information systems, especially in human-computer interaction, usability and e-commerce, recognized as a fundamental aspect in web interface design (Lee & Koubek, 2010; Wu *et al.*, 2008; Coursaris *et al.*, 2008; Kang & Corbitt, 2001). Research has found color to be an important factor in e-commerce, influencing website aesthetics (Agarwal and Hedge, 2008; Coursaris *et al.*, 2008; Schmidt & Liu, 2005) e-retailer perceptions (Agarwal & Hedge, 2008), user preference for e-commerce web sites (Lee & Koubek, 2010). With the interaction with the system, as well as the aesthetics of the interface being also of similar importance in the context of e-learning, it can be assumed that these works can imply a link between color and memorization and trust in e-learning,

Color is one of the components that take part in the physical consistency (Rhee et al. 2005). The physical consistency is part of the interface consistency of the e-learning platform. The latest is also composed of communicational and conceptual consistency.

- Physical consistency is the consistency of the graphical appearance or the visual characteristics of an interface feature;
- Communicational consistency is the consistency of the input and the output of the interface;
- Conceptual consistency is the consistency of metaphor applied to an interface feature or an action that is embodied within a feature.

Among the modes of visual recognition of the human being, color is at the origin of a significant number of work on the acquisition of knowledge, allowing to improve the development of printed materials (Moore & Dwyer, 1997). For example, we know that the structure of the text will affect the memorising of some part of it (Kintsch, 1970; Meyer, 1975; Meyer et al, 1980). Color allows the learner to distinguish the learning object (Gibson, 1966). It is also a major paralinguistic organizer, which means that a presentation can be graphical or temporal and allows presenting the displayed information in a structured manner. This effect is even more strengthened when it is associated with a color that allows either to focus on some of the text, or to categorize the information (Dwyer & Lamberski, 1982-83). The color should also be used as a learning strategy. The choice of colors used to raise specific issues or to prioritize the importance of information to be integrated into the thinking at the design stage is thus very important to take into consideration. This is to limit the potential risk of overloading. The concepts of learning and memory are thus directly related since the content of e-learning platforms and the colors used are key factors for the readability of information.

The graphic of the e-learning platform website, which represents the whole graphical presentation, contains basically two colors: the foreground color, also called tonic or dynamic color by webmasters, and the background color, which is the dominant one. This association reveals the contrast, which corresponds to a strong difference between a foreground and a background color, as defined by the W3C (Accessiweb, 2008). Its main function is to enhance the readability of the displayed information, and a fortiori the memorization and the fact of passing the examination in the case of an e-learning platform. It seems therefore interesting to know what information is stored and what can be the successful completion of an examination related to a previously learned content through such a platform.

In any case, research about the impact of color on the effectiveness of online learning does not concern the use of color that much for an entire text, although the use of black text on white backgrounds seem preferred to transfer knowledge (Hall & Hanna, 2004). As part of an experiment the purpose of which was to measure how much and what quality of information was retained by the consumer when he visited an e-commerce website, interesting results have emerged. They show that the memory score grows when colors of foreground and background are chromatic, i.e. different than black and white and their intermediate greys (Pelet, 2008). Interaction effects of hue and brightness operate on free recall, especially when background and foreground colors, i.e. green and yellow in the context of the evocated experimentation, are used.

Memorization is a very important factor for the large number of information-based websites that currently exist. It is important for e-learning applications, since the user goal is usually to retain the information beyond the time the page is being read. In order to understand the effects of color on consumer memorization we have to take into account the quality and quantity of information a user has memorized while visiting an e-learning website. We posit that memorization varies according to the colors of the website, and especially according to the contrast between the dominant and dynamic colors, in agreement with the work of Hall and Hanna (2004). In general, information is stored according to an encoding process enabling one to sort out information thanks to criteria which will then allow one to retrieve this information. The role of these criteria is to connect a piece of information to other similar information already stored (Ladwein, 1999).

4. E-LEARNING AND TRUST

Trust has traditionally been a complex, multidimensional and context-specific phenomenon (Lewis & Weigert, 1985; Butler, 1991; Barber, 1983). It has been the topic of numerous research studies in various disciplines, including psychology, sociology, economics and marketing. It has long been identified as a key factor in any kind of relationship and has been shown to be of particular importance in traditional as well as in electronic settings, such as e-commerce (Gefen *et al.*, 2003; Papadopoulou, 2007).

As is evidenced in the literature, trust is characterized by a lack of consensus regarding its definition. It is a multi-faceted concept, resulting in a collection of multiple, diverse definitions of trust, which is evidenced across all disciplines where trust has been studied. Divided in two dominant conceptualizations, trust is viewed by many as a belief or expectation implying a notion of confidence and trustworthiness stemming from the partner's expertise, reliability, or intentions (Blau, 1964; Pruitt, 1981). Under this stream of research, the concept of trust has also been widely studied under the notion of beliefs about trust relevant attributes of the trustee, largely referring to the perceived benevolence, competence, integrity and predictability of the trustee (McKnight *et al.*, 1998). From another perspective, trust is seen as a behavioral intention or willingness to be vulnerable and rely on another party, accepting the risk and uncertainty emanating from the assumption that the party will behave within accepted norms (Williamson, 1975; Coleman, 1990; Deutsch, 1960). Researchers adopting this view have separated beliefs from trust conceptualizations and defined them as antecedents of trust (Mayer *et al.*, 1995). While discriminating between beliefs and behavioral intention, a third research stream has argued that both are necessary components of trust (Moorman *et al.*, 1992; McAllister, 1995; McKnight *et al.*, 1998).

Trust is an important factor in the decision of people when they choose others with whom to interact (McKnight *et al.* 1998, Zaheer *et al.* 1998). In the case of an e-learning platform, the learner, in order to use it, decides if the interface can be trusted or not or if the lecturer can be trusted or not depending on whether the content displayed on the screen is acceptable for him. Several factors may explain why high initial levels of trusting beliefs and intentions might be observed (Stewart, 2003). These include individuals' disposition to trust (Rotter, 1967), the existence of assurance mechanisms (Zucker, 1986), security felt in the situation in which the trust target is encountered (Lewis and Weigart, 1985), calculations regarding the incentives and penalties to the target of acting in a trustworthy manner (Lewicki and Bunker, 1996), and cognitive processes such as stereotyping and categorization (McKnight *et al.*, 1998).

Trust has recently been identified as being an important parameter in e-learning. Anwar and Greer (2008) characterise trust as a cornerstone of safe and engaging e-learning environment. Orton-Johnson (2009) has shown a high tendency of students not to use e-learning material attributing it to their trust in traditional texts as authentic academic knowledge and an instrumental and strategic approach to study. Arguably, in the context of e-learning, trust can be an important variable for e-learner behavior. As e-learning involves the use of an electronic, usually online, means, it changes the learning process and creates new challenges, with trust being one of them.

Trust in the context of e-learning can include several aspects that affect the success of an e-learning system. In order to effectively adopt and use an e-learning system, users should have trust in the e-learning system or platform. This refers to the interface and the characteristics offered by the platform. In addition, users should have trust in the content provided in the e-learning platform. This trust aspect is very important for e-learning since if users do not believe that the content of an e-learning system is reliable and trustworthy, they will not use the e-learning system or they will not exploit its advantages to a maximum. Trust is also related to the actors involved in the e-learning platform, i.e. students, educators, moderators/administrators, content providers. The trust of the users of an e-learning system may vary among the actors according to their role and their content provision. For instance, students of an e-learning platform may have more trust in the content provided by an educator than in the content provided by other students.

Color, as an important interface element, is expected to influence user perceptions regarding the trustworthiness of the e-learning system. A previous study has shown that trust was considered as an important variable to consider when accomplishing a research on the effects of color interfaces on users. In particular, color proved to be directly linked to the trustworthiness felt when facing an interface where the association of background and foreground colors reinforces readability and memorization of the content. Indeed, the contrast reinforces the buying intention when it is accurate (Pelet & Papadopoulou, 2010). In a similar vein, color can be associated with the trust that a user creates in an e-learning environment. The effect of color on trust can be attributed to the colors used as well as to the contrast of foreground and background colors. The colors, in terms of hue, saturation and brightness as well as the contrast can provide interface characteristics, such as ease-of-use, which can be perceived by the user as signals of trustworthiness. The colors used in an e-learning platform could affect user trust in the e-learning content, as well as trust in the educator/tutor of the platform or in other actors contributing content to the e-learning platform. In this direction, an investigation of the effect of color on trust perceptions of an e-learning user will facilitate our understanding regarding the effect of color on e-learning. This study will focus on trust in an e-learning system largely referring to trust in the content as well as the teacher of the e-learning platform.

5. THE EFFECT OF COLOR ON ONLINE TRUST: A QUALITATIVE STUDY

An exploratory qualitative study was conducted in order to understand the relationship between color and trust (Pelet & Papadopoulou, 2010). The objective of the study was to investigate how user trust in a website is influenced by the colors of the interface. The study was based on the literature and the former studies examining these websites. We wanted to understand what made the respondents feel that they could trust a website they already have used. Our data collection aimed at identifying and classifying the elements which referred to the trust felt by users during a website visit, as a result of the colors.

The study was conducted using semi-structured interviews with 21 persons, regular users and skilled ones, close to web designers in terms of use of the web. We asked interviewees to speak about past visits to websites of their choice. The interview guide was structured and open and allowed us to collect data related to the subjects experience in websites. We adopted a neutral attitude with regard to them so as not to influence them in the way they answered. Interviews were conducted using what interviewees recalled from their past visit on websites. To ensure that the respondents only used their memory to provide information about their browsing websites, it was not possible for them to look at a screen showing a website interface during the interview (Jallais, 2006). The reason for which we investigated the memory can be understood by the following example. When people learn in traditional or electronic classrooms by relying on what they stored in their brain regarding courses information (date, equation, theoretical concept, etc), they rely on what they memorized. Indeed, in everyday life there is no trigger helping learners recall the content of an e-learning website they visited or compare it with another course. Most of the time, learners can not visit the e-learning website where they found relevant information while they walk in the street. In case they can try and compare courses information for example, they can only trust what they stored in their long term memory. It seemed therefore relevant to conduct our interviews by respecting the same conditions as real life ones.

The objectives of the qualitative study were pursued on the basis of the following questions:

- Is color one of the atmospheric elements of websites that is important for Internet users?
- Do Internet users feel particular emotions while using a website? If yes, does color affect the emotions of the users?
- Do Internet users trust a website thanks to the colors as a variable of the website atmosphere?
- Does a website's color affect user perceptions of a website? How is trust influenced?
- Does color affect users trust and perceptions about a website in a positive or in a negative way?

5.1. Respondents Characteristics

The exploratory qualitative study took part in France with French respondents. A total of 21 subjects participated in the study, 43% of which were female. The age of the sample population ranged from 31 to 57 years old. The majority of

Table 1. Qualitative analysis results summary

Respondent No	Respondent expertise	Trust Color Effect
1	RU	rich website, interactivity, cold, austere, minimalism of the interface
2	RU	Reassuring, facilitate the link visibility, content is put to the fore, convenience of website, usability procures help
3	RU	delimitate zones, ergonomics (search engines), zones facilitate tracking, harmonious, ergonomics helps the user
4	W	fluid navigation, clear, playful, harmonious, playability to use the whole website
5	RU	delimitate zones, reassuring, sober, facilitate tracking of links, clarity of the presentation
6	RU	rapidity, communicate, practical, enlighten, fluidity of actions
7	W	feeling of speed, cold, beautiful, not practical website, high level of information aspect
8	RU	enhances the content, boring, clarity, beautiful, sober
9	RU	Vivid, speed, convivial, procure advice, help information is simple
10	RU	make dream, deceive, lack of trust, poor information, poverty of information deceives
11	RU	clarity, aggressive, conviviality, no respect of the consumer, violent colors aggress users
12	RU	trusted website, secure website, competence, feeling of speed, fluidity reassures
13	RU	absorbed, professionalism of the website, ergonomics, feeling to save time, information structure serves information content
14	W	pleasure, ease of use, practicality, ease of finding information, ease of use thanks to color contrasts
15	W	distrust, poverty of information, conviviality, cheap look, vivid colors don't enhance trust
16	RU	absorbed, give advice, feeling of saving time, reliability, conciseness of information helps the user
17	W	richness of information, deception, rapidity of finding information, indecision, too much information doesn't serve the website
18	RU	rapidity to find information, attractive prices, stress, boredom, simplicity and aesthetic aspect
19	W	lot of choice in the catalog, suspicious, practicality, beautiful, aesthetic, richness of information and right choice of color family
20	RU	indecision, sober, easy-to-learn website, fluidity of information, clear zones of explanations help users
21	RU	purchase without pressure, conviviality, complex way of buying, espionage, conviviality and simplicity

the participants (15 persons) were regular Internet users, while 6 of them were webmasters. The representation of webmasters and regular users according to the users numbers used in the study is summarized in Table 1. The interviews were based on websites related to different activities according to the interviewed persons and their habits. Different websites related to everything about selling were also used for description since selling products and courses can be compared according to us. Websites dedicated to downloads for webmasters also came to the discussion with the interviewed webmasters, such as Flashloaded for example.

5.2. Data Analysis

Data were analyzed following the different stages of content analysis suggested by Vernette and Giannelloni (1994). Interviews can be analyzed in multiple ways following the research objectives (Evrard, Pras and Roux, 2003), which in our case was to explore the importance of color in trust and in what ways it was important. The responses were coded and the preliminary readings (Bardin, 1996) allowed us to identify the items. A cluster by topic and frequency of appearance of the variables followed. We respected a certain number of stages in identifying the subjects and the useful semantic units, as in any content analysis. The qualitative data were analyzed with a table summarizing all the results of our respondents, where each construct was assigned a '1' value for every instance. For each respondent, we analyzed the interview and we extracted instances of trust that were mentioned and that were related to color. We then grouped the 'result' columns of all our respondents in the final table in order to understand precisely what the answers meant. This corresponded to a total of 22 columns, with cells having either a '1' value or being empty. Rows in this table indicated the different themes issued from the interview guide. Topics and words related to a precise field appeared in a 'result' column, showing the frequency of appearance of each construct. The results are summarized in Table 1.

This exploratory qualitative analysis enabled us to note that color was actually an integral part of the websites interfaces. A number of elements, presented in Table 1, were revealed as important to user perceptions, as a result of the colors used. The elements, which appear essential to the interface due to the impact of color, can be mainly grouped into:

- elements related to usage - putting the organization of the site as a main factor, thanks to its clarity and the readability of its tree structure: this serves in favor of building trust
- elements allowing a rapid navigation within the site, by the provision of search engines in particular: this plays in favor of memorization since no time is wasted.

6. THE CASE OF KNOWLEDGE MANAGEMENT AND CONTENT MANAGEMENT SYSTEM KMCMS

In this section we attempt to investigate how the interface of an e-learning environment and particularly the use of the colors can influence memorization and trust. A 6-years experience (between 2003 and 2009) using a knowledge management platform and content management system, is presented. Our description and pictures are extracted from an actually in use e-learning platform, Knowledge Management and Content Management System (KMCMS). KMCMS is an online learning tool, in the form of a wiki, designed to help students to prepare their major projects in different courses. The wiki was built in 2003 and today, around 1500 contributions dedicated to very precise topics are available inn it. Project-related material, home works or major projects such as dissertations, has to be easily publishable and accessible for editing by any student logged on the platform. As time passes, more and more students publish their work (syntheses of academic papers, journalistic ones or reviews of literature on a particular topic) making it available for the overall community of students. The phenomenon is thus becoming autonomous since students get help from the contribution of other students, and also help the others to find easily what they could be interested in by working on it, adding their *"knowledge"*. One of the key factors of success of such a tool lies in the search engine mechanism that is embedded into the system to accelerate search and results found. This is one of our findings since any student working on this platform refers to this search engine first. As for any User Generated Content website, the role of the lecturer is to moderate each new contribution by asking the student to modify it if necessary.

This platform aims at sharing knowledge issued from reading papers (academic publications, student projects, books, managerial cases, etc), on a social network platform. The possibility to create pages with synthesis of academic readings or work experience projects, to communicate with other students or to present works from the platform to the wall of a room during "live" lessons, by using PowerPoint presentations for example, contributes to share this way of learning. The platform is therefore a powerful e-learning tool built on the restitution of readings from different levels (students, professors) and in different topics related to marketing, information systems and design themes, focusing on consumer behavior, ergonomics or usability on the Internet. These topics represent the main themes evocated by authors during lectures or across their academic publications. The same principles as those already used in the e-business industry are followed in the educational sector, with ratings and log files so as to understand the user behavior more precisely. Log files can show what content is read first, what page is the most popular and so on, exactly like a teacher does with an e-learning platform. (Figure 1).

Figure 1. Log file from kmcms.net

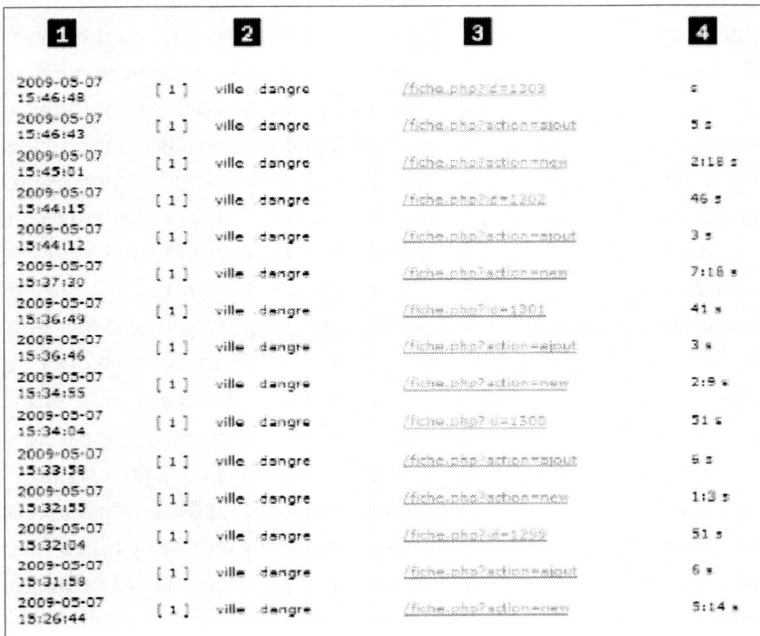

1		2	3	4
2009-05-07 15:46:48	[1]	ville dangre	/fiche.php?id=1303	s
2009-05-07 15:46:43	[1]	ville dangre	/fiche.php?action=ajout	5 s
2009-05-07 15:45:01	[1]	ville dangre	/fiche.php?action=new	2:18 s
2009-05-07 15:44:15	[1]	ville dangre	/fiche.php?id=1302	46 s
2009-05-07 15:44:12	[1]	ville dangre	/fiche.php?action=ajout	3 s
2009-05-07 15:37:30	[1]	ville dangre	/fiche.php?action=new	7:18 s
2009-05-07 15:36:49	[1]	ville dangre	/fiche.php?id=1301	41 s
2009-05-07 15:36:46	[1]	ville dangre	/fiche.php?action=ajout	3 s
2009-05-07 15:34:55	[1]	ville dangre	/fiche.php?action=new	2:9 s
2009-05-07 15:34:04	[1]	ville dangre	/fiche.php?id=1300	51 s
2009-05-07 15:33:58	[1]	ville dangre	/fiche.php?action=ajout	6 s
2009-05-07 15:32:55	[1]	ville dangre	/fiche.php?action=new	1:3 s
2009-05-07 15:32:04	[1]	ville dangre	/fiche.php?id=1299	51 s
2009-05-07 15:31:58	[1]	ville dangre	/fiche.php?action=ajout	6 s
2009-05-07 15:26:44	[1]	ville dangre	/fiche.php?action=new	5:14 s

Legend:

1: date/hour the page has been viewed

2: name of the user (student in this case)

3: viewed page

4: time spent on the page

Figure 2. Homepage of the e-learning platform

6.1. Exploring Memorization and Trust

On the home page of this e-learning platform, students can easily see what is new about a lecture or a topic and what they have to do for their lesson. Once logged onto the e-learning platform, the trust of students in the e-learning platform increases thanks to the facility to view each lecture and the provided URL allowing to go to the page of a lesson, case or literature in one click. Colors of schools or Institution logotypes reinforce the feeling of trust that students need to have when logging on this type of website. (Figure 2)

Like on any e-learning platform, a search engine helps the student to look for any type of content. (Figure 3)

After having typed their request, students can filter the search results by title, abstracts, keywords, authors, date of creation and content. In order to enhance the trust of the student in the results found, the word typed in the search engine is highlighted with a fluorescent yellow color on the results page. It helps the student to identify if the word comes from the title, abstract or any other part of the offered results. For example, when typing «color», in the search engine, 223 pages have been found in the database. As Maruyama and Akahori (2008) have shown, the use of color to highlight certain words in a text is effective in memorizing when one word only was

Figure 3. Search engine of the e-learning platform

highlighted. The first results appear on the pages as shown below (Figure 4).

The possibility to ask for the "search in context" also exists; its function is to avoid for the student to open each page obtained among the results of the "results page". This feature can be very useful as it could indeed be dramatically long to check 223 pages to look for an information in this case. In this way, facilitating their search and reducing the time required for locating results, students trust as well as memorization are increased. The search engine offers answers containing the abstract, titles and keywords from pages of the databases. By providing paragraphs in the answer, containing the 200 characters before and the 200 characters after the word typed in the search engine, it is then unnecessary to open the page linked to the results page, since the student understands in which domain the typed word belongs to, in order to appear on the e-learning platform. The fact that there is no need to open unuseful pages enhances students trust towards the e-learning system. The system saves the students time, and enhances their trust in proceeding with their next request. It also reduces students effort to retaining the content found. Here is an example of the *"search in context"* result, when the word "humeurs" has been typed in the search engine (Figure 5).

This e–learning platform is usable for old or low vision students. An option offers the possibility to enlarge and decrease the size of the fonts used on the e-learning platform. This is especially relevant when the student has to prepare a presentation and then present it in the classroom using a video transmitter. There is no need to bring the presentation on a USB stick, since it can be viewed directly from this platform and projected to a white board. It is simple to increase the size of the font by pressing several times on the links,

Figure 4. Results of the search engine request appear thanks to a highlighted word

Figure 5. Results in context save students time by showing the result pages without opening them

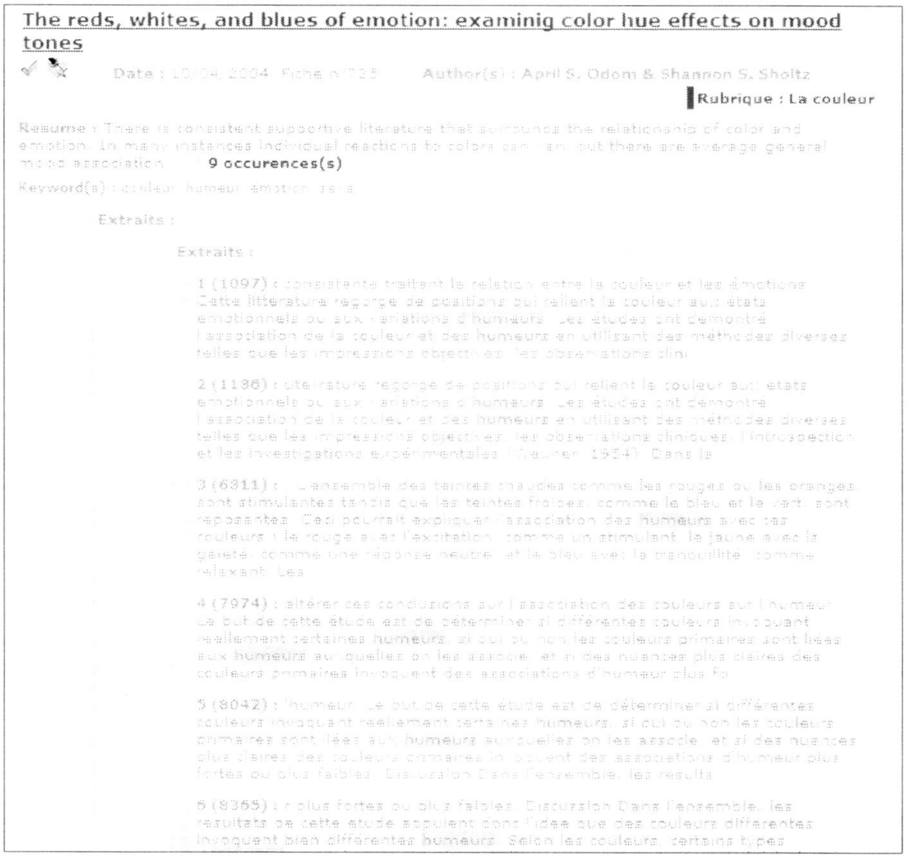

dedicated to this functionality (A, a), and the presentation thus appears in a proper manner on the screen. (Figure 6).

There is also a possibility to change both the foreground (text) and background colors, by using a scroll bar that modifies the hue, brightness and saturation levels of both of them. This functionality improves the quality of the readability. The system appears to be particularly relevant for colorblind students (8% of males and 0,5% of females), studying in their proper conditions of colors when setting the colored appearance of the screen thanks to this functionality. It therefore contributes in enhancing students memorization of the content as well as their trust in the e-learning platform. (Figure 6).

Here is an example of the platform after being modified on the color level (Figure 7).

The customization of e-learning systems for computing novices appears like a way to bridge the gap in education equity according to Rhee, Moon and Choe (2005). The possibility offered to modify the aspect of the interface can also contribute to play in favor of the sustainable development since learners can use proper tools in this direction. Dark screens effectively participate to save the electric energy of the screen. Thereby, e-learning contributes to protect the environment by the use of a function aiming at modifying the colors of the interface by preserving a good readability to stay concentrated.

A vertical navigation bar is also provided as part of the KMCMS interface. The description of

Figure 6. An e-learning platforms option allows to modify the size of the text, and the colors of both text and background

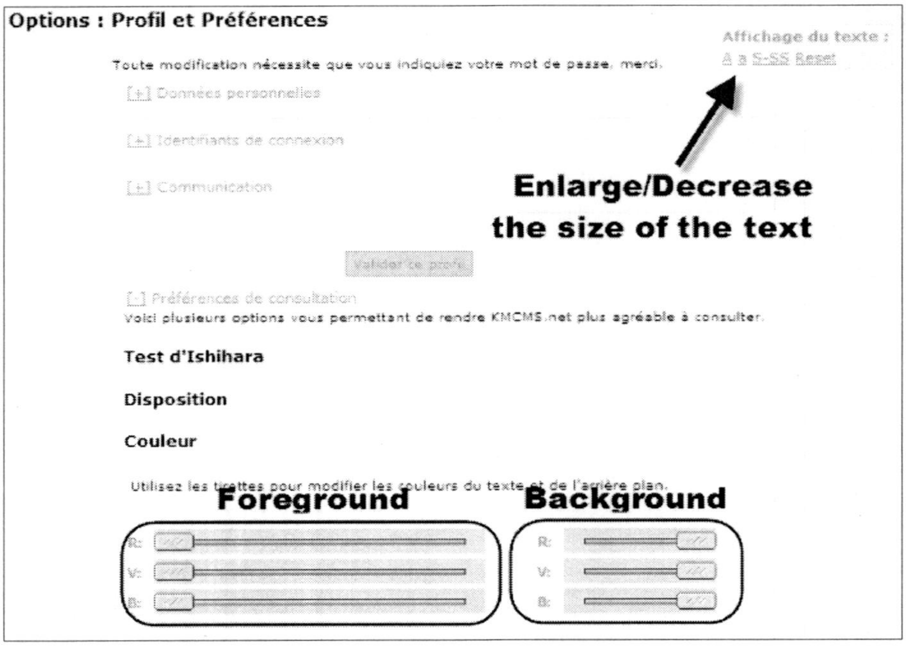

Figure 7. Appearance of the e-learning platform once its colors have been modified

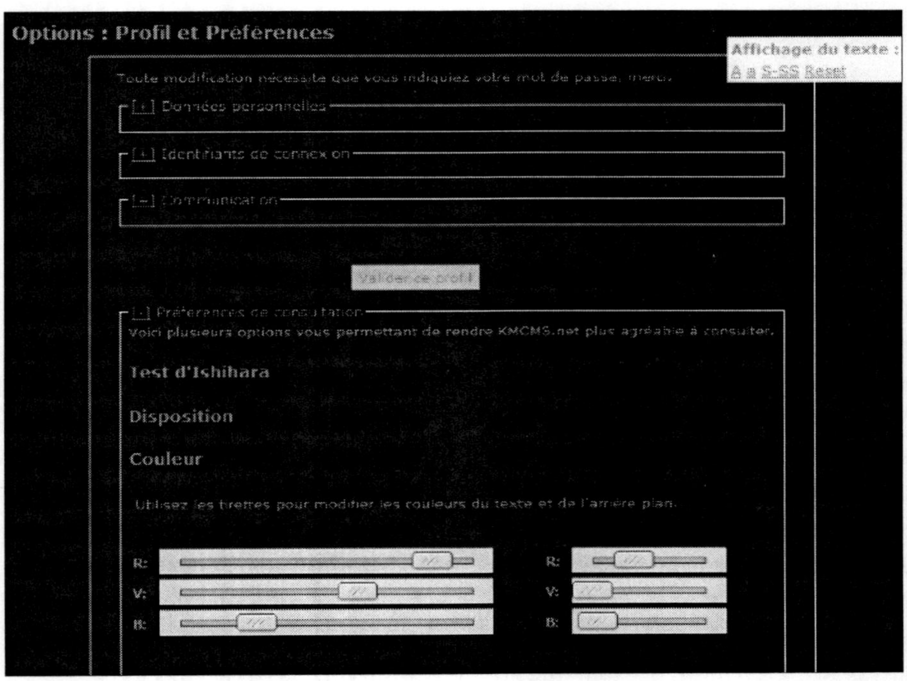

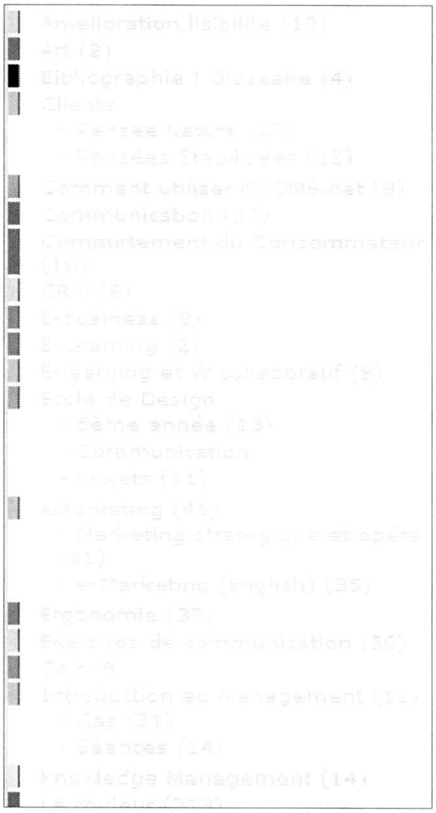

Figure 8. Each lecture owns its proper color, they reveal the different lectures that are available on the e-learning platforms at once

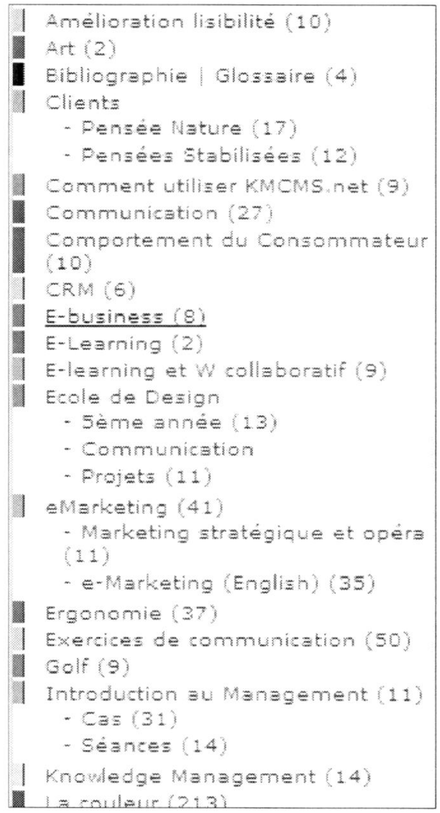

Figure 9. On rollover, the vertical navigation bar has a bigger contrast to get a better readability

the possibilities offered is explained between the two images (Figures 8 & 9).

In the vertical navigation bar, each domain, lecture, field etc. corresponds to a proper color, enabling the students to look for a particular content more easily, by recognizing the color dedicated to it.

The figures into brackets provide information to the student: the number of pages available for the lecture/field etc. This is another way to enhance student memorization of content and trust towards the whole system.

One can notice that in order to make the content easier to read, the navigation bar uses a light grey font: a pale color making it discrete during an "unused period" (Figure 8), whereas once the cursors is on rollover, on the e-business section on Figure 9, the grey becomes darker. It is supposed to help the student to read the menu of the navigation bar more easily thanks to a better contrast.

The same colors also appear when the student looks for content which is available on the platform. The domain of the content is then easier to recognize since the student approximately knows what he is looking for on the system. This can be even more relevant if he knows exactly which color corresponds to the lecture he has to read, learn and memorize. It is also another way to enhance his trust towards the system. It is noticeable that the number of pictures contained in each page is written in another color in order to provide information that can help the student, especially if he looks for particular content. (Figure 10)

Figure 10. Each page offers an aspect that is differentiated thanks to colors

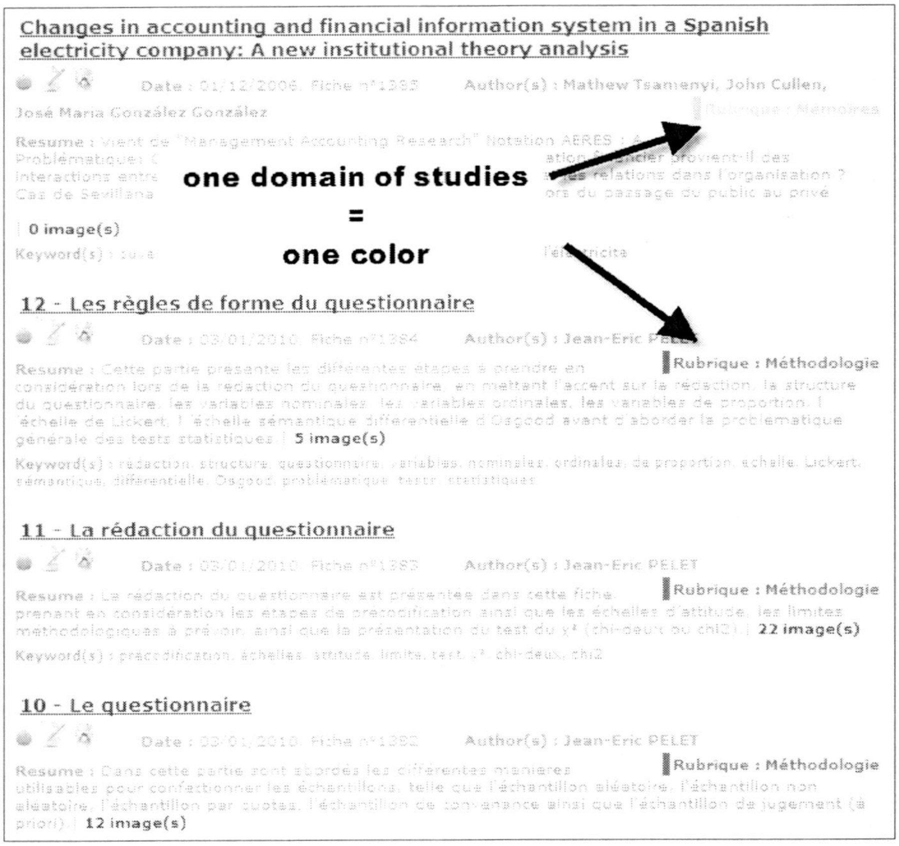

When a page is opened, it offers a common visual aspect in order to help students know where the information is on the interface. The content of the box (n°1 in Figure 11) uses the same background color as the title of the navigation bar already presented, and the same color of the course 'color'. This box - green on the following picture - always provides information such as:

- the date the page has been created,
- the domain it takes part of on the platform,
- the name(s) of the author(s) of the already read publication, in the case of a publication,
- the abstract of the page and the keywords chosen by the editor.

By preserving this consistency in the interface, the students are facilitated in memorizing the content and in trusting the platform.

A control bar (n°2 in Figure 11) is also placed on the right hand-side of the page, in order to facilitate the most repeated actions on the e-learning platform. This control bar will be described below. Finally, the content (n°3 in Figure 11) always appear in a proper format making reading easy, thanks to the chosen font, length of each line, size of the underline, and of course, use of styles such as bold, italic, underline and colors, following ergonomics guidelines (Figure 11). All these characteristics and their combination are important for increasing students memorization of the content and also their trust in the e-learning system.

Figure 11. Each page has a common interface in order to facilitate the memorization of the content

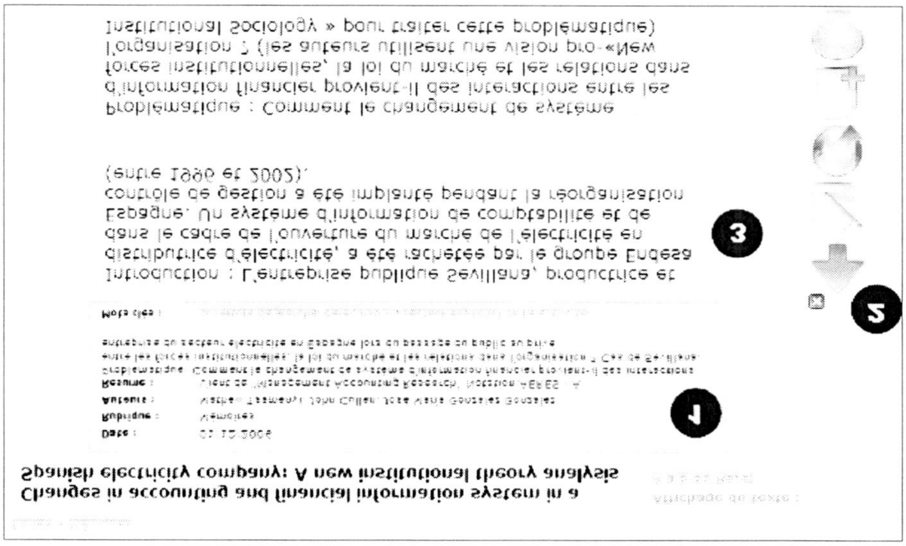

Figure 12.The control helps the student to gain time by accessing quickly to the main functionalities of the platform during a session on the platform

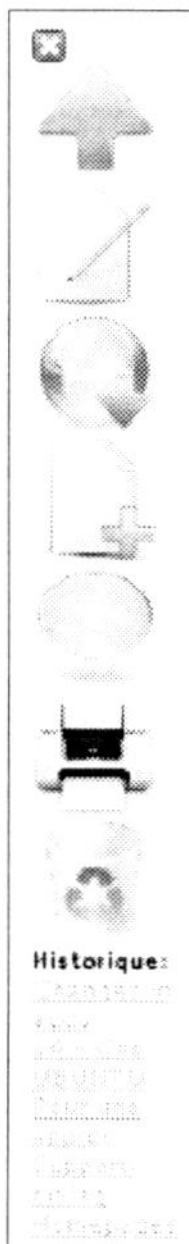

1 1: the **green arrow** is a *"super elevator"*, to quickly attain the top of page.

2 2: the **editor page** offers the possibility to modify a page built by the logged in student

3 3: the **earth link** indicates that there is a URL available to look for other resources

4 4: the **blank page** permits to create a new page

5 5: the **dialog bull** offers access to posted comments

6 6: the **printing machine** enables the student to print the page in a easy to read on paper (different than the Internet one) format

7 7: the **bin** deletes the page (available according to the profile of the student

8 8: these **green links** are shortcuts showing the 5 last visited pages.

The control bar appearing on each page of the e-learning platform is a tool which is particularly relevant when the time spent on the platform is important. Each button has effectively a proper role that will activate the navigation by making it more fluid. The control bar is very important

Figure 13. Each part of the content of each page of the e-learning platform can be modified easily thanks to a wiki system

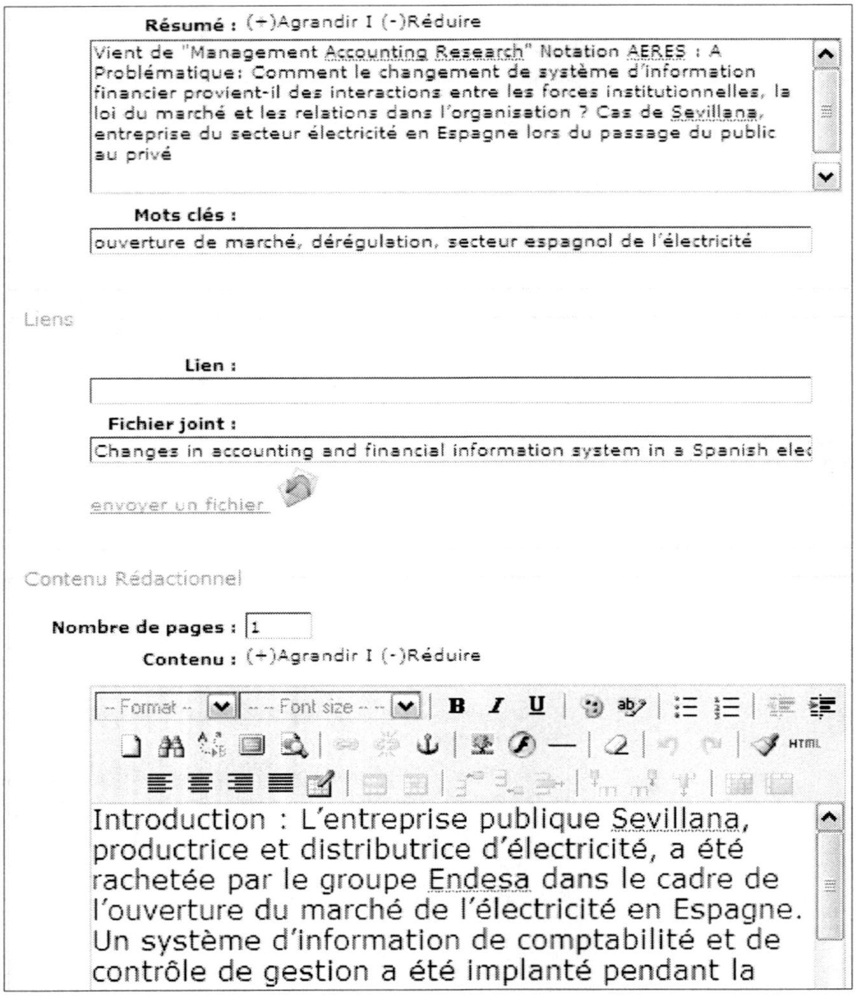

for students trust and memorization, particularly in navigating and using the interface (Figure 12).

Like in any content management system, a wiki platform enables the student to edit the content in a simple way (Figure 13). Each cell of the edition page is modifiable, and offers the common tools available in the Web 2.0 Internet community. Thus, a student can choose the font, its size and color; insert a link, an image, a flash or video content, and bullets and so on. This is very common nowadays, but it contributes to a better appropriation of the content by students when they decide to modify any part of it on their own. This is useful in enhancing trust in the content, based on personalized, student by student, content management.

In order to further augment student trust in the e-learning platform as well as in the e-learning content, the system shows when the page has been updated (Figure 14).

An e-learning platform, like a social network, can help students develop trust and stay loyal by offering them the possibility to comment what the professor says, or what their colleagues say.

Figure 14. The use of connectors serves the content since it provides information such as the date it has been updated among other functionnalities

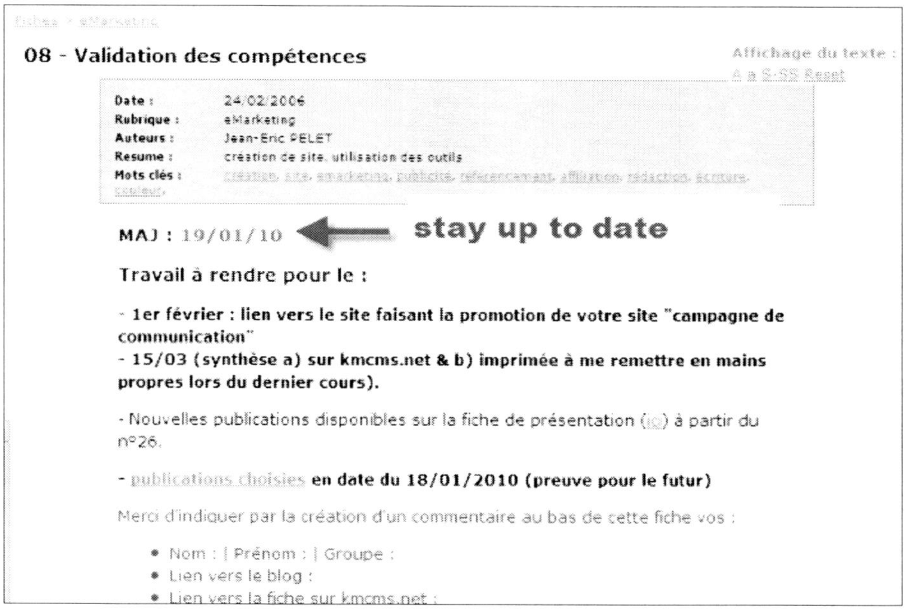

It obviously stimulates the dialog and enhances the trust towards the platform, especially if students are solicited to comment the content of pages. It also helps in memorizing the sequence of activities related to specific content. These comments can be described after the content of a lecture. An administration of these comments has not been useful for the last 6 years since students need to log in, being therefore identifiable by the professor (Figure 15).

Exams or whole class projects can also take part of the content of the e-learning platform, permitting the lecturer to follow the class progress in a single view. This feature facilitates not only students trust but also the educator trust in the e-learning platform. It also aids in keeping in memory the status of a project and the progress of students. It is then possible to think larger and work on major projects involving a whole group of students for example (Figure 16).

7. CONCLUSION

The design of an e-learning platform is of paramount importance for e-learning success, influencing learner interaction and behavior. This is in line with a large stream of research in information systems which have studied web design and have shown that web site quality and web site characteristics influence online consumer behavior (Torkzadeh & Dhillon, 2002; Singh *et al.*, 2005; Hampton-Sosa & Koufaris, 2005). Color constitutes an important variable for the design of e-learning platforms, as reported in information system studies, especially in usability, human-computer interaction and e-learning literature (Lee & Koubek, 2010; Coursaris *et al.*, 2008; Agarwal & Hedge 2008; McKracken *et al.*, 2004).

Memorization is a key goal of an e-learning system, ultimately aiming to make students learn and retain the educational content available. At the same time, trust in the e-learning platform is necessary in order for users to effectively adopt e-learning. The advanced usability of an e-learning

Figure 15. Comments of the page enrich its content, and enhance trust towards the platform

platform offered by a careful selection of colors can enhance students trust in the e-learning platform and the available content. Our conclusion follows Rhee, Moon and Choe (2005) results, showing that skilled students are generally more sensitive to interface consistency than novice students. The trust felt thanks to a proper assemblage of colors can enhance the appropriation of the e-learning platforms and help novices to become skilled, in order to perform more quickly and easily to exams. This study extends previous research on color in the context of e-learning and website design and contributes to existing information systems literature offering findings with important implications. Our study is the first, to our knowledge, that examines memorization in the context of e-learning. We introduce trust as a factor influencing memorization, contributing to the existing body of literature on antecedents of memorization in e-learning. We also examined two factors influencing memorization, color and trust.

The contribution of our study to information systems and more specifically e-learning literature mainly lies in providing findings on color that can inform and guide the design of websites in e-learning so that they are effective for attracting learner interest. By looking into trust, as a variable, which is an antecedent of retention and is influenced by color, our study further contributes in information systems. The research implications are most important for trust, as it is proposed as a new factor predicting memorization of online learners.

Figure 16. An entire class can work on a single page easily, for big projects for example

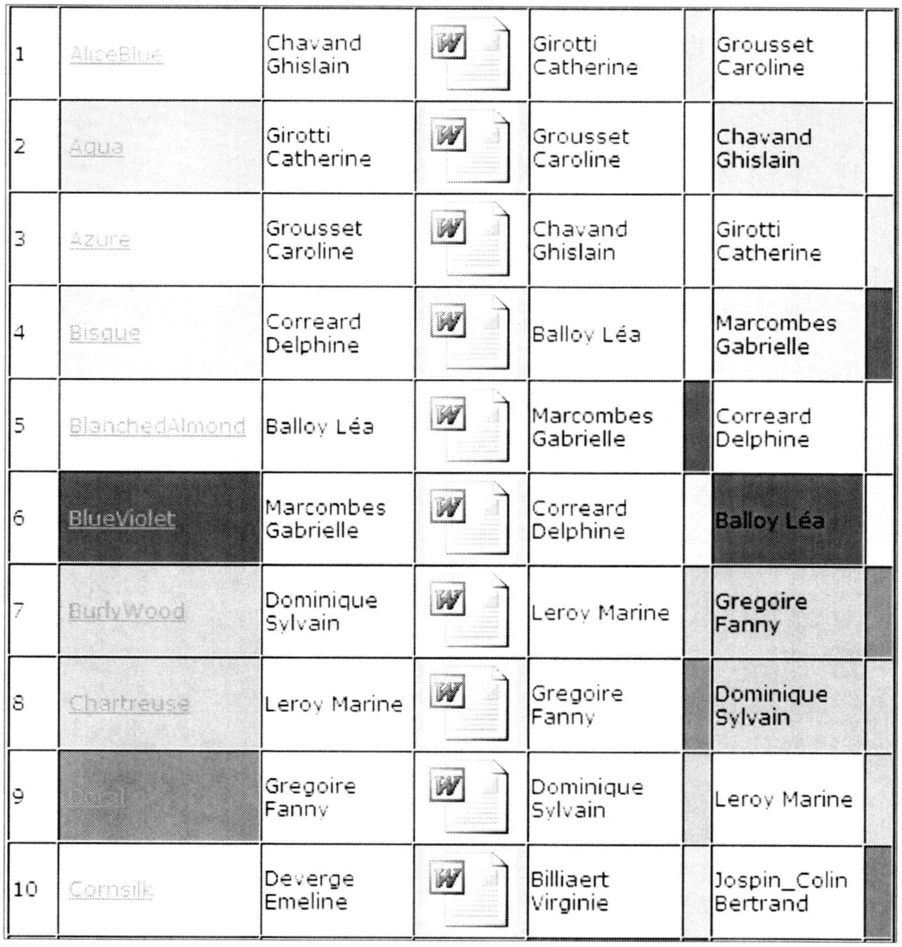

1	AliceBlue	Chavand Ghislain		Girotti Catherine		Grousset Caroline	
2	Aqua	Girotti Catherine		Grousset Caroline		Chavand Ghislain	
3	Azure	Grousset Caroline		Chavand Ghislain		Girotti Catherine	
4	Bisque	Correard Delphine		Balloy Léa		Marcombes Gabrielle	
5	BlanchedAlmond	Balloy Léa		Marcombes Gabrielle		Correard Delphine	
6	BlueViolet	Marcombes Gabrielle		Correard Delphine		Balloy Léa	
7	BurlyWood	Dominique Sylvain		Leroy Marine		Gregoire Fanny	
8	Chartreuse	Leroy Marine		Gregoire Fanny		Dominique Sylvain	
9	Coral	Gregoire Fanny		Dominique Sylvain		Leroy Marine	
10	Cornsilk	Deverge Emeline		Billiaert Virginie		Jospin_Colin Bertrand	

Our results on the effect of color on trust and memorization are not only relevant to e-learning web design but to web interface design in general and are also of value for research in web aesthetics, usability and human-computer interaction. Hence, our findings are relevant and important to information systems researchers, particularly to those active in the areas of usability, human-computer interaction and e-learning or the intersection of these areas.

These results must be related to the studies conducted by Silverstein (1987) who noticed that monochrome screens entailed more eyestrain and overall tiredness. Therefore, e-lecturers should be aware of this and choose carefully the colors that they will use on their site so as to adjust them to their target. They should also take into account the aesthetic and functional impact of those colors: their contrast facilitates finding information on a webpage. Learners recall information more easily when they had difficulty reading on an e-learning website.

REFERENCES

Accessiweb. (2008). Du référentiel AccessiWeb créé par le consortium W3C. Retrieved from http://www.accessiweb.org/

Agarwal, A., & Hedge, A. (2008). The impact of Web page usability guideline implementation on aesthetics and perceptions of the e-retailer. *Proceedings of the Human Factors and Ergonomics Society*, (pp. 528-532).

Ando, M., & Ueno, M. (2008). Effect of pointer presentation on multimedia e-learning materials. In *Proceedings of World Conference on Educational Multimedia, Hypermedia and Telecommunications*, (pp. 5549-5559). Chesapeake, VA: AACE.

Anwar, M., & Greer, J. (2008). *Enabling reputation-based trust in privacy-enhanced learning systems*. (5091 LNCS), (pp. 681-683).

Aspillaga, M. (1991). Screen design: Location of information and its effects on learning. *Journal of Computer-Based Instruction, 18*(3), 89–92.

Barber, B. (1983). *The logics and limits of trust*. New Brunswick, NJ: Rutgers University Press.

Blau, P. (1964). *Exchange and power in social life*. New York: John Wiley and Sons.

Butler, J. K. (1991). Toward understanding and measuring conditions of trust: Evolution of the conditions of trust inventory. *Journal of Management, 17*, 643–663. doi:10.1177/014920639101700307

Buzhardt, J., Abbott, M., Greenwood, C., & Tapia, Y. (2005). Usability testing of the class-wide peer tutoring-learning management system. *Journal of Special Education Technology, 20*(1), 19–29.

Cantoni, V., Cellario, M., & Porta, M. (2004). Perspectives and challenges in e-learning: Towards natural interaction paradigms. *Journal of Visual Languages and Computing, 15*, 333–345. doi:10.1016/j.jvlc.2003.10.002

Chang, J. E., Simpson, T. W., Rangaswamy, A., & Tekchadaney, J. R. (2002). *A good website can convey the wrong brand image! A preliminary report*. Working Paper, E-Business Research Center (EBRC), Université de Pennsylvanie.

Chou, C. (2003). Interactivity and interactive functions in Web-based learning systems: A technical framework for designers. *British Journal of Educational Technology, 34*(3), 265–279. doi:10.1111/1467-8535.00326

Chu, L. F., & Chan, B. K. (1998). Evolution of website design: Implications for medical education on the Internet. *Computers in Biology and Medicine, 28*, 460–472. doi:10.1016/S0010-4825(98)00027-4

Clark, J. M., & Paivio, A. (1991). Dual coding theory and education. *Educational Psychology Review, 3*, 149–210. doi:10.1007/BF01320076

Coleman, J. S. (1990). *Foundations of social theory*. Cambridge, MA: Harvard University Press.

Coursaris, C., Swierenga, S., & Watrall, E. (2008). An empirical investigation of color, temperature, and gender effects on Web aesthetics. *Journal of Usability Studies, 3*(3), 103–117.

Crowther, M. S., Keller, C. C., & Waddoups, G. L. (2004). Improving the quality and effective of computer-mediated instruction through usability evaluations. *British Journal of Educational Technology, 35*(3), 289–303. doi:10.1111/j.0007-1013.2004.00390.x

Dailey, L. (2004). Navigational Web atmospherics explaining the influence of restrictive navigation cues. *Journal of Business Research, 57*, 795–803. doi:10.1016/S0148-2963(02)00364-8

De Groot, A. (1965). *Thought and choice in chess*. The Hague, Netherlands: Mouton.

Deutsch, M. (1960). The effect of motivational orientation upon trust and suspicion. *Human Relations, 13*, 123–139. doi:10.1177/001872676001300202

Divard, R., & Urien, B. (2001). *Le consumer vit dans un monde en couleurs* (pp. 3–24).

Dwyer, F. M., & Lamberski, R. J. (1982-83). A review of the research on the effects of the use of colour in the teaching-learning process. *International Journal of Instructional Media, 10*(4), 303–328.

Ellis, K., & Blashki, K. (2004). Toddler techies: A study of young children's interaction with computers. *Information Technology in Childhood Education Annual, 2004*, 77–96.

Ettinger, A., Holton, V., & Blass, E. (2006). E-learner experiences: What is the future for e-learning? *Industrial and Commercial Training, 38*(4), 208–212. doi:10.1108/00197850610671991

Gauss, B., & Urbas, L. (2003). Individual differences in navigation between sharable content objects–an evaluation study of a learning module design. *British Journal of Educational Technology, 34*(4), 499–509. doi:10.1111/1467-8535.00346

Gefen, D., Karahanna, E., & Straub, D. W. (2003). Trust and TAM in online shopping: An integrated model. *Management Information Systems Quarterly, 27*(1), 51–90.

Gibson, J. J. (1977). The theory of affordances. In Shaw, R., & Bransford, J. (Eds.), *Perceiving, acting and knowing*. Hillsdale, NJ: Erlbaum.

Hall, R. H., & Hanna, P. (2004). The impact of Web page text-background colour combinations on readability, retention, aesthetics, and behavioral intention. *Behaviour & Information Technology, 23*(3), 183–195. doi:10.1080/01449290410001669932

Hampton-Sosa, W., & Koufaris, M. (2005). The effect of website perceptions on initial trust in the owner company. *International Journal of Electronic Commerce, 10*(1), 55–81.

Hinostroza, J. E., & Mellar, H. (2001). Pedagogy embedded in educational software design: Report of a case study. *Computers & Education, 37*, 27–40. doi:10.1016/S0360-1315(01)00032-X

Itten, J. (1970). *The elements of color*. New York: Van Nostrand Reinhold Company.

Jarmon, L., Traphagan, T., Mayrath, M., & Trivedi, A. (2009). Virtual world teaching, experiential learning, and assessment: An interdisciplinary communication course in Second Life. *Computers & Education, 53*, 169–182. doi:10.1016/j.compedu.2009.01.010

Jonassen, D. H., & Wang, S. (1993). Acquiring factual knowledge from semantically structured hypertext. *Journal of Computer-Based Instruction, 20*(1), 1–8.

Kang, K.-S., & Corbitt, B. (2001). Effectiveness of graphical components in website e-commerce application-a cultural perspective. *Electronic Journal on Information Systems in Developing Countries, 7*(2), 1–6.

Kintsch, W. (1970). Models for free recall and recognition. In Norman, D. A. (Ed.), *Models of human memory*. New York: Academic Press.

Kobayashi, S. (1986). Theoretical issues concerning superiority of pictures over words and sentences in memory. *Perceptual and Motor Skills, 63*, 783–792.

Lee, S., & Koubek, R. J. (2010). The effects of usability and web design attributes on user preference for e-commerce websites. *Computers in Industry, 61*, 329–341. doi:10.1016/j.compind.2009.12.004

Lemoine, J.-F. (2008). Atmosphère des sites web marchands et réactions des apprenants. *Revue Française de Marketing, 217*, 2/5.

Lewicki, R. J., & Bunker, B. B. (1996). Developing and maintaining trust in work relationships. In Kramer, R. M., & Tyler, T. (Eds.), *Trust in organizations: Frontiers of theory and research* (pp. 114–139). Thousand Oaks, CA: Sage Publications.

Lewis, J. D., & Weigert, A. J. (1985). Trust as a social reality. *Social Forces, 63*(4), 967–985. doi:10.2307/2578601

Livingston, L. A. (1991). The effect of colour on performance in an instructional gaming environment. *Journal of Research of Computing in Education, 24*(2), 246–253.

Maruyama, Y., & Akahori, K. (2008). Effect of bichrome prints on memory in the middle-school educational environment. In K. McFerrin, et al. (Eds.), *Proceedings of Society for Information Technology and Teacher Education International Conference,* (pp. 4650-4655). Chesapeake, VA: AACE.

Mattelart, A. (1996). *The invention of communication*. London, Minneapolis: University Minnesota Press.

Mayer, R. C., Davis, J. H., & Schoorman, F. D. (1995). An integrative model of organizational trust. *Academy of Management Review, 20*(3), 709–734. doi:10.2307/258792

Mayer, R. E. (2001). *Multimedia learning*. New York: Cambridge University Press.

Mayer, R. E., & Gallini, J. K. (1990). When is an illustration worth ten thousand words? *Journal of Educational Psychology, 82*, 715–726. doi:10.1037/0022-0663.82.4.715

McAllister, D. J. (1995). Affect- and cognition-based trust as foundations for interpersonal cooperation in organizations. *Academy of Management Journal, 38*(1), 24–59. doi:10.2307/256727

McCracken, D. D., & Wolfe, R. J. (2004). *User-centered website development: A human-computer interaction approach*. Upper Saddle River, NJ: Pearson Prentice Hall Inc.

McKnight, D. H., Cummings, L. L., & Chervany, N. L. (1998). Initial trust formation in new organizational relationships. *Academy of Management Review, 23*(3), 473–490. doi:10.2307/259290

Mehrabian, A., & Russell, J. A. (1974). *An approach to environmental psychology*. Cambridge, MA: MIT Press.

Meyer, B. J. F. (1975). *The organization of prose and its effect on memory*. Amsterdam: North Holland Publishing Company.

Meyer, B. J. F., Brandt, D. M., & Bluth, G. J. (1980). Use of top-level structure in text: Key for reading comprehension of ninth-grade students. *Reading Research Quarterly, 16*, 72–101. doi:10.2307/747349

Moore, D. M., & Dwyer, F. M. (1997). Effect of colour-coding on locus of control. *International Journal of Instructional Media, 24*, 145–151.

Moorman, C., Zaltman, G., & Deshpande, R. (1992). Relationships between providers and users of market research: The dynamics of trust within and between organizations. *JMR, Journal of Marketing Research, 29*, 314–328. doi:10.2307/3172742

Orton-Johnson, K. (2009). I've stuck to the path, I'm afraid: Exploring student non-use of blended learning. *British Journal of Educational Technology, 40*(5), 837–847. doi:10.1111/j.1467-8535.2008.00860.x

Ozok, A. A., & Salvendy, G. (2003). The effect of language inconsistency on performance and satisfaction in using the Web: Results from three experiments. *Behaviour & Information Technology, 22*(3), 155–163. doi:10.1080/0144929031000107081

Pastoor, S. (1990). Legibility and subjective preference for color combinations in text. *Human Factors, 32*, 157–171.

Pelet, J.-É. (2008). *Effets de la couleur des sites marchands sur la mémorisation et sur l'intention de l'internaute.* Thèse de doctorat en Sciences de Gestion, Université de Nantes

Pelet, J.-É. (2010a). Using Web 2.0 social computing technologies to enhance the use of Information Systems in organizations. In Papadopoulou, P., Kanellis, P., & Martakos, D. (Eds.), *Social computing theory and practice: Interdisciplinary approaches*. Hershey, PA: IGI Global. doi:10.4018/9781616929046.ch007

Pelet, J.-É. (2010b). Effets de la couleur des sites web marchands sur la mémorisation et sur l'intention d'achat. *Systèmes d'Information et Management*, *15*(1), 97–131.

Pelet, J.-É., & Papadopoulou, P. (in press). The effect of e-commerce websites colors on customer trust. *International. Journal of Business Research.*

Pruitt, D. G. (1981). *Negotiation Behavior*. New York: Academic Press.

Reisman, S. (2003). *Electronic learning communities*. Connecticut: Information Age Publishing.

Rhee, C., Moon, J., & Choe, Y. (2005). Web interface consistency in e-learning. *Online Information Review*, *30*(1), 53–69. doi:10.1108/14684520610650309

Richard, M. O. (2005). Modeling the impact of internet atmospherics on surfer behavior. *Journal of Business Research*, *58*, 1632–1642. doi:10.1016/j.jbusres.2004.07.009

Rotter, J. (1967). A new scale for the measurement of interpersonal trust. *Journal of Personality*, 651–665. doi:10.1111/j.1467-6494.1967.tb01454.x

Rubens, P., & Krull, R. (1985). Application of research on document design to online displays. *Technical Communication*, *32*(4), 29–34.

Schmidt, K. E., & Liu, Y. (2005). *Design of consumer product Web pages: Experimental investigations of aesthetic and performance factors* (pp. 1743–1746). Human Factors and Ergonomics Society.

Scott, B., Shurville, S., Maclean, P., & Cong, C. (2007). Cybernetic principles for learning design. *Kybernetes*, *36*(9/10), 1497–1514. doi:10.1108/03684920710827445

Shieh, K. K., Chen, M. T., & Chuang, J. H. (1997). Effects of color combination and typography on identification of characters briefly presented on VDTs. *International Journal of Human-Computer Interaction*, *9*(2), 169–181. doi:10.1207/s15327590ijhc0902_5

Silverstein, L. D. (1987). Human factor for color CRT display systems: Concepts, methods and research. In Durrett, H. J. (Ed.), *Color and the computer* (pp. 27–61). Orlando: Academic.

Singh, S. N., Dalal, N., & Spears, N. (2005). Understanding Web home page perception. *European Journal of Information Systems*, *14*, 288–302. doi:10.1057/palgrave.ejis.3000525

Siqueira, S. W. M., Braz, M. H. L. B., & Melo, R. N. (2007). Modeling e-learning content. *International Journal of Web Information Systems*, *3*(1/2), 140–152. doi:10.1108/17440080710829261

Soulier, J. (1988). *The design and development of computer based instruction*. Toronto, ON: Allyn and Bacon.

Steinberg, E. R. (1991). *Computer-assisted instruction: A synthesis of theory. Practice, and technology*. Hillsdale, NJ: Lawrence Erlbaum.

Stewart, K. J. (2003). Trust transfer on the World Wide Web. *Organization Science*, *14*(1), 5–17. doi:10.1287/orsc.14.1.5.12810

Sweller, J. (1999). *Instructional design in technical areas*. Melbourne, Australia: Acer Press.

Sweller, J. (2002). *Visualisation and instructional design*. International Workshop on Dynamic Visualizations and Learning, July, 18.-19., Knowledge Media Research Center (KMRC), Tübingen, Germany.

Szabo, M., & Poohkay, B. (1994). So what if it's in color and moves? A critique of multimedia. *Proceedings of the World Conference on Ed-Media/EdTelecomm 94*. Vancouver, Canada. Retrieved on June 29, 2006, from http://www.quasar.ualberta.ca/IT/research/Szabo/poohkay.html

Torkzadeh, G., & Dhillon, G. (2002). Measuring factors that influence success of Internet commerce. *Information Systems Research*, *13*(2), 187–204. doi:10.1287/isre.13.2.187.87

Tricot, A. & Plégat-Soutjis, F. (2003). Pour une approche ergonomique de la conception d'un dispositif de formation à distance utilisant les TIC. *Sciences et Technologies de l'Information et de la Communication pour l'Éducation et la Formation, 10*.

Weinreich, H., Obendort, H., Herder, E., & Mayer, M. (2006). Off the beaten tracks: Exploring three aspects of Web navigation. *Proceedings of the 15th international conference on World Wide Web table of contents*, Edinburgh, Scotland, (pp. 133-142).

Williamson, O. E. (1975). *Markets and hierarchies, analysis and anti-trust implications*. New York: Free Press.

Wright, P., Lickorish, A., & Milory, R. (2000). Route choices, anticipated forgetting, and interface design for online reference documents. *Journal of Experimental Psychology. Applied*, *6*(2), 158–167. doi:10.1037/1076-898X.6.2.158

Wu, C.-S., Cheng, F.-F., & Yen, D. C. (2008). The atmospheric factors of online storefront environment design: An empirical experiment in Taiwan. *Information & Management*, *45*, 493–498. doi:10.1016/j.im.2008.07.004

Zaheer, A., McEvily, B., & Perrone, V. (1998). Does trust matter? Exploring the effects of interorganizational and interpersonal trust on performance. *Organization Science*, *9*, 141–159. doi:10.1287/orsc.9.2.141

Zucker, L. G. (1986). Production of trust: Institutional sources of economic structure. In Staw, B. M., & Cummings, L. L. (Eds.), *Research in organizational behavior* (pp. 53–111). Thousand Oaks, CA: Sage Publications.

Chapter 5
Technology, Trust and B2B Relationships:
A Banking Perspective

Raechel Johns
University of Canberra, Australia

ABSTRACT

Technology has altered business processes, and the use of self-service technologies changes the nature of service delivery. The importance of developing and fostering relationships with customers has long been regarded as important within services marketing (Berry, 1983) and also within B2B relationships (Ford, 1990). In the 1980s and 90s, a shift in marketing focus has seen an increased emphasis on relationship marketing (Morgan & Hunt, 1994). This chapter reviews the results and implications of recent exploratory research conducted with a small sample of Australian business bank customers (n=20). This research makes a contribution toward answering the research question "what impact does the use of self service technologies have on relationships within a business-to-business context?" Utilising a banking context, this research explores how the use of Internet banking impacts on relationships between the bank and its business customer. In exploring this research area, a greater understanding of banking relationships and business banking customers has been identified.

DOI: 10.4018/978-1-60960-501-8.ch005

INTRODUCTION

Technology has dramatically altered the way businesses operate in a Business-to-Business (B2B) context and has had profound influences on services, altering the way services are delivered (Bitner, Ostrom & Meuter, 2002). The increased use of self-service technologies could have a great impact on B2B relationships, yet there is a shortage of research in the area. Electronic banking is one technology that has streamlined business transactions, encouraged by banks as a way to reduce service delivery costs and improve service quality for customers (Australian Bankers Association, 2000). Banking has attempted to move what used to be a physical transaction to one that is virtual, having implications for relationships. Despite this, there is little understanding about how well received this is – particularly when considering business customers.

The importance of developing and fostering relationships with customers has long been regarded as important within services marketing (Berry, 1983) and also within B2B relationships (Ford, 1990). In the 1980s and 90s, a shift in marketing focus has seen an increased emphasis on relationship marketing (Morgan and Hunt, 1994). This chapter examines the Australian banking industry, and discusses relationship marketing in the context of banking. The chapter reviews the results and implications of recent exploratory research conducted with a small sample of Australian business bank customers (n=20).

Through analysing the literature, it is evident that there are several key reasons self-service technologies have been introduced in banking. They are cost savings, switching costs and relationships from the bank perspective; relationships and convenience from the customer perspective. This chapter provides an overview for the reasons self-service technologies has been used in banking. It will highlight key relationship marketing theory, and a model of the expected relationship will be provided. Qualitative research will be discussed to highlight the impact of technology on business-to-business relationships. This chapter will indicate that the old rules still apply in banking, however banking is conducted differently than it was in the past.

REASONS FOR INTRODUCTION OF TECHNOLOGY IN A BANKING CONTEXT

Bank Perspective: Cost Savings

Competitive pressures from other banks and building societies have forced banks to reconsider the way they do business. Over the past several decades, distribution of banking has been seen as a means of providing customer service and reducing costs (Sathye, 1999; Hughes & Hughes, 2004) and differentiating themselves from competitors (Easingwood & Storey, 1996). Self-service delivery has become a driving force in banking since the mid 1990s (Pikkarainen, Pikkarainen, Kajaluoto & Pahnila, 2004), and, in general, the use of Self Service Technologies in the industry is increasing. Automatic teller machines (ATMs) (Howcroft, 1993; and Carson, Gilmore & Walsh, 2004), telephone banking, and ultimately internet banking were introduced to enhance distribution, improve banking processes and make cost savings (Hughes & Hughes, 2004). This rapid increase in the use of Internet banking within financial services (Eriksson & Marquardt, 2001) highlights it as an appropriate industry for investigation, due to the removal, or at least, reduction, of face-to-face relationships.

Whilst cost savings are essential, pressure to increase profitability has meant that banks are required to focus on developing and maintaining long terms relationships with customers (Kandampully & Duddy, 1999); particularly in a B2B context, where relationships are key to keeping customers (Adamson, Chan & Handford, 2003; Hawke & Heffernan, 2005a). Regardless of whether technology is utilised or not, customers still seek quality when adopting services (Bitner, 2001). Generally customers benchmark their service delivery against dependable outcomes, easy access, responsive systems, flexible response to customer needs and apologies if a service delivery goes wrong (Bitner, 2001). Therefore it is imperative that service delivery is appropriate for business customers due to the importance in relationship building. Cost savings benefit the bank, while relationship building provides a benefit to both the bank and business customer.

Mutual Perspective: Relationship Focus

From a business-to-business perspective, banking is also interesting as both standard products and tailor-made products are utilised. Long-term relationships are often likely within banking (Eriksson & Marquardt, 2001). Though there is a considerable amount of research on the impact of technology on banking relationships, this is drawn from a personal customer perspective, which is different to business customers. There is evidence to suggest that personal banking customers and business banking customers utilize different banking activities.

Table 1 indicates how personal and business banking customers users utilise online banking.

Though there could be differences in reporting by the authors, it is important to note that business bankers utilise internet banking for more specialised activities, such as invoicing and receipt creation (Johns & Perrott, 2008). This indicates the importance of examining businesses separately from personal banking.

Although literature suggests that relationship activities in banking include the physical evidence, process activities and people activities (Carson, Gilmore & Walsh, 2004) and product, distribution and pricing activities are considered transaction activities (Carson, Gilmore & Walsh, 2004), it is essential to consider the way technology has altered the emphasis on transactions. It is assumed that internet banking is a transaction marketing activity because of the nature of the activity. The customer does not engage with staff on an interpersonal level, but rather, conducts their relationship at a distance. If internet banking is a transaction activity, it is therefore focused on distribution and pricing, rather than a relationship focus.

Table 1. Usage of online banking

Activity	Percentage of Australian online bankers who undertake activity	Percentage of online Australian business banking customer who undertake activity
Use financial services	28	-
Check account balances	90	100
Transfer funds between accounts	78	-
Schedule payments	48	90
Transfer money to third parties	47	100
Withdrawls	-	40
Creation of Invoices	-	60
Receipt creation	-	60
Stopping cheques	-	60
Refunding customers	-	50

Source: Personal banking data adapted by Greenspan, 2003; Business banking data adapted by Johns and Perrott, 2008

Because customers highly value banking relationships (Colgate & Alexander, 1998; Barnes 1997) it is important to consider relationship marketing in the context of this study. Strong relationships with business customers foster increased profit, improved communication and an increase in satisfaction, creating loyalty (Petersen & Rajan, 1994). Within banking, it is suggested that interpersonal liking could impact on the likelihood of maintaining a business relationship – and factors such as personality, trust, professionalism, communication and similarities appear to influence interpersonal liking (Hawke & Heffernan, 2005b). This interpersonal liking leads to the outcomes of commitment; cooperation and business referrals (Hawke & Heffernan, 2005a) as the outcome. What impact does the removal of face to face conduct have on relationships and what is the benefit to the customer when technology is used?

Customer Perspective: Convenience

While cost savings have motivated banks to introduce self-service technologies, the main attraction of internet banking from a customer perspective is convenience. Customers tend to seek convenience in products where they have no interest, perceive the product as a commodity, and trust the supplier (Boyes, 2003). It is evident that banking would fit in with these categories, and therefore it would be expected than consumers would seek convenience in their banking habits. For business customers, convenience is necessary to ensure that business processes are managed efficiently.

Bank Perspective: Switching Costs

Apart from offering customers convenience, and providing cost savings for the bank, there is a very important reason for banks to encourage customers to utilise Internet banking. This is because customers who utilise Internet banking perceive the switching costs of swapping banks as being higher (Pyun, Scruggs & Nam, 2002). As a result, banks see Internet banking as a way to retain customers and increase market share (Gardener, Howcroft & Williams, 1999). This has obvious benefits for banks in terms of fostering commitment, important in a business-to-business context (Morgan & Hunt, 1994).

Within commercial banking, it has been noted that brand switching generally occurs because competitors' offers were better, and current provider's fees were too high (Bogomolova & Romaniuk, 2005; Colgate, Stewart & Kinsella, 1996; Colgate and Hedge, 2001). This is in direct contrast to that of consumer markets, where price dissatisfaction was not the major reason for leaving a bank (Keaveney, 1995). These differences highlight the need for evaluating the impact of self service technologies on business relationships, as Business to Consumer studies prove insufficient in this context.

SATISFACTION AND DISSATISFACTION WITH SELF-SERVICES

Satisfaction with service delivery is greater where the customer interacts directly with technology, compared to when they do not have direct access (Dabholkar, 1994). As an example, Internet banking allows the customer direct access while telephone banking does not (Jospeh, McClure and Joseph, 1999). Previous research within a Business to Consumer context indicates that sources of satisfaction in banking include: attentiveness, responsiveness, care, availability, reliability, integrity, friendliness, courtesy, communication, competence and functionality (Johnston, 1995), however this study is limited in that it looks at face to face relationships (not Self Service Technologies) and a Business to Consumer context (not business-to-business). Within the Business to Consumer sector, research indicates it is highly important that electronic banking provides records that all transactions have taken place, and guarantees that all transactions have taken place. It is also important that the bank is able to satisfy any complaints within 24 hours (Joseph, McClure & Joseph, 1999). Research needs to consider important satisfaction issues from a Business-to-Business perspective.

Previous research has found that the number of contacts with a bank decreases when internet banking is used, however, users of Internet banking are more likely to try new financial products than those that do not utilize Internet banking (Eriksson & Marquardt, 2001). Ultimately it seems that the use of technology enhances relationships between banks and their customers, and indicates the importance of Internet Banking to banks. Eriksson and Marquardt's study (2001) was conducted with personal bank customers, rather than in a B2B context, and does not provide an understanding about the impact this plays on the customer's perception of value in the relationship. Therefore, it is essential to understand relationship-marketing principles in this study.

RELATIONSHIP MARKETING

Prior to understanding the impact of technology on relationships, it is essential to clearly identify what relationship marketing is. This is more problematic than it may seem with over fifty published definitions (Dann & Dann, 2001). There has been many attempts at defining relationship marketing over the years and yet there are commonalities between them. In 1999, Harker undertook a content analysis of definitions, and this paper is a good starting point for defining relationship marketing. As a single definition needs to be determined and used throughout the research, it is important to determine one to use. The definition Harker (1999) settled on after the content analysis of twenty-six definitions encompasses all necessary key words judged by a series of academic evaluators. This definition is as follows:

"Relationship marketing includes tasks undertaken to "Identify and establish, maintain and enhance and, when necessary, terminate relationships with customers and other stakeholders, at a profit, so that the objectives of all parties involved are met; and this is done by mutual exchange the fulfillment of promises" (Grönroos, 1994: 9).

This definition indicates the importance of interaction. Relationship marketing has mutual benefit to both parties. Furthermore, the long term, commitment focus is indicated. Commitment and trust are part of the 'emotional content' of RM definitions. Although commitment and trust do not appear to be highlighted in many definitions of relationship marketing (Harker, 1999), several authors (Morgan & Hunt, 1994; and O'Malley & Tynan, 2000) discuss the importance in their papers in terms of relationship marketing theory (Harker, 1999). Furthermore, Bennett (1996) does highlight commitment and trust in his definition of relationship marketing. Trust and Commitment are difficult to test as likelihood to stay in a relationship may be due to switching costs and other factors, not simply due to satisfaction.

Relationship marketing is particularly relevant in a services context. Services are based on interactive marketing, and, therefore, direct contact is necessary (Grönroos, 1994) relying on relationship marketing principles. In fact, services marketing was the discipline that first introduced relationship marketing in the 1970s after debates that the marketing mix was insufficient for services (Möller & Halinen, 2000) requiring more interactive processes. In services, customers are active participants in the service delivery process, and trust and ultimately commitment are essential constructs (Kelley & Davis, 1994). Relationship marketing is, therefore, viewed as essential in business-to-business relationships, and particularly within services. While the establishment of trust can build relationships, there is little understanding as to how the use of information technology can impact on relationships.

TRUST AND COMPUTER MEDIATED ENVIRONMENTS

The intangible nature of services makes them difficult to evaluate prior to service, making decision making more complicated for a customer. However, if a customer trusts a service provider, they are likely to go back (Berry, 2002) and therefore trust and commitment is important. Building relationships with customers can enhance trust and ultimately commitment (Morgan & Hunt, 1994), and the way organizations recover from service failures has huge implications for customer evaluation of a firm (Tax, Brown & Chandrashekaran, 1998; Reichheld, 1993).

Trust is a key area of relationship marketing, particularly important within services marketing due to the intangibility of services (Berry & Parasuraman, 1991). This means that perceived risk of the purchase is increased, relative to goods (Murray & Schlacter, 1990). Trust has been heavily researched in the past, however this is mostly from a Business to Consumer (B2C) approach (Jevons & Gabbott, 2000), rather than a B2B approach. Relationship marketing is most essential in a B2B context due to the fragile nature and importance of inter-firm relationships (Dibb & Meadows, 2004) and therefore it is essential to consider trust in a business to business context.

Trust exists when one party has confidence in the reliability of the exchange partner (Morgan & Hunt, 1994; and Dwyer & Tanner, 2002). Relationship commitment, defined as "an exchange partner believing that an ongoing relationship with another is so important as to warrant maximum efforts at maintaining it" (Morgan & Hunt, 1994:23) is viewed as essential for both relationship marketing and Services Marketing (Berry & Parasuraman, 1991). However, commitment is only fostered through trust, because trust leads to repeat patronage by customers (Berry, 2002), and this loyalty can foster commitment.

Trust can be developed through shared interactions over time (Young & Wilkinson, 1989), and is related to a number of elements such as competitive advantage and satisfaction (Ratnasingam & Pavlou, 2003). The literature indicates there are many types of trust, but in business relationships, there are two key types of trust– trading partner trust and technology trust (Ratnasingam & Pavlou, 2003). Trading partner trust is about trust between the network participants. When purchasing from other businesses, either for use within the business, or for resale, issues such as product quality, delivery issues and prices are concerns for buyers, while sellers are concerned with payment (Violino, 2002). Technology trust indicates a trust in the technology infrastructure to conduct interfirm exchanges (Ratnasingam & Pavlou, 2003). Obviously when considering the impact of technology on relationships, the concept of technology trust in particular needs consideration.

What impact does the use of IT have on business relationships? Literature indicates the importance of "getting it right" with technology (Stone & Woodcock, 1997; Joseph, 1998; Grimm, 1999; and Lang & Colgate, 2003). Customers dissatisfied with the amount of technology use in their relationships (that is, they would prefer to use more or less) perceive their relationship to be weaker than those satisfied with the amount of technology use (Lang & Colgate 2003). This research, undertaken in a B2C context, is an important finding, indicating the impact of IT use on relationships but does not provide an overall understanding of contexts where it is appropriate and how much is "enough". Doesn't it matter at all, is it simply considered a value added service? It is, therefore, essential for organisations to consider customer satisfaction with IT as ultimately this will contribute toward satisfaction with the relationship and no doubt commitment to the relationship – all key constructs in relationship marketing.

Services such as automated package shipping, internet banking and airline log in could easily be perceived as complementary services which enhance relationships. From a business-to-business context, additional services such as goods ordering and tracking systems could also be viewed as value adding. However, if this became the only choice, customers can become frustrated. Furthermore, if something went wrong with these services, customers require human beings to assist (Bitner, 2001). The literature indicates that too much technology or too little is problematic in relationships (Lang & Colgate, 2003). In business today, it appears some technology is necessary. Indeed, even in the implementation of relationship marketing, technology is a requirement. In addition to traditional direct contact with customers, customer orientated service systems and a database is necessary (Grönroos, 1994). With the understanding, therefore, that technology is necessary; the question for every marketing practitioner becomes "how much is enough?" Is there some kind of ratio marketers can work toward with a balance between face to face contact and self-service? Does it differ depending on the industry or demographics of the customer? Additionally, do different customers have different requirements?

Relationship marketing, from a services marketing perspective, emphasises the importance of personal relationships between a marketer and customer. This is in direct contrast from that of database marketing where relationships are distant and more often carried out through mass communication (Möller & Halinen, 2000). Are self-service technologies to be considered similar to database marketing, or are they more similar to personal interactions? If direct contact with customers is required in relationship marketing, does this mean that relationships conducted through self-service cannot be deemed to be "relationship marketing"? It would appear so, yet the literature also suggests that relationship marketing is necessary in services marketing (Morgan & Hunt 1994; Berry & Parasuraman, 1991) in order for an organisation to be successful. Perhaps self-service technologies fall into their own category entirely (Johns, Low & Blackman, 2009).

While there is some suggestion that IT use within relationships could have an impact (Jones, 1996), there is little discussion of the direction of the impact. There is evidence to suggest that IT has a positive impact on business relationships (Stone & Woodcock, 1997; Joseph, 1998; Grimm, 1999), but this appears to be when used as "complementary" to a face to face relationship, rather than when self-service technologies are forced. Furthermore, there is some suggestion in the literature that that IT does not inhibit the creation of trust, satisfaction or commitment central to relationship marketing (Comer, Mehta & Holmes, 1998). While the literature suggests that complementary IT services can add value to a relationship (Stone & Woodcock, 1997; Joseph, 1998; Grimm, 1999), there seems to be a requirement for face-to face contact to ensure trust, and ultimately relationship commitment, something all marketers are striving for. This study examines this.

Because customer perceptions of quality are often drawn from their satisfaction with their relationship with the marketer (Berry & Parasuram, 1993; Möller & Halinen, 2000) it is imperative that marketers consider relationships between themselves and their customers. In services, customers are active participants in the service delivery process, increasing the importance of trust and commitment (Kelley & Davis, 1994), essential areas for relationship marketing (Morgan & Hunt, 1994).

To summarise - technology could have a negligible impact on firms relationships with customers, but there has been very little research done in this area (Lang & Colgate 2003). The research undertaken seems to indicate that when an additional service is offered, there is little or no impact on relationships, however, there is no discussion of the impact on relationships when technology forms a major part of the relationship, nor when the relationship is a key business-to-business relationship, for example, in the case of business banking. The exploratory study was conducted to determine the impact of technology on B2B relationships.

The Exploratory Study

As introduced before, there are four key reasons for the introduction of self-service technologies in banking. Cost savings and the reduction of switching costs from the bank's perspective all relate to profit making, and are less focused on the customer. The relationship benefits and convenience which may come from the use of self-service technologies are more customer-centric, but there is limited understanding about the impact of technology on relationships. Banks seeking to co-create and collaborate with customers could introduce proprietary software with the business customer. To offer value, banks appear to see it in terms of the bank's perspective. Rather than using technology to build relationships, banks utilise their staff. Finally, banks strive for customer-centric relationships, to build trust and commitment. Table 2 indicates how each of these relate to the reason for introduction of Self-Service Technologies.

Due to the nature of banking and how it fits in with the four key areas of relationship marketing, it is ideal to explore the research objectives in this industry. From a thorough analysis of the literature, three objectives for the study have been determined:

- *Objective 1*: To examine relationship marketing theory in the context of self-service technologies
- *Objective 2*: To understand how the reduction of face-to-face contact can impact on relationships in a B2B context
- *Objective 3*: To understand how the use of technology impacts on trust and relationship commitment in a B2B context

Methodology

The previous sections of this chapter outlined a gap in the literature and three research objectives. To explore this study, a series of twenty qualitative interviews were conducted with business customers. This section of the chapter will discuss the research findings from the study. The questions asked sought to explore the question "what impact do internet banking have on relationships within a B2B context".

Prior to addressing the sample for this study, it is important to note that in business to business research there are two distinct groups in the research – the marketer or "focal firm" (Rowley, 1997) and the customer. A marketer may supply goods or services to one organisation, and be a customer of another. In a bank to business customer

Table 2. Self-service technology introduction and relationship to relationship marketing and service dominant logic

Reason for introduction	Co-creation and collaboration	Value	Resource Focus	Relationship Requirements
Cost saving		a		
Relationships		a	a	a
Convenience	a			
Switching costs	a			

Source: Developed for this study

perspective, it is possible to clearly define the role of the marketer versus the customer. As mentioned previously, the user often differs from the decision maker in business (Zaltman, Dunaca & Holbeck, 1973; Leonard Barton & Deschamps, 1988), both users and the decision maker were investigated in this study. Further, samples of focal firms (Banks) were also investigated as their decision-making is expected to influence the business relationship.

The size of the sample is less important than data richness (Patton, 1990). When research provides homogenous results, as few as ten interviews is viewed as sufficient (Sandelowski, 1995). Therefore, in this study a reasonably small sample was utilised with a total of twenty interviews, or "cases" (Patton, 1990). If there are vast differences in interviews, and new information is continually revealed, it will be more important to keep interviewing to best understand the issues relating to business use of Self Service Technologies. An appropriate sample size in qualitative research is generally a matter of judgement in terms of how useful the information is (Sandelowski, 1995) and in this study, it was deemed that twenty was appropriate. Perhaps more important than the number of subjects in qualitative research is the information that is received from the subjects (Patton, 1990). As a result, the research has been carried out until it is believed enough information has been achieved to successfully analyse the results or achieve "theoretical saturation" (Eisenhardt, 1989). Furthermore, in qualitative research,

Table 3. Overview of interview respondents

Case #	Industry	Decision maker or user	Gender of interviewee	Estimated age group of interviewee	Length in role; length with bank	Org size
1	Insurance	Decision maker	Male	50-60	30; 30	Medium
2	Non profit	Decision maker	Male	30-40	3; 10	Medium
3	Warranties	User	Female	20-30	3; 5	Small
4	Insurance	User	Female	30-40	2; 30	Medium
5	Retail	User	Female	30-40	4; 10	Small
6	Investments	Decision Maker	Male	50-60		Small
7	Retail	User	Female	20-30	5; 10	Small
8	IT	Decision maker	Male	20-30	6; 6	Small
9	Medical	User	Female	20-30	4;4	Small
10	Investments	Decision Maker	Female	30-40	4; 7	Small
11	Property	Decision Maker	Male	30 – 40	15; 20	Medium
12	Services – Tours	Decision Maker	Male	30-40	3; 0.5	Small
13	Corporate Gifts	Decision Maker	Female	30 – 40	2; 10	Small
14	Telecommunications	User	Female	40 – 50	3;6	Large
15	Building	Decision maker	Male	40-50	6; 6	Small
16	Fitness	User	Female	30-40	1; 5	Small
17	Building	Decision Marker	Female	30 - 40	5; 5	Small
18	Golf – Services	Decision Maker	Male	30-40	6; 6	Medium
19	Casino – Services	User	M ale	30-40	3; 10	Large
20	Retail	Decision Maker	Female	30-40	2; 4	Small

Organisation size: Small < 10 employees; Medium 11- 99 employees; Large >100 employees
Source: Developed for this study

sample sizes can be too large, not allowing for sufficient analysis of each case (Sandelowski, 1995). As a result, it was determined that twenty customers formed an appropriate sample size for this exploratory study.

Table 3 provides an overview of the respondents.

RESULTS

The Use of Internet Banking

To commence the interviews, respondents were asked to discuss their experiences with internet banking (IB) within the organisation. Most of the respondents found their experience with using IB had been relatively positive, within the business. All but one of the respondents utilised internet banking for their personal (non business) accounts, as well. This indicates the experience level of respondents, to better understand how that relates to the use of IB within the organisation.

In order to explore the way IB is used, respondents were asked how the organisation used the technology in their normal business activities. This was important, because it gave a greater understanding as to the importance of tasks, and therefore how much the organisation relies on internet banking. It was obvious through the interviews that internet banking is useful in keeping up to date on account details, often on a daily basis. One statement which typifies responses to the question was: "*in the morning, I do the account balance, which means I print up a list of all the transactions we've done over the last day or two, and balance those against our records. So we're able to see when cheques go through and all of that basically as it happens. We can also stop cheques, do fund transfers to customers who want refunds... and things like that*".

In order to understand why businesses utilise IB, their motivations for adopting it were examined. This was elicited by asking informants how and why they came to adopt IB, although this was not necessary in all cases as a number informants discussed this without prompting. Media reports had previously indicated that organisations may have felt compelled to adopt IB, however this was not the case. Instead, feelings tended to be quite positive and there were clearly justified reasoning as to why IB was in their organisation's best interests. For example, it is perceived as "*a lot easier and quicker... because you're able to do it yourself, you're not relying on the bank to do it for you. You can just get online in the morning and do what you need to do.*"

One respondent, a volunteer treasurer for a non-profit organisation, felt there were very clear reasons to utilise IB. There seemed to be different issues from a volunteer organisation to other organisations. "*Being a volunteer non profit organisation, it's very hard to get everyone together in one spot to sign cheques and that sort of thing. So time was a crucial factor... A lot of suppliers were insisting in fact that we move to electronic payment. The change (from using traditional banking) was ... mainly my time commitments. I didn't want to spend my life being an unpaid accountant for this organisation and I could see it happening very quickly. And it was very difficult to get people to meet, because they were just unwilling to do so at the one place and one time to sign cheques. And I would find I was doing all the running around for them and not the other way around, so I thought 'enough's enough'*". The respondents felt that convenience and time saving were major advantages of utilising internet banking within the business, therefore benefiting the organisation. Despite this, there was some degree of feeling forced to use IB. " *(I do feel forced), because it's the easiest alternative. There's not too many other ways that you could do it that's practical and efficient.*" This was in direct contrast to another interviewee, who did not feel forced at all: *I don't have to (use internet banking), it just makes my life... easier. It just means one trip to the bank, not several. No telephone banking... it's easier.*"

The research also indicates that IB does not remove all face-to-face contact with the bank, as it depends on the organisation structure. This is illustrated by the following quote "*It's easier for everyone just to have electronic payment... oh, many of our tenants are still using cheques. I still have to go to the bank.*"

Loyalty

In order to best understand how positive the relationship was, loyalty to the relationship was explored. Through the interviews, it was obvious that there was a feeling of some degree of loyalty. Is it loyalty, however, or simply are switching costs too high? This was indirectly discussed by an accounts keeper who appeared less certain of the loyalty: "*to an extent (we are loyal), because it would be pretty hard to change banks – just because we've already got all our records in there, and all of our daily direct debits are in there... But I don't know if that is really a loyalty so much as a convenience of using the bank that we've been using. (It would be hard to switch banks) because... well actually, (the bank we're with just) changed their system, and I found that hard to just learn using their new system let alone changing the bank and discovering a whole new system.*"

Similarly, the treasurer for the non profit organisation felt little loyalty. He explained that it was switching costs that kept him with his bank, and that we would love to switch banks: "*because the way it's set up... it's not user friendly. It's been based from the bank's point of view which is fair enough, because that's where the finance system needs to operate from ... security, trust... and confidentiality of the system. But for this organisation, and... a lot of other organisations, ... it makes the assumption that ... they have someone in the office that can just walk over and enter their log in, user name and password details...*". He explained that this was particularly relevant in a non profit organisation. In contrast, interviewee three, an accountant and business owner felt a great degree of loyalty, as illustrated by this quote: "*Yes, I'm loyal. We've been with them over thirty years, so...*"

Perhaps decision makers are always more certain of their loyalty. For instance, one respondent felt that loyalty was 'inherited' from her employer, which is an interesting finding. As she explained "*I think my boss is (loyal)! I don't have a strong opinion. They're (banks) all the same, I guess*". This potentially raises the question about whether or not this is common place in terms of banking loyalty, that is, my boss trusts the bank, and therefore I will, too.

Not all decision makers are loyal and trusting, though. A young small business owner, when asked if he was loyal to his bank replied "*loyal? No. They're not loyal to me when they're introducing this fee, or that fee, and that's okay... they're a profit making business, but so am I. That's what we're both loyal to*".

Loyalty, or commitment to the relationship, is essential to understand – it indicates how likely it is that relationships will be maintained, and also the potential impact self-service technologies could have on these business relationships. As indicated in the literature review, relationships in a B2B context are essential. Further, relationships within banking are viewed as very important; therefore a sense of commitment, or loyalty, to the bank is important to understand. Through the interviews, there is evidence of a strong loyalty to the banks, or at minimum, a perception of high switching costs.

Trust and Value

As trust is such an important aspect of relationship marketing, the level of trust of the bank was examined to determine whether IB changed their trust level. All respondents trusted their bank, with one surprised that there would be any reason not to. "*Yeah, absolutely (I trust them). But it never occurred to me not to. I think banks in general...*

(are trustworthy)". It was interesting to see how naturally trust for the bank came to the respondents.

The respondents' perception of "value" was also explored, and value meant different things for different respondents. This is the most troubling issue for banks, as there is no one solution on what value is to business customers. Some customers felt that relationships created value. *"I like having someone to call if I need to organise a loan. I like that he knows who I am,"* a respondent with strong banking relationships stated. In contrast, another respondent believed that all that was important was the transaction side of banking *"I would rather not get mail from them, not have to meet up with them, and just get on with my business"*. The concept of value for this organisation was simply transactions that worked. In the twenty interviews conducted, three interviewees had a similar concept of value – transactions that work. The similarities with these three were that they were all small business owners, with a positive cash flow, who did not borrow funds from the bank. The three organisations were all service orientated businesses, relatively young (under 45 years of age) and confident with technology and felt that if they could do it online, rather than dealing with a staff member, it was preferable. Perhaps this is a subset of the market, or perhaps this group of customers will increase as consumers gain confidence using the technology. Perhaps relationships in business banking are less important that the literature posits.

Problems with Internet Banking

Finally, in order to better understand whether there were any problems with the technology or use of IB within the business, respondents were asked about any issues. This was a concluding question designed to complete the interview, but provided more insight about the perceptions of the technology and even the service delivery from the bank.

The respondent from the non-profit organisation had difficulty with the logistics of organisational internet banking. He felt that non-profit organisations, or at least volunteer organisations, had different issues. The volunteers within this organisation are not the same place, at the same time, unlike most traditional organisations. Their bank requires two internet banking log ins, as their account is a two signatory account. This is best illustrated by the following quote: "*Internet banking is difficult (for us)t because the banks require you to be at the same place at the same time. Unless you have their log in details and password – which you're not meant to do, because banks tell you not to give out passwords and log in details. ... to other people – for obvious reasons."*

The question regarding issues was asked towards the end of another interview. This particular interviewee has been relatively quiet throughout his entire interview, only answering the bare minimum. This interviewee required the most probing overall, however, he became much more responsive when discussing issues with his internet banking system. Therefore, it is concluded that this is a relatively major issue for him, and has impacted on his satisfaction with the system. This was regarding a change implemented by the bank which was not well received by the organisation, as indicated in the following quote "*Well I'm still using the old (system)... if I can, because the new one's hopeless. I mean they've actually had issues ... I've seen it in the press... they've been involved in court action with the people who did it, and ... so it hasn't really been a success for them, I think. From what I know of it." He* also struggles to get help for the new system: "T*he helpdesk with Quickline was good. I've found the ... that I've used the helpdesk with, what is it? Commbus... I haven't used it very much, but in the initial stages when I was having problems with it, I used it a fair bit. I found them to be not very helpful. ... Actually going back to your question about being forced to use internet banking... I don't feel forced to use internet banking, but I certainly was forced to use ... commbus."*

Perhaps the most important part of this interview is the fact that this respondent was most interested in talking about issues, despite discussing how loyal and committed he was to his bank just minutes prior. He seemed to see the issues with IB as separate from that of his banking relationship. Perhaps this is because the bank had another organisation develop the technology, and is obviously doing something about it. In the interim, he is sometimes being able to use his preferred system. The helpdesk is the only thing that could be solved at this point by the bank and appears to have had little investigation in terms of service delivery. This respondent, therefore, feels a great deal of loyalty to his bank, having been with them for over thirty years, but doubts the service delivery and feels forced to use a particular system. It has not yet influenced his view of the bank, rather just his perception of the technology.

Implications for Banks

The research indicated limited evidence to suggest that the removal of face-to-face contact impacts on B2B relationships in any way, despite the literature indicating that face-to-face relationships were important in a B2B context. It is important to understand how the use of technology impacts on trust and relationship commitment in a B2B context. While there was little evidence to suggest that technology impacts on trust or commitment to the relationship, it was evident that trust of the bank and commitment to the bank actually impacts on the perception of the technology, which was unexpected and interesting. Nevertheless, there were some limitations of the research, which must be discussed.

This research was aimed at exploring the following research question: "what impact does the use of business internet banking have on relationships within a B2B context?" It was apparent through the research that internet banking has had little impact on relationships, but that the existing relationship quality can impact on the perception of the technologies. This is an important finding, which requires further investigation to determine how important the pre-existing relationship is at influencing the perception of technology quality. This research was conducted because it was expected that technology would impact on the relationships between bankers and commercial customers. After conducting exploratory research, it appears that this is not the case – in fact, it is the relationship, which impacts on the perception of the technology. Therefore, the old rules still apply for bank managers – relationship marketing is still essential in the day of technology. Surprisingly, face-to-face relationships were not seen as that important for building loyalty, but all interviewees had an established relationship prior to the use of internet banking. How this will change when the younger generation are establishing business relationships is still to be determined. It would be interesting to research this further with Generation Y business people. Quantitative research will be utilised to empirically test these areas. As banks seek ways to be more competitive, it is essential to understand the way in which the use of self-service technologies could impact on business relationships.

As technology is continually improved and utilised in organisations, it is essential to have an understanding of how technology impacts on relationships. Whilst a number of studies indicate how products are adopted or accepted, few examine B2B relationships and self-service technologies. Self-service technologies have altered the relationship between the service deliverer and customer, however, customers still demand outcomes that are dependable, provide easy access, flexibility and compensation when problems arise (Bitner, 2001).

Overall, it appears that the level trust has a large impact on the perception of the relationship, and the perception of the relationship influences how respondents see the self-service technologies. There were no obvious decision maker attributes influencing trust, nor were there any obvious user attributes influencing trust.

Limitations

The major limitation for this research was that a number of businesses did not wish to be involved. As banking is considered to be a private matter, and security concerns are heavily promoted, particularly in terms of Internet banking, many people contacted for interviews were wary. It is expected that there was some self-selection bias, where only those that felt they could contribute to the topic participated. Furthermore, it was anticipated that the use of Internet banking would feel like a forced decision, but it appears that this was not the case, changing the direction of the research and the discussion. Lessons learnt from this, however, is to allow the respondent to talk more openly about their banking relationship, as this is where more insights were provided, such as problems with the system, or feeling the help desk did not fulfil their needs. Consequently, there are some implications for practice and further research. The three major objectives for the research were explored while attempting to answer the research question: "what impact does the use of business internet banking have on relationships within a B2B context?"

SUMMARY

Whilst a number of studies indicate how products are adopted or accepted, few examine B2B relationships and self-service technologies. Self-service technologies have altered the relationship between the service deliverer and customers, however, customers' still demand outcomes that are dependable, provide easy access, flexibility and compensation when problems arise (Bitner, 2001). The level trust has a large impact on the perception of the relationship, and the perception of the relationship influences how respondents see the self-service technologies. Overall, the most insight gleaned from the interviews was that the stronger the relationship, the more likely the respondents were to think positively of the technology, or at the very least, not think negatively of Internet Banking. This was in direct contrast to the expected relationship, where it was anticipated that the respondents' perception of the technology would impact on the perception of the relationship. In the case where there was a negative relationship with the bank, the respondent did not find the system useful or easy to use, therefore perception of the technology was negative. Further, while there was little evidence that the quality of the relationship impacted on relationship length, there was evidence to suggest that where there was a weaker relationship, respondents were more likely to want to leave the relationship. Another interesting finding was that there appears to be a subset of customers who do not wish to have any kind of relationship with their bank, perceiving value from transactions that "work". This needs further investigation in research. The majority of customers, however, did perceive value from a relationship with the bank, but the perception of value differed depending on whether the respondent was a business owner/manager or banking user.

The objectives of this chapter were:

- *Objective 1:* To examine relationship marketing theory in the context of self-service technologies. This was achieved, and examined areas such as trust and commitment when using self-service technologies; however further research is required to examine this further, particularly in terms of commitment.
- *Objective 2:* To understand how the reduction of face-to-face contact can impact on relationships in a B2B context. There was little evidence to suggest that the removal of face-to-face contact impacts on B2B relationships in any way, despite the literature indicating that face-to-face relationships were important in a B2B context.

- *Objective 3:* To understand how the use of technology impacts on trust and relationship commitment in a B2B context. There was little evidence to suggest that technology impacts on trust or commitment to the relationship, but trust of the bank and commitment to the bank actually impacts on the perception of the technology, which was unexpected and interesting.

As businesses seek ways to be more competitive, it is essential to understand the way in which the use of self-service technologies could impact on business relationships.

REFERENCES

Adamson, I., Chan, K., & Handford, D. (2003). Relationship marketing: Customer commitment and trust as a strategy for the smaller Hong Kong corporate banking sector. *International Journal of Bank Marketing*, *21*(6/7). doi:10.1108/02652320310498492

Australian Bankers Association. (2000). *The banking industry and overview.*

Barnes, J. (1997). *Exploring the importance of closeness in customer relationships*. (pp. 227-40). Paper presented at American Marketing Association Conference, Dublin.

Bennett, R. (1996). Relationship formation and governance in consumer markets: Transactional analysis versus the behaviourist approach. *Journal of Marketing Management*, *12*(12), 417–436. doi :10.1080/0267257X.1996.9964425

Berry, L. (1983). *Relationship marketing*. Chicago: AMA.

Berry, L. (2002). Relationship marketing of services–perspectives from 1983 and 2000. *Journal of Relationship Marketing*, 1.

Berry, L., & Parasuraman, A. (1991). *Marketing services—competing through quality*. New York: Free Press.

Bitner, M. (2001). Service and Technology: Opportunities and paradoxes. *Managing Service Quality*, *11*(6). doi:10.1108/09604520110410584

Bitner, M., Ostrom, A., & Meuter, M. (2002). *Implementing successful self-service technologies*. Paper presented at Academy of Management Executive, (pp. 96-119).

Bogomolova, S., & Romaniuk, J. (2005). *Why do they leave? An examination of the reasons for customer defection in the business banking industry*. Paper presented at ANZAMC 2005.

Boyes, G. (2003). E-Business opportunities in financial services. *Journal of Financial Services Marketing, 8*.

Carson, D., Gilmore, A., & Walsh, S. (2004). Balancing transaction and relationship marketing in retail banking. *Journal of Marketing Management*, *20*(3/4), 431–455. doi:10.1362/026725704323080489

Colgate, M., & Alexander, N. (1998). Banks, retailers and their customers: A relationship marketing perspective. *International Journal of Bank Marketing*, *16*(4). doi:10.1108/02652329810220693

Colgate, M., & Hedge, R. (2001). An investigation into the switching process in retail banking services. *International Journal of Bank Marketing*, *19*(5). doi:10.1108/02652320110400888

Colgate, M., Stewart, K., & Kinsella, R. (1996). Customer defection: A study of the student market in Ireland. *International Journal of Bank Marketing*, *14*(3). doi:10.1108/02652329610113144

Comer, J. M., Mehta, R., & Holmes, T. L. (1998). Information Technology: Retail users versus nonusers. *Journal of Interactive Marketing*, *12*(2), 49–62. doi:10.1002/(SICI)1520-6653(199821)12:2<49::AID-DIR5>3.0.CO;2-D

Dabholkar, P. (1994). Incorporating choice into an attitudinal framework: Analysing model of mental comparison process. *The Journal of Consumer Research, 21*, 100–118. doi:10.1086/209385

Dann, S., & Dann, S. (2001). *Strategic Internet marketing*. Milton, Australia: John Wiley and Sons.

Dibb, S., & Meadows, M. (2004). Relationship marketing and CRM: A financial services case study. *Journal of Strategic Marketing, 12*, 111–125. doi:10.1080/0965254042000215177

Dwyer, F., & Tanner, J. (2002). *Business marketing: Connecting strategies. Relationship and learning*. Boston: McGraw Hill.

Easingwood, C., & Storey, C. (1996). The value of multi-distribution systems in the financial services sector. *Service Industries Journal, 16*(2), 223–241. doi:10.1080/02642069600000023

Eisenhardt, K. M. (1989). Building theories from case study research. *Academy of Management Review, 14*, 532–550. doi:10.2307/258557

Eriksson, K., & Marquardt, R. (2001) *Is relationship marketing theory applicable to Internet bank relationships?* Paper presented at the 17th IMP-conference in Oslo, Norway.

Ford, D. (1990). *Understanding business markets: Interaction, relationships and networks. The IMP Group*. Academic Press Limited.

Gardener, E., Howcroft, B. & Williams, J. (1999). The new retail banking revolution. *The Service Industries Journal, 19*(2).

Greenspan, R. (2003). *What they're doing down under*. Retrieved on April 15, 2004, from http://www.clickz.com/stats/big_picture/geographics/article.php/5911_3090361

Grimm, M. (1999). Ford connects: Hatches massive relationship program via Web. *Brandweek, 40*(1), 19–20.

Grönroos, C. (1994). From marketing mix to relationship marketing-towards a paradigm shift in marketing. *Management Decision, 32*(2). doi:10.1108/00251749410054774

Harker, M. (1999). Relationship marketing defined? An examination of current relationship marketing definitions. *Marketing Intelligence & Planning, 17*(1). doi:10.1108/02634509910253768

Hawke, A., & Heffernan, T. (2005a). *The outcomes of interpersonal liking in lender-customer relationships in the Australian banking sector.* Paper presented at the ANZMAC Conference, 2005.

Hawke, A., & Heffernan, T. (2005b). *Developing interpersonal liking in lender-customer relationships in the Australian banking sector*. Paper presented at the British Academy of Marketing Conference Proceedings, Dublin.

Howcroft, B. (1993). Branch networks and alternative distribution channels. *International Journal of Bank Marketing, 11*(6). doi:10.1108/02652329310045738

Hughes, R., & Hughes, A. (2004). *The adoption of Internet banking services by young Australian consumers*. Paper presented at the AusWeb Conference, Gold Coast, Australia.

Jevons, C., & Gabbott, M. (2000). Trust, brand equity and brand reality in Internet business relationships: An interdisciplinary approach. *Journal of Marketing Management, 16*(6), 619–634. doi:10.1362/026725700785045967

Johns, R., Low, D., & Blackman, D. (2009). Information Technology and relationship marketing in an inter-firm context: Implications for research. Paper presented at the ANZMAC Conference, Melbourne.

Johns, R., & Perrott, B. (2008). The impact of internet banking on business-customer relationships. Are you being self-served? *International Journal of Bank Marketing, 26*(7), 465–482. doi:10.1108/02652320810913846

Johnston, R. (1995). The determinants of service quality: Satisfiers and dissatisfiers. *International Journal of Service Industry Management*, *6*(5), 53–71. doi:10.1108/09564239510101536

Jones, N. (1996). Talking pages. *Marketing Week*, *19*(18), 37–40.

Joseph, M., McClure, C., & Joseph, B. (1999). Service quality in the banking sector: The impact of technology on service delivery. *International Journal of Bank Marketing*, *17*(4). doi:10.1108/02652329910278879

Joseph, R. (1998). *Customer orientation and the creation of new technology based services*. Electronic Media and Executive Education.

Kandampully, J., & Duddy, R. (1999). Competitive advantage through anticipation, innovation and relationships. *Management Decision*, *37*(1), 51–56. doi:10.1108/00251749910252021

Keaveney, S. (1995). Customer switching behaviour in service industries: An exploratory study. *Journal of Marketing*, 59.

Kelley, S., & Davis, M. (1994). Antecedents to customer expectations for service recovery. *Journal of the Academy of Marketing Science*, *22*(1), 52–61. doi:10.1177/0092070394221005

Lang, B., & Colgate, M. (2003). Relationship quality, on-line banking and the Information Technology gap. *International Journal of Bank Marketing*, *21*(1), 29–37. doi:10.1108/02652320310457785

Leonard-Barton, D., & Deschamps, I. (1988). Managerial influence in the implementation of new technology. *Management Science*, *34*(10), 1252–1265. doi:10.1287/mnsc.34.10.1252

Möller, K., & Halinen, A. (2000). Relationship marketing theory: Its roots and direction. *Journal of Marketing Management*, *16*(1), 29–54. doi:10.1362/026725700785100460

Morgan, R., & Hunt, S. (1994). The commitment-trust theory of relationship marketing. *Journal of Marketing*, 58.

Murray, K., & Schlacter, J. (1990). The impact of services versus goods on consumers' assessment of perceived risk and variability. *Journal of the Academy of Marketing Science*, *18*(1), 51–65. doi:10.1007/BF02729762

O'Malley, L., & Tynan, C. (2000). Relationship marketing in consumer markets. Rhetoric or reality? *European Journal of Marketing*, *34*(7), 797–815. doi:10.1108/03090560010331225

Patton, M. Q. (1990). *Qualitative evaluation and research methods* (2nd ed.). Newbury Park, CA: Sage.

Petersen, M., & Rajan, R. (1994). The benefits of lending relationships: Evidence from small business data. *The Journal of Finance*, *49*(1). doi:10.2307/2329133

Pikkarainen, T., Pikkarainen, K., Karjaluoto, H., & Pahnila, S. (2004). Consumer acceptance of online banking: An extension of the technology acceptance model. *Internet Research*, *14*(3). doi:10.1108/10662240410542652

Pyun, C., Scruggs, L., & Nam, K. (2002). *Internet banking in the US, Japan and Europe*. Multinational Business Review.

Ratnasingam, P., & Pavlou, P. (2002). *Technology trust: The next value creator in B2B electronic commerce*. Paper presented at the International Resource Management Association Conference: Seattle, Washington, May 21-23.

Reichheld, F. (1993). Loyalty-based management. *Harvard Business Review*, *71*(2), 64–73.

Rowley, T. (1997). Moving beyond dyadic ties: A network theory of stakeholder influences. *Academy of Management Review*, *22*(4), 887–911. doi:10.2307/259248

Sandelowski, M. (1995). Sample size in qualitative research. *Research in Nursing & Health, 18*, 179–183. doi:10.1002/nur.4770180211

Sathye, M. (1999). Adoption of Internet banking by Australian consumers: An empirical investigation. *International Journal of Bank Marketing, 17*(7). doi:10.1108/02652329910305689

Stone, M., & Woodcock, N. (1997). *Winning new customers in financial services: Using relationship marketing and Information Technology in consumer financial services*. London: Pitman Publishing.

Tax, S., Brown, S., & Chandrashekaran, N. (1998). Customer evaluations of service complaint experiences: Implications for relationship marketing. *Journal of Marketing, 62*(2), 60–76. doi:10.2307/1252161

Violino, B. (2002). Building B2B trust. *Computerworld, 36*(25), 32.

Young, L., & Wilkinson, I. (1989). The role of trust and cooperation in marketing channels: A preliminary study. *European Journal of Marketing, 23*, 109–122. doi:10.1108/EUM0000000000550

Zaltman, G., Duncan, R., & Holbeck, J. (1973). *Innovations and organisations*. New York: Wiley and Sons.

KEY TERMS AND DEFINITIONS

Business to Business: Marketing conducted by one business, and to another.

Commercial Banking: Banking relationship conducted between an organisation and a bank.

Relationship Marketing: Provides a mutual benefit, usually from a B2B perspective.

Value: Based on the perception of the customer. What they seek out of the relationship.

Chapter 6
E-Banking Development and Applications:
Current Issues and Challenges in Rural Areas and Emerging Nations

Jiaqin Yang
Georgia College & State University, USA

ABSTRACT

This chapter explores the recent e-banking application and development issues, focusing on two main areas: (1) the small to medium sized banks in remote rural areas, and (2) a comparative investigation to examine how different cultural, economic, and political factors influence the e-banking systems in different nations. Given the conditions with less population and financial resources, the research objective of the first section is to investigate how smaller and community banks located in rural areas have attempted to catch up their counterparts in larger cities in terms of the application of e-banking, focusing on emerging issues and challenges, such as Internet accessibility, culture, and tradition playing a vital role in the development of e-banking industry among different nations.

1. INTRODUCTION

Electronic finance has been continuously growing as a new industry during the last decade. The banking industry has led this trend in recent years, and called e-banking referring to all banking transactions completing through internet applications (Boss, et al., 2000; Costanzo, 2000; Ebling, 2001; Feinman, et al., 1999). E-Banking has revolutionized the way business is transacted by globalizing the business enterprise. All businesses, including small and medium sized firms, no matter their geographical locations, are all beneficiaries of e-banking. It encompasses all kind of commercial transaction that is conducted on an electronic medium, mostly through the Internet. It allows companies to make new business contacts from different global business alliance, test new products and services, and make market research and other enquiries all at a minimal cost

DOI: 10.4018/978-1-60960-501-8.ch006

both financial and otherwise (Barnes & Corbitt, 2003; Laukkanen & Lauronen, 2005). E-banking services currently available include online inquiry, e-payments, e-transfer, online 24/7 banking services greatly reduce the labor cost, extend the service edge, and increase the banking efficiency and effectiveness (Gonzalez et al., 2008; Smith, 2008). In recent years, mobile banking service is emerging as another new direction of electronic banking (Gerpott & Kornmeier, 2009). Mobile banking (M-banking) is that customers use wireless devices (cell phone, PDA, laptop, etc.) to access the internet and conduct the online services at any time from any place. M-banking is viewed as a powerful tool to complement regular e-banking with new developed services (Scornavacca & Hoehle, 2007).

While the e-banking application has advanced significantly during recent years, especially in the big cities among developed nations, in comparison, the development of e-banking application has been far behind in many rural areas (within the developed nations), and in most developing nations, although many banks in those areas and nations have been implementing E-banking during recent years. Therefore, the primary purpose of this chapter is to promote and disseminate research and studies in the e-banking application and development in the rural areas and those in emerging developing nations.

In the following discussion, the following section presents a literature review about the recent e-banking application and development in rural areas and the current issues and challenges facing the e-banking industry in these emerging nations. Section three addresses the major e-banking issues and challenges for the banks in the rural less-populated areas. Following section describes a comparative study investigating the issues in the current e-banking services in those emerging nations under different cultures and traditions. Finally, the last section discusses the potential research directions for the e-banking industry in the rural areas and those emerging nations respectively.

2. LITERATURE REVIEW

2.1 E-Banking in Rural Areas

It has been reported that the banks leading the current wave of e-banking application are those large (or super larger) national banks with most of their headquarters are located in larger cities and urban areas (Harris, 2000). And most published e-banking research works have been carried out primarily on the e-banking applications of larger banks (Shin, 2008). In comparison, the issues of e-banking applications for those smaller and community banks located in rural and remote areas have not been fully explored in the published research. Especially there are special or unique challenges facing those local smaller and community banks in terms of their e-banking applications – how to provide easy and accessible e-banking services to their local customers while having adequate operational funding to provide those services. In summary, the lack of finance resource and weak in technical expertise have been cited currently as two major obstacles in developing and maintaining e-banking services for many local smaller community banks located in remote and rural cities or counties. As such, research effort, especially an empirical study on the development of e-banking application for those local smaller community banks is apparently in need, and is the primary focus of this section.

Facing extremely intensive competition from non-banking sector, the banking industry has adopted a more aggressive approach in the development of new e-banking services. For instance, some large banks have started to install advanced software to process all consumer loan applications on-line, a new paperless e-loan process (Li, 2002; Brown et al., 2004). Empirical studies from the consumer side of e-banking have been reported recently targeting mortgage lending, consumer lending, and small business loan products (Hitt, et al., 2002; Kaynak & Harcar, 2005; Halperin 2001; Rao, 2004). It has been suggested that for smaller community banks, the addition of e-banking services can increase their capacity to offer more banking products and reduce their dependence on core deposits (Stamoulis, 2000). But in practice, larger banks located in a centralized urban area tend to have the greatest incentive to adopt e-banking services, while in comparison, smaller community banks tend to have a high initial technological cost in developing e-banking services (Treadwell, 2001). Looking forward, most small banks are motivated to develop e-banking services for potential future cost savings and gaining a competitive edge in the competition (Timmons, 2000). In summary, the effort of developing and maintaining e-banking services for many local smaller community banks located in remote and rural areas is currently lag behind their counterparts in larger city and urban areas. As such, an empirical study to investigate the recent development of e-banking application for those local smaller community banks is described in next section.

2.2 E-Banking in Economic Emerging Nations

There are several factors which may impact the development of e-banking industry, including social (cultural, tradition, education, etc.), economic (economic system, average income level), and technological (industrial infrastructure, technologic background) considerations. As such, there is a huge gap between the developed nations (i.e. U.S. and European countries) and those emerging developing nations (i.e. China and Russia) in terms of the development of e-banking industry (Gao & Owolabi, 2008). Unlike most developed countries, e-banking has just been emerged in recent years in most developing countries, including those economic emerging nations like China (Yin, 2001). Most banks in China so far have only adopted the incumbent bank model to take advantage of their numerous branches and ATMs across the nation. Considering those specific factors which have a large impact on the development of e-banking industry, like popularity of Internet usage, educational level of general public, average income level, cultural and traditions, and specific governmental regulations, Table 1 shows the differences among those factors between China and United States, for comparison purposes. (Note: The data for Table 1 are summarized from Internet World Statistic, Indexmundi.com, United States Census Bureau, and National Bureau of Statistics of China.)

In 2003, 91% of U.S. households held bank accounts and 93 percent of those used at least one electronic transfer of funds option with their account (Kolodinsky & Hogarth, 2004). However, Fest (2007) points out that only 40% of US households took advantage of e-banking service, while over 50% of households that had not been attracted yet to e-banking because those customers might have had a bad experience on a self-service site (Swann, 2008). The winners in e-banking industry are those banks which are able to successfully enhance their offerings while simultaneously enhancing security measures and getting customers to believe in them (Rombel, 2006). In addition, for all electronic banking customers, customer satisfaction is not only affected by banks' service quality, but also by their cultural features (Levesque and McDougall, 1996).

E-banking in developing countries grows rapidly in the past decade (Akinci et al., 2004). Their research indicates that for consumers' attitudes and adoption toward e-banking, there were significant differences between the two groups, e-banking users and non-e-banking users, with respect to demographic profiles, attitudinal properties, and preferences for service delivery channels. For instance, in China, there were only 6,000 computers connected to the Internet with 40,000 Internet users in 1995, but there were 10.2

Table 1. A comparison between China and United States

	China				USA			
Internet Penetration (% Population)	253,000,000 Internet users as of June/08, 19.0% penetration, per CNNIC				220,141,969 Internet users as of June/08, 72.5% penetration, Nielsen			
Education	Only 3.6% post secondary through advanced degree 11.1% complete secondary				Less Than 9th Grad: 3.10% High School: 35.89% College:61.01%			
Income Level	Per Capita Annual Disposable Income of Urban Households: RMB11759.5 ($1680) Per Capita Annual Disposable Income of Rural Households: RMB3587.0 ($512.42)				Median money income: $50,233			
Age Structure	**Age**	**Population**	**% Population**	**Total**	**Age**	**Population**	**% Population**	**Total**
	15-19	105,023	8.81%		15-19	21,324	7.12%	
	20-24	76,160	6.39%		20-24	21,111	7.05%	
	25-29	74,110	6.21%		25-29	20,709	6.92%	
	30-34	93,398	7.83%		30-34	19,706	6.58%	
	35-39	113,952	9.55%		35-39	21,186	7.08%	
History	Around 10 years.				More than 30 years.			
Core Value of Culture	Moderate, Harmony of spirituality and reality				Freedom, Democracy, Competition, Independent.			
Payment Method	Cash, sometimes use debit card				Credit Card			
Banking Industry	Most banks are fully or partially government owned				Banks compete in all fields.			
Laws and Regulation	Laws and Regulation falls behind the development				A set of laws and regulations protect clients.			

(Resource from: Internet World Statistic & indexmundi.com, United States Census Bureau, National Bureau of Statistics of China)

million Internet-connected computers and 26.5 million Internet users nationwide by the end of June 2001 (Zhao, 2002). Lu et al. (2005) reveal that one of key strategic responses of banks in China before joining WTO was to develop e-banking to a more competitive environment, even under the current condition of lack a practical customer credit system. In another research, Lafore and Li (2005) examine the extent of e-banking and m-banking in China by investigating its market status, identifying the target customers, the demographic characteristics of users and non-users, and comparing their attitudes towards electronic banking adoption. They conclude that there was a low awareness of such services in China, due to security concerns, perceived risks, low computer skills, and a Chinese tradition of cash-carry banking.

The following section in this chapter describes a comparative study to empirically investigate the attitude, acceptance, usage frequency, concerns, and satisfaction degree toward the current e-banking services among the young consumers between two nations: China vs. U.S. All related factors, such as different cultures, traditions, and industry development backgrounds are considered in the study.

3. AN EMPIRICAL INVESTIGATION OF E-BANKING IN RURAL AREAS

3.1 Research Methodology and Data Collection

The primary data for this study are collected through a Web-based survey along with on-site interviews. To validate the survey questions, the first two early drafts were sent and reviewed by the selected local bank officials – who are in charge of their e-banking operations. Based on the feedback, the survey questionnaire was revised both in wording and choices accordingly. As the targeted participating banks are small and local community banks, an Internet search was first conducted to identify banks that are located in relative small cities and counties with a population less than 50,000. To ensure a large sample size and valid responses, about 120 small and local banks were identified and their Websites and e-mail contact information (of their e-banking managers) were obtained. After first inquiry e-mail asking for their participation, the second e-mail with the online survey link and the instruction was sent out to those with a positive response. The survey consists of four sections: (a) demographic information, (b) questions about the e-banking website, (c) e-banking services currently provided, and (d) the future direction of the e-banking industry.

3.2 Results Analysis

Among 120-targeted banks, only 30 completed the survey. This section presents some important results from this study. First, all surveyed banks can either be categorized as small (50%) with less than $100 million in total asset, or relative small with a total asset over $100 million (50%). That is, comparing to their large city counterparts (most of them have a total assets over $500 million or more), all surveyed banks are those local small community banks. In addition, the majority of those small local community banks (about 57%) have a low e-banking customer rate of less than 30%, and only a few banks (about 7%) have a 60% or higher of their current customers using their e-banking services. While a third (36%) has only 30% to 60% of their customers using e-banking services and none has a proportion over 80%. That is, there is a sizeable potential for those small and local community banks to improve their e-banking operations in the near future. When asked about the awareness of their e-banking services among their customers, an overwhelming majority (93%) of surveyed banks have a positive response. The low percentages of current e-banking customers for those participating local community banks is not a result of unawareness about the available e-banking services, but rather the lack of effective promotion efforts by those banks. Additionally, a large majority (72%) of participating banks had outsourced their e-banking website design to outside IT professionals, only one fifth (21%) had their e-banking websites designed by in-house IT professionals. It has been argued that having the e-banking website designed by an in-house IT team can help the banks to gain the capacity of upgrading and maintaining their e-banking services on a timely fashion, and in comparison the banks that depending on outside IT professionals for their e-banking website design will be slow in responding any emerging customer requests (Shin, 2008). For most of those small local community banks, over a quarter (27%) starting their online banking services less than a year, about 60% within two years, and only one fifth (20%) have started their e-banking services over three years or longer. A further research about this issue – the relationships among the potential e-banking users and the competitive edge of those banks providing their e-banking services, should provide more managerial insights for the banks, especially those small and local community banks.

The current available e-banking services provided by the surveyed banks are listed in Table 2. It can be seen that those basic online services are now available by almost all survey participants, such as: online account inquiry (100%), online bill payment (93%), online check order option (73%), and Internet deposit account (64%). Some relative newer e-banking options, however, are only available by a few banks surveyed, such as: online mortgage application (27%), online small business loans (20%), online trust funds (13%), online investment brokerage service (13%), and online Travelers Check ordering (13%). Certain popular new e-banking services provided by many large city and urban banks, including: online money order purchasing, e-signature, online insurance and financing options, and online purchasing of CDs and IRAs, are not available at all in the current e-banking service packages provided by those surveyed banks. The possible managerial implications for those small and local community banks are twofold: (1) their early e-banking development should focus on those basic banking services to have a good start and gain necessary experiences for further expansion on those advanced e-banking options; and (2) there is a huge room for improvement in their e-banking operations to attain new competitive edges in the market by developing those new advanced e-banking options into their online banking strategy.

For the impact of the current e-banking services on banks' operational performance and the related policies adopted by the participating banks, as shown in Table 3, over 54% of the banks have provided e-banking brochures, and 60% making e-banking demo available on their websites. But only 20% have prepared necessary e-banking training sessions to their customers. Other listed assistance includes: personal help in bank or over phone, e-mailed instructions, and "Help" option placed on their websites. It should strongly suggest to those small and local community banks that for many of their customers (especially those senior citizens lived in a remote location), certain assistance might be a necessary for them to be able to enjoy the benefits and convenience from using e-banking services. A very encouraging sign can be highlighted here – when asked if those banks believe that the new e-banking services will enhance their competitive position in the market, all surveyed banks (100%) returned with a positive response. Similar results are shown with regard to the belief that the addition of e-banking services has increased customers' satis-

Table 2. E-banking services currently provided

Questions		Survey Results
Does your bank offer any of the following services?	Online Inquiry and Transfers	100%
	Online Bill-Pay	93%
	Online Mortgage Application	27%
	Online Small Business loans	20%
	Online Trust Funds	13%
	Online Investment and Options Management	13%
	Online deposit accounts	64%
Does your bank currently offer any of the following new services?	Online Check Order	73%
	Online Order of Travelers Checks	13%
	Online Money Order	0%
	E-Signature	0%
What kind of help or assistance is offered to your e-banking customers?	Brochures	54%
	Training	20%
	Online Banking Demo	60%
	Others (e.g., Employee Assistance, Personal service over the telephone, Emailed instructions, Website "Help" capabilities)	13%

Table 3. Operational impact of e-banking services

Questions		Survey Results
Does your bank believe that e-banking services will enhance your competitive position in the market?	Strongly Agree	46%
	Agree	54%
	Disagree	0%
	Strongly Disagree	0%
Has your e-banking service increased the degree of customer satisfaction?	Strongly Agree	14%
	Agree	86%
	Disagree	0%
	Strongly Disagree	0%
Does your bank believe your e-banking services have helped to reduce your banks daily operating cost?	Strongly Agree	7%
	Agree	73%
	Disagree	20%
	Strongly Disagree	0%
Does your bank believe more customer training/ customer education is needed in promoting e-banking services?	Strongly Agree	13%
	Agree	80%
	Disagree	7%
	Strongly Disagree	0%
Does your bank believe that the customer personal information security is better now than it was before?	Strongly Agree	22%
	Agree	70%
	Disagree	8%
	Strongly Disagree	0%
Does your bank believe that senior citizens or rural residents represent recognizable growth potential for e-banking services?	Strongly Agree	7%
	Agree	53%
	Disagree	40%
	Strongly Disagree	0%
Has your bank adopted (or plans to adopt) any new approach that had increased your e-banking users?	Promotion with new customers	34%
	On every monthly statement	66%
	Mail out brochures to customers that don't use e-banking	27%
	Other (please specify)	0%
Has your bank adopted (or plans to adopt) any new technique or promotional effort that had increased your e-banking users?	Cell phone Internet access	0%
	Service outlets in public areas e.g. library, super market	0%
	Other (please specify)	33%
	None of the above	47%

faction to certain degree. In addition, a majority of surveyed banks (80%) agree that the addition of new e-banking service has reduced their daily operating cost. A further research investigating this issue is in need to bring some light and insight for those small and local community banks. For a key concern about customer personal information security, an overwhelming majority (92%) agreed that this issue has been improved greatly and no longer a real threat to the e-banking industry. A presumption made for this study is that those senior citizens or rural residents living in a remote location should represent a critical growth potential for the e-banking services of those surveyed small and local community banks. While a small majority (60%) of those banks believes so, however, a significant proportion (40%) has a different view – which certainly generates a quite surprise and a warrant for further research. Finally, when asked about using some most recent new technology or promotional effort to increase their current e-banking services, a significant gap between those small and local community banks and their counterparts in large city and urban area is demonstrated here. While almost half (47%) of the surveyed banks have answered "none", about a third (33%) answered with "Others" but none giving specifics, and none among those banks has ever had a concept or a plan for the newest m-banking (i.e., wireless Internet banking), or utilizing public Internet capacity to promote their e-banking services.

The results about the future direction of their e-banking are presented in Table 4. For instance, for the major challenges and difficulties for their current e-banking operations, many surveyed banks (29%) still argued with the lack of in-house IT professionals for the delays and slow responses from outside IT technological support - when those small banks outsourced the development of their e-banking service operations. In an Internet era, customers would be very unhappy about the slow response to their e-request and expect a more timely result. There are also similar responses to the need for more training in IT technology (28%), more innovative design of e-banking services

Table 4. Future directions of e-banking

Questions		Survey Results
What major challenges and difficulties has your bank faced when dealing with E-Banking?	Lack of In-house IT professionals	29%
	Lack of interest from customers	21%
	Extra workload from processing online banking services	21%
	Need for employment training in IT technology	29%
	The innovative nature of E-Banking	29%
	Others (expenses, customers' computer skill, extra work load, user friendliness, user passwords updates)	14%
In which direction does your bank believe it will shift to (next 10 years)?	Traditional Banking will still dominate (e.g. more than 50%)	65%
	E-banking will dominate (greater than 50%)	35%
How frequently does your bank update your services and websites?	Daily	15%
	Weekly	0%
	Bi-Weekly	8%
	Monthly	8%
	As Needed	69%
	None of the above	0%
Has your bank established e-banking into your banks future Strategic planning?	Strongly Agree	22%
	Agree	78%
	Disagree	0%
	Strongly Disagree	0%

(28%), increased extra workload due to the task of processing all those online requests (21%), and the lack of customer interest (21%). But for a critical issue by which many e-banking customers have complained about – the timely updating of banks' website, the response received is not optimistic at all. Only a few banks (15%) answered that it updates on a daily basis, while bi-weekly and monthly updating each acclaim the same 8%, and a majority (69%) of surveyed banks answered with the "As Needed" option. That is, there are banks (16% in this study) that wouldn't keep their e-banking websites updated until the end of every other week or the end of each month. As a result, customers' complaints about the useless and slowness for their e-banking experiences have been received. Finally, for the future direction of the e-banking industry over a foreseeable future, a majority of surveyed banks (64%) believe that traditional banking services will still dominate over e-banking services, but all participating banks (100%) indicated that they have placed their e-banking into their banks' future strategic planning. This is really interesting- are there any similar studies that confirm your points as well- or vice versa contradict?

3.3 Conclusion and Future Research

While the large city and urban area banks have been leading the way in the recent application and development of e-banking, many small and local community banks are catching up in this trend and becoming more interested in the e-banking services to gain competitive edges in the marketplace. This empirical study provides several managerial insights in evaluating the performance of e-banking operations for those small and local community banks. For example, it reveals that the e-banking is still in its infant stage for most small and local community banks, and further research about the relationships among the potential e-banking users and the competitive edge from providing the e-banking services is important to provide more insights. Finally, this study confirms that these new e-banking services can reduce banks' operating cost and increase the degree of customer satisfaction, but those small and local community banks should speed up their effort in using some recent new technologies and promotional effort to increase their e-banking services.

4. E-BANKING IN U.S. VS. CHINA: A COMPARATIVE STUDY

It is believed that the younger generation of a nation will become the majority of potential customers for online banking service (Akinci, et al., 2004). The Gallup survey finds that in the developed nations the typical profile of an e-banking user is someone who is young, prosperous, and familiar with computer and Internet. Similarly, in developing countries, the e-banking users are young or mid-aged, male, more technology-oriented and convenience-minded consumers (Akinci et al., 2004). However, there has been very limited research in the published reports to focus on the youth consumers in examining their consumption attitude and satisfaction in terms of e-banking development. It is the primary criterion to select the youth consumers in this study about the future development of e-banking.

This section describes a comparative study about the issues in the current e-banking services among the young consumers between two nations: China vs. U.S. It is believed that the attitude and usage of young consumers will be a good indicator for the trend of e-banking service in the future, and different cultures and traditions will play a role in the development of E-banking industry among different nations. As such, this comparative study provides insightful guidelines for the development of E-banking industry in both nations and worldwide.

4.1 Research Methodology and Data Collection

The data for this study are collected through a questionnaire survey in two languages (i.e., English and Chinese). The primary research questions are focused on the issues about e-banking service from the consumer's perspective. Specifically, the survey focused on two perspectives of e-banking service: current acceptance and awareness toward the current e-banking services and customers' perception for the future development of e-banking industry.

E-banking in China is still in its early development stage. Chinese banks are lack of experience to respond to customers complains and problems, which may eventually hurt customers' willingness to use online banking service. In addition, the long-time tradition of Chinese customers to do banking by going to a local branch is relatively hard to persuade (especially the elders) to change. Low average income also limits many families in China to using e-banking, as computers and Internet connection fees are still belong to luxury items in their daily life. Lack of education is another factor for many Chinese customers in using online banking services. All of the above concerns are designed into the survey questionnaire.

All respondents of the survey are grouped into two categories: e-banking users (the ones who have used e-banking before) and non-users (the ones who have not used e-banking so far). Those non-users are then offered a chance to rank 10 most commonly selected reasons (that prevent one from using e-banking) with 5-point scale. For those e-banking users, the questions about the length, frequency of their e-banking usage and the awareness and experiences of available e-banking services are followed. The respondents' history and frequency of using e-banking service will be very indicative for their higher acceptance and awareness of e-banking. Also, the information about the available e-banking services and their popularity among customers will be reflective on the current gap between the e-banking development stages between the two nations and provide insight for future direction of e-banking improvements for both nations. The customers' beliefs and perceptions about the future development of e-banking industry are also considered in the study, especially those e-banking services that have been available U. S. but not in China yet, such as online inquiry & transfer, online payments, online loan/mortgage application, online loan for small businesses, online investment, and online trust fund. Based on the information collected, a comparative analysis is conducted to examine the current acceptance and awareness of e-banking among customers of two nations, and to identify the most influential factors and considerations about the future development of e-banking from the perspective of customers.

4.2 Result Analysis

In terms of demographic information, the results show that more than half of the sampled are males (54.81% in China, and 51.49% in U.S.) in both China and the United States. In terms of age distribution, it is interesting to notice the difference between China and U.S. - that almost all survey participants (98%) in China are very young, between age of 18 to 25, but in comparison, only 81% in U.S. are in this age group, with over 10% are older than 30, assured that all survey participants in this research are familiar with computer and Internet applications.

Regarding the past e-banking experience, the majority of the survey respondents in United States indicate that they have used the e-banking services (82%). In contrast, less than half (46%) of survey respondents in China have any e-banking experience, a clear empirical evidence to confirm the expected gap between two nations in terms of e-banking industry development. One of the key research questions for this study is to understand - what are the real reasons that prevent those non-users from using e-banking services.

For this purpose, after surveying the literature and interviewing with bank managers, about 10 most recognized reasons are selected in this study. As seen in Table 5, for all items – the respondents in China scores a higher mean than their counterpart in U.S. This certainly reconfirms a presumption that with more advanced e-banking technology and long exposure to the available e-banking services, the young consumers in U.S. are less likely to be prevented from using e-banking services, comparing to their counterpart in China. More specifically, among the few with higher mean scores, one is that many customers still prefer face-to-face communication while conducting banking service, even among young consumers in U.S. (3.39 vs. 3.41). For the banks in both nations, this result suggests that for many customers, e-banking service is currently only viewed as an expansion of traditional banking service, and a more powerful campaign to promote e-banking as the future banking service has its merit. While it has been recognized that the issue of security concern has been addressed and solved with the advanced IT technology, it is clearly not the belief among young consumers in two nations – with the highest mean scores of all (4.01 in China and 3.29 in U.S.). An extremely high mean score of 4.01 in China deserves a further discussion. In fact, due to the lack of a well-developed legal system in China, online fraud stories have been reported daily in Chinese news media involving many banking incidents (reference??), this may be one factor which contributes to the low acceptance of Chinese consumers in using e-banking services. In addition, several key website design features have also may have played a role in discouraging potential e-banking users, such as out-of-date online information (China: 2.45, United States: 2.27), user unfriendly(China: 2.67, United States: 2.83), and lack of supportive service (China: 2.68, United States: 2.70). As discussed earlier, most mean scores from the U.S. sample are lower than those from their Chinese counterpart, implying that some concerns – which may not be a big deal in U.S., but are quite problematic in China. In addition, the survey result confirms that currently banks in U.S. provide more e-banking services than those in China. Most banks in China are owned, regulated, and controlled by governments, either by central government or local government. As such, there is more cooperation than competition among banks in the industry. A mean of 3.78 from "Uncertainty about additional charge" in China shows that people take this into account in their decision about whether they will use the e-banking services or not. In comparison, the

Table 5. Major factors that preventing customers from using e-banking

	China			United States		
	Mean	**Std. Dev.**	**N**	**Mean**	**Std. Dev.**	**N**
Unaware of service	2.31	0.72	54	2.21	1.23	37
Face-to-face preference	3.39	0.90	64	3.41	1.19	35
Clueless of operating process	3.69	0.76	67	2.34	1.07	37
Security concerns	4.01	0.73	67	3.29	1.09	37
Unavailability of required service	3.12	0.93	59	2.47	1.02	29
Out of date information	2.45	0.80	58	2.27	1.10	35
User unfriendly	2.67	1.00	58	2.83	1.06	33
Insufficient customer support	2.68	0.91	60	2.70	0.88	27
Poor internet connection	3.78	0.83	59	2.79	1.01	29
Uncertainty about additional charge	3.78	0.94	63	2.88	1.21	15

banking industry is basically a free market in U. S. and all banks compete independently with each other. Consequentially, for all banks, e-banking is a weapon to gain competitive advantages in the competition.

For those respondents who have prior e-banking experience, the data about their e-banking experience are displayed in Table 6. As expected, the differences of e-banking usage experience and frequency are quite significant between the two countries. For example, almost 80% of respondents in U. S. indicate that they have e-banking experience over a year; while in contrast, only a quarter (25%) respondents in China fall in this category. Similarly, the respondents in U. S. indicate a more frequent use of e-banking services (near 70% used the service once a week) than their Chinese counterpart (less than 10%). In addition, a correlation analysis shows that the longer the customers has used e-banking service, the more frequently they will use the service for the respondents from the U.S. sample, but the same correlation is nonsexist among the respondents from the Chinese sample. One implication here is for the banks in China, if their current e-banking promotional effort continues, there will be the time that more customers will use their e-banking services more frequently.

The banking systems in two nations are significantly different in terms of banks' ownership, governmental regulation policy, and the degree of the competition among the banks. As a result, the current available e-banking services provided in two nations are not exactly the same. It's not surprising to see that both Online Inquiry & Transfer and Online Payments are top two e-banking services most well known and popularly among the respondents in both countries. In comparison, other two more advanced e-banking services (Third-Party Deposit and Online Investment) are still in their early development stage, much less popular (less than 30% in both nations). It is also noteworthy about the "gap" between the "awareness" of an e-banking service and the actual "using" of that service. Based on the respondents from China, the average gap between the "awareness" and actual "using" of the same service is larger than 20%, while the same "gap" in U. S. is only about 10% in average. Obviously, a larger gap indicates a higher unwillingness among the respondents to use that e-banking service. This certainly implies that there are more opportunities and challenges for the banks in China to attract more potential customers for their e-banking services through further improvement of their current e-banking systems and a more aggressive promotional effort.

Table 6. E-banking usage history and frequency distribution

		China		United States	
		No.	%	No.	%
Usage History	Less than One month.	11	15.94%	1	1.23%
	One month to Half a Year	15	21.74%	5	6.17%
	Half a year to One Year	22	31.88%	8	9.88%
	One Year to Three Years	16	23.19%	39	48.15%
	More than Three Years	1	1.45%	28	34.57%
Frequency	Once a Week	5	7.25%	55	67.90%
	Two or Three time per Month	12	17.39%	13	16.05%
	Once a Month	14	20.29%	9	11.11%
	Less than once a month	34	49.28%	3	3.70%

The survey respondents of both two nations were asked to rate their e-banking service performance, and the results are summarized in Table 7. On a 5-ponit scale, U.S. respondents are much satisfied with their current e-banking services than their Chinese counterpart, with all ratings are higher than 3.5 (two even higher than 4.0) while most ratings of Chinese respondents are below 3.0 (except one). That is, in general most Chinese e-banking customers have a negative perception about the current e-banking service's performance. Specifically, for U.S. respondents, the highest two rating scores are from "Reliable Secure Protecting" (4.11) and "Simple Operating Process" (4.07). In comparison, for Chinese respondents, the highest rating score is from "Continual Customer Training" (3.61), the only rating score higher than 3.0, followed by "Reliable Secure Protecting" (2.80), "Fast Internet Accessibility" (2.74), and "Various Customized Service" (2.51). Clearly, there is a long way to go for the banks in China to continuously improve their e-banking services up to the level of the advanced stage. It is not surprised to see that Chinese customers expressed such as strong negative perception about their current e-banking service performance. Possible explanations may include: (1) Technology factor: less advanced IT and Internet network technology generates less convenient and less reliable e-banking services. (2) Ownership: since all banks in China are owned by government, there is less competition among banks, which result in less interest for banks to improve their e-banking performance.

For the customers' perception about e-banking in the future, about 85% of customers in both nations indicate that they will continue to use e-banking services in the future. But it is that 15% who indicate that they will not use e-banking services in the future deserve a further examination. When measured by - if they would recommend e-banking to others, however, the positive response from customers in China is declined significantly (from 85% to 64%), comparing only a slight decrease from customers in U.S. (still at 80%). It just reconfirms the fact that a large proportion of Chinese customers currently have quite a negative perception about their e-banking service performance.

4.3 Future Research

This section describes a comparative study about the issues of current available e-banking services among the young consumers between China and. U.S. The findings of this study provide several important and insightful guidelines for the future development of e-banking services in both nations and worldwide. For example, the gap between two nations about the awareness and usage of e-banking services is quite significant, as less

Table 7. Rating of e-banking service performance

	China			United States		
	Mean	Std. Dev.	No.	Mean	Std. Dev.	No.
Simple Operating Process	2.28	0.91	69	4.07	0.70	81
Sufficient and available information which updates frequently	2.37	0.83	68	3.82	0.83	78
Quick Searching Method	2.40	0.87	67	3.64	0.93	76
User Friendly Interface.	2.37	0.74	65	3.74	0.92	78
Fast Internet Accessibility	2.74	0.97	65	3.94	0.80	78
Reliable Secure Protecting	2.80	0.98	66	4.11	0.86	79
Various Customerized Service	2.51	0.95	65	3.57	0.92	77
Continual Customer Training and Notice	3.61	1.22	64	3.29	1.07	70

available services and lower service quality are two critical problems, due to a low competitive banking industry in China and lack of nationwide credit system. It is also interesting to see that Chinese customers are more willing and open to new available services (both in E-banking service and M-banking service) of which United States customers are less aware and more cautious to, due to the different cultures and traditions.

Several important managerial implications can be derived for the banks in both nations. Especially for those U.S. banks recently opened their banking business in China; they certainly should use their advanced e-banking technology and services to establish competitive advantages among young Chinese consumers. In comparison, for those banks in China; there are several emerging tasks to be targeted before the development of e-banking industry can really upgrade to a more advanced level, including: (a) developing a national individual credit system for a large scaled nationwide effort of personal check account and individual credit card; (b) encouraging those state-owned banks for more market competition – for the purpose of providing a better service to all consumers; (c) improving current e-banking services step-by-step to provide a better and more reliable banking channel; and (d) promoting all new electronic banking services (including m-banking) more aggressively to increase the general awareness of e-banking services, especially among all consumer groups.

5. CONCLUSION

This chapter describes the related research issues in the recent e-banking application and development, focusing on two specific areas: (1) the relative small to medium sized banks in rural areas, and (2) a comparative investigation to examine – how the different cultural, economic, and political factors influencing the e-banking system in different nations. The chapter provides the updated related literature and identifies the major issues and challenges of e-banking development in those two areas. Specifically, with less population and financial resources, the research objective of the first study is to investigate – how those smaller and community banks located in rural areas have attempted to catch up their counterparts in larger cities in terms of the application of e-banking, focusing on emerging issues and challenges. In the later study, a comparative study about the issues in the current e-banking services among consumers between two nations are conducted to show that the different cultures and traditions will play a critical role in the development of E-banking industry among different nations. Managerial implications are discussed along with suggestions for future research.

REFERENCES

Akinci, S., Aksoy, S., & Atilgan, E. (2004). Adoption of Internet banking among sophisticated consumer segments in an advanced developing country. *International Journal of Bank Marketing*, *22*(2/3), 212–219. doi:10.1108/02652320410530322

Barnes, S. J., & Corbitt, B. (2003). Mobile banking: Concept and potential. *International Journal of Mobile Communications*, *1*(3), 273–288. doi:10.1504/IJMC.2003.003494

Boss, S., McGranahan, D., & Mehta, A. (2000). Will the banks control online banking? *The McKinsey Quarterly*, 70–77.

Brown, A., Remenyi, D., & Bajomo, A. (2004). E-banking for SMEs: A case study of Fortis Bank UK. *International Journal of Electronic Business*, *2*(4), 366–382. doi:10.1504/IJEB.2004.005639

Costanzo, C. (2000). *Pioneer Internet-only bank has a new strategy*. American Banker.

Ebling, T. (2001). The economics of online banking. *Target Marketing*, 67.

Feinman, T., Goldman, D., Wang, R., & Cooper, N. (1999). Security basics: A Whitepaper. Retrieved from http:/ /www. pwc.com

Gao, P., & Owolabi, O. (2008). Consumer adoption of Internet banking in Nigeria. *International Journal of Electronic Finance*, *2*(3), 284–299. doi:10.1504/IJEF.2008.020598

Gerpott, T. J., & Kornmeier, K. (2009). Determinants of customer acceptance of mobile payment systems. *International Journal of Electronic Finance*, *3*(1), 1–30. doi:10.1504/IJEF.2009.024267

González, M. E., Mueller, R. D., & Mack, R. W. (2008). An alternative approach in service quality: An e-banking case study. *The Quality Management Journal*, *15*(1), 41–60.

Halperin, K. (2001). *Balancing act*. Company Business and Marketing.

Harris, G. (2000). The big four banks: Dealing with the online challenge. *Journal of Financial Services Marketing*, *4*(4), 296–305.

Hughes, T. (2003). Marketing challenges in e-banking standalone or integrated? *Journal of Marketing Management*, *19*, 1067–1085. doi:10.1362/026725703770558321

Hwang, H. G., Chen, R. F., & Lee, J. M. (2007). Measuring customer satisfaction with Internet banking: An exploratory study. *International Journal of Electronic Finance*, *1*(3), 321–335. doi:10.1504/IJEF.2007.011502

Kaynak, E., & Harcar, T. D. (2005). Consumer value creation in mobile banking services. *International Journal of Technology Marketing*, *1*(1), 62–78. doi:10.1504/IJTMKT.2005.008125

Kolodinsky, J. M., Hogarth, J. M., & Hilgert, M. A. (2004). The adoption of electronic banking technologies by US consumers. *International Journal of Bank Marketing*, *22*(4), 233–241. doi:10.1108/02652320410542536

Laforet, S., & Li, X. Y. (2005). Consumers' attitudes towards online and mobile banking in China. *International Journal of Bank Marketing*, *23*(4/5), 362–381. doi:10.1108/02652320510629250

Laukkanen, T., & Lauronen, J. (2005). Consumer value creation in mobile banking services. *International Journal of Mobile Communications*, *3*(4), 325–338. doi:10.1504/IJMC.2005.007021

Levesque, T., & McDougall, G. H. (1996). Determinants of customer satisfaction in retail banking. *International Journal of Bank Marketing*, *14*(7), 12–20. doi:10.1108/02652329610151340

Li, F. (2002). Internet banking: From new distribution channel to new business models. *International Journal of Business Performance Management*, *4*(2-4), 136–160. doi:10.1504/IJBPM.2002.000112

Lu, M. T., Liu, C. H., Jing, J., & Huang, L. J. (2005). Internet banking: Strategic responses to the accession of WTO by Chinese banks. *Industrial Management & Data Systems*, *105*(3/4), 429–443. doi:10.1108/02635570510592343

Rao, S. (2004). Building inter-firm relationships online: Perspectives from the Australian Internet banking industry. *International Journal of Internet Marketing and Advertising*, *1*(4), 371–387. doi:10.1504/IJIMA.2004.006330

Rombel, A. (2006). Banks build on twin pillars of security and service. *Global Finance*, *20*(8), 39–43.

Scornavacca, E., & Hoehle, H. (2007). Mobile banking in Germany: A strategic perspective. *International Journal of Electronic Finance*, *1*(3), 304–320. doi:10.1504/IJEF.2007.011501

Servon, L. J., & Kaestner, R. (2008). Consumer financial literacy and the ompact of online banking on the financial behavior of lower-income bank customers. *The Journal of Consumer Affairs*, *42*(2), 271–305. doi:10.1111/j.1745-6606.2008.00108.x

Shin, K. H. (2008). Is e-banking a competitive weapon? A causal analysis. *International Journal of Electronic Finance, 2*(2), 180–196. doi:10.1504/IJEF.2008.017539

Smith, A. D. (2008). Online payment service providers and customer relationship management. *International Journal of Electronic Finance, 2*(3), 257–283. doi:10.1504/IJEF.2008.020597

Stamoulis, D. S. (2000). How banks fit in an Internet commerce business activities model. *Journal of Internet Banking & Commerce, 5*(1).

Swann, J. (2008). Catering to consumer expectations online. *Community Banker., 17*(2), 55–56.

Sweeney, A., & Morrison, M. (2004). Clicks vs. bricks: Internet-facilitated relationships in financial services. *International Journal of Internet Marketing and Advertising, 1*(4), 350–370. doi:10.1504/IJIMA.2004.006329

Timmons, H. (2000). Online banks can't go it alone. *Business Week Online.*

Yin, L. (2001). An empirical study about development and monitoring issues in online banking development in China. *Finance Research, 1*(1), 1–7.

Zhao, H. (2002). Rapid Internet development in China: A discussion of opportunities and constraints on future growth. *Thunderbird International Business Review, 44*(1), 119–138. doi:10.1002/tie.1042

ADDITIONAL READING

Anguelov, C. E., Marianne, A. H., & Hogarth, J. M. (2004). "U.S. Consumers and Electronic Banking", 1995-2003. *Federal Reserve Bulletin, 90*(Winter), 1–18.

Byers, R. E., & Lederer, P. L. (2001). Retail Banking Service Strategy: A Model of Traditional, Electronic, and Mixed Distribution Choices. *Journal of Management Information Systems, 18*(2), 133–156.

Dandapani, K., Karels, G. V., & Lawrence, E. R. (2008). Internet banking services and credit union performance. *Managerial Finance, 34*(6), 437–447. doi:10.1108/03074350810872804

Giannakoudi, S. (1999). Internet Banking: the Digital Voyage of Banking and Money in Cyberspace. *Information & Communications Technology Law, 8*(3), 205–243. doi:10.1080/13600834.1999.9965811

Herington, C., & Weaven, S. (2007). Can banks improve customer relationships with high quality online services. *Managing Service Quality, 17*(4), 404–414. doi:10.1108/09604520710760544

Howcroft, B., Hamilton, R., & Hewer, P. (2002). Consumer attitude and the usage and adoption of home-based banking in the United Kingdom. *International Journal of Bank Marketing, 20*(3), 111–121. doi:10.1108/02652320210424205

Jenkins, H. (2007). Adopting internet banking services in a small island state: assurance of bank service quality. *Managing Service Quality, 17*(5), 523–534. doi:10.1108/09604520710817343

Liu, C. C. (2008). The Relationship between Digital Capital of Internet Banking and Business Performance. *International Journal of Electronic Finance, 2*(1), 18–30. doi:10.1504/IJEF.2008.016882

Murphy, P. A. (2007). Digital DIVIDE. *Independent Banker, 57*(11), 57.

Rose, P. S., & Hudgins, S. C. (2005). *Bank Management and Financial Services.* New York: McGraw-Hill Irwin.

Saleh, M. N. and Andrea, S. (2002) "Challenges of the 'E-Banking Revolution'", A quarterly magazine of International Monetary Fund (IMF), Vol. 39, No. 3, pp.34-56.

Sathye, M. (1999). Adoption of Internet Banking by Australian Consumers: An Empirical Investigation. *International Journal of Bank Marketing*, *17*(7), 324–334. doi:10.1108/02652329910305689

Shi, W. H., & Lee, C. F. (2008). Does quality of alternatives matter for internet banking? *International Journal of Electronic Finance*, *2*(2), 162–179. doi:10.1504/IJEF.2008.017538

Witman, P. D., & Poust, T. L. (2008). Balances and accounts of online banking users: a study of two US financial institutions. *International Journal of Electronic Finance*, *2*(2), 197–210. doi:10.1504/IJEF.2008.017540

Ziqi, L. and Michael, T. C. (2003) "Challenges to Internet e-banking". Communications of the ACM archive: Mobile computing opportunities and challenges. Vol. 46, Issue 12, pp. 248-250.

KEY TERMS AND DEFINITIONS

E-Banking: referring to all banking transactions completing through internet applications.

Electronic Finance: Internet application by the finance function with an organization or government agency.

Emerging Nations: Referring to those emerging nations with rapid economic growth (like China, India, Russia, and Brazil).

E-Payment: A payment made through an electronic channel.

E-Transfer: A transfer transaction between two or more parties made through an electronic channel.

M-Banking: A banking transaction made through a mobile (wireless) devices through accessing the internet and conduct the online services at any time from any place.

Rural Areas: The remote areas outside the major city and urban boundaries.

Chapter 7
ICTs and Social Inclusion: The Case of Microfinance in Developing Countries

Ana Marr
University of Greenwich Business School, UK

Lin Yan
University of Greenwich Business School, UK

ABSTRACT

Microfinance, defined as the provision of small-scale financial services for low-income populations, has widely been regarded in alleviating poverty and facilitating social inclusion. While much has been debated on the impact of information and communication technologies (ICTs) on social inclusion, paucity remains on how ICTs contribute to microfinance in developing countries. Social inclusion, particularly in the sense of increasing access to microfinance, is important to entrepreneurs in developing countries, especially among women entrepreneurs in rural areas. A major challenge is to understand how ICTs contribute to microfinance, both in terms of reaching to a large population, and in providing efficient and effective services. This chapter investigates the role ICTs play in facilitating microfinance in developing countries. To do so, the current literature on ICTs and social inclusion and the literature on microfinance are reviewed in order to provide an integrated conceptual framework on how ICTs contribute to microfinance in enhancing social inclusion in developing countries.

DOI: 10.4018/978-1-60960-501-8.ch007

INTRODUCTION

The United Nations Population Fund predicted that the population of developing countries will rise from 5.6 billion in 2009 to 7.9 billion in 2050, constituting 88% of the world's population (UNPF, 2009). Contrasting the sheer size of the population is the low level of living standard. Many people in most developing countries lack basic resources such as clean water, adequate housing, healthcare, and education. To enhance social and economic development, many scholars have looked into the role of ICT, and established a clear link between ICT penetration and development (e.g. Avgerou, 2003; Mosse & Sahay, 2003; Krishna & Walsham, 2005; Walsham et al., 2007). In a recent review, Walsham et al (2007: 317) argued:

'The question has now become not whether, but how ICTs can benefit development. ICTs have high potential value across both public and private enterprises; and at multiple levels, for example from software business in urban areas to health delivery in rural villages.'

Yet challenges remain in understanding how ICTs contribute to development. Mixed findings have been reported as to the mutual shaping of ICTs and social change in developing countries. In a similar vein, microfinance has often been credited as contributing to social inclusion and poverty alleviation in developing countries (e.g. Ivatury, 2009; Wiezorek-Zeul, 2009). 5,000 microfinance institutions worldwide serve approximately 50 million low-income individuals and their families (Frederick, 2009).

A major challenge is to understand how ICTs, such as mobile phones, Automated Teller Machines (ATMs), and Point of Sale terminals (PoS), contribute to microfinance, both in terms of reaching to a large population, and in providing efficient and effective services (Firpo, 2009). We therefore find it necessary to explore the inter-section between these two areas, namely the role ICTs play in delivering and transforming microfinance, and the subsequent social impact in developing countries.

To this end, we will first review the literature on ICTs and social inclusion in developing countries, followed by a review on the current trends of using ICTs in microfinance. We will then integrate the two bodies of literatures and propose a conceptual framework and future research questions to explore the interplay among ICTs, microfinance, and social inclusion.

MICROFINANCE

In the 1980s and 1990s, widespread disillusion with government programmes for poverty reduction, including financial schemes, led to the mushrooming of non-governmental organisations (NGOs) as channels for donors' assistance. Building on informal financial practices such as ROSCAs,[1] NGOs experimented with savings and credit groups following a bottom-up, demand-led approach. In Bangladesh, a university project that resulted in the now well-known Grameen Bank, also worked with groups for the delivery of financial services. Modern microfinance, as we understand it today, was hence born in the image of those enduring informal financial arrangements – and its fundamental aim was to achieve the social development goals of helping reduce poverty.

Such microfinance practices also encouraged further theoretical analysis of finance for the poor. As one of the first and most influential studies, Stiglitz (1990) contended that joint-liability lending schemes – such as the Grameen Bank groups where members share responsibility for loan repayment – can induce peer monitoring that leads to the overcoming of some of the main constraints to finance for the poor. By making group members liable for one another's loan defaults, they have the incentive to monitor each other's loan usage to ensure that default risks are minimised.

Information asymmetries are hence reduced by tapping information that exists between group members and by allowing them to act upon that information in order to ensure loan repayment. In addition, through groups it is possible to minimise transaction costs inherently high due to the small size of loans. This line of thinking spawned, in the 1990s and 2000s, a large number of studies that look closely at how joint-liability schemes work and how they can ensure loan repayment and hence envision microfinance as a potentially profitable enterprise (Besley & Coate 1995; Conning 2000; Ghatak *et al.,* 2005).

Thus, academic investigation, practitioner discovery and political ideology converge to deem microfinance as a poverty tool that can be run as a financially sustainable business. An international political agenda supportive of market-based instruments and the prospect of declining subsidies persuaded microfinance initiatives to state as an important goal (together with the existing poverty-reduction goal) that of attaining financial sustainability. The rationale was that, in order to make lasting and great impacts on poverty, microfinance institutions needed to become financially sustainable – which would enable them to stay in the financial market in the long run, independent from subsidies that might stop in the short term. Academic and empirical understanding of the design of microfinance contracts and institutions appeared to support those arguments and hence in the last decade or so, microfinance has embraced the twin objectives (those of trying to reduce poverty while becoming financially sustainable) as the main purpose of its existence today.

The expectation that microfinance should attain both of its twin goals, therefore, appears to have given rise to two distinctive schools of thought: the *institutionalist* or financial systems approach and the *welfarist* or poverty reduction approach (Morduch, 2000; Robinson, 2001; Hermes & Lensink, 2007). The so-called *institutionalists* support the view that financial sustainability is of paramount importance and that subsidies should be phased out quickly, while the *welfarists* assert that microfinance ought to concentrate on helping communities overcome poverty even if this means maintaining subsidies for longer. The significance of these ostensibly contrasting approaches is not only that they clearly place more emphasis on one goal over the other – but also that the debate seems to uncover a fundamental suspicion that there might be trade-offs between attempting to achieve both goals simultaneously.

Academic studies that explicitly explore whether there are trade-offs between poverty reduction versus the strife for financial sustainability are still scarce. However, since the mid-1990s, there has been some evidence to suggest that trade-offs exist (von Pischke, 1996; Montgomery, 1996) while more recent work has been mainly case-study driven (Khandker, 2005; Lafourcade *et al.,* 2005; Marr, 2006). Navajas *et al* (2000) for example, show that microfinance institutions in Bolivia that have been successful in achieving financial sustainability are only serving those individuals who are just above or just below the poverty line, i.e. the marginally poor, suggesting that these institutions encounter difficulties in reaching poorer customers or that their pursuit for financial sustainability have driven them away from the original target.

Most recently, Balkenhol (2007), Cull *et al* (2007) and Mersland and Strom (2008) have attempted to address the issue of potential trade-offs more overtly and by use of larger samples. Balkenhol (2007) looks at 45 microfinance institutions (MFIs) in 24 countries and finds that the evidence on trade-offs is mixed. Some MFIs perform well in achieving one goal but not the other, suggesting that there is proof in support of existing trade-offs between trying to reduce poverty while becoming financially sustainable. Yet again, some other MFIs perform well on both accounts while others perform badly in each. The reasons for this mixed result are related to the differing internal management of MFIs but also, and importantly, to contextual factors such as the diverse level of ICT development that MFIs face in their own localities.

Cull *et al* (2007), meanwhile, based their analysis on 124 MFIs in 49 countries to find evidence of a trade-off between the depth of outreach to the poor and the pursuit of profitability by MFIs. As such, this study focuses on a specific aspect of poverty reduction, i.e. depth of outreach, which represents the ability of the MFI to reach people at the bottom end of the poverty spectrum. For measurement purposes, the authors use a simple proxy, i.e. average loan size, which in turn assumes that very small loan sizes are held only by very poor people and large-size loans by wealthier customers. The proxy variable is evidently a rough estimate of depth of outreach and highlights the current difficulties in identifying quantitative indicators to measure poverty-reduction outcomes accurately, particularly for cross-country analyses.

Measuring financial returns is relatively straightforward. Thus, in this study, profitability is proxied by the commonly used financial self-sufficiency ratio,[2] which indicates the MFI's ability to generate sufficient revenue to cover its costs without ongoing subsidies including soft loans and grants. An important aspect of the research is that it examines the relevance of 'institutional design' with respect to the trade-off between financial performance and depth of outreach of MFIs. Institutional design, in this case, is understood by the type of loan-delivery methods that the MFI employs. These can be group-based (whereby loans are delivered through groups); individual-based; or village banking (where teams are larger than group-based models and participants take on more managerial responsibilities).

The findings are generally in line with existing anecdotal evidence and common opinion amongst microfinance practitioners. That is, individual-based MFIs appear to perform better in terms of financial profitability but least well on measures of outreach, compared to group-based institutions. Table 1 shows some striking differences: while individual-based MFIs have the highest level of financial self-sufficiency, they extend loans that are eight times larger than those offered by village banks (i.e. US$1,220 versus US$148), which implies that they serve wealthier people than those catered by village banks. By contrast, village banks show high dependency on subsidies with 30% of their loan portfolios comprise of donations but appear to be able to reach poorer customers and a very large fraction (88%) of women who, in developing countries, are considered amongst the poorest of the poor.

Another interesting finding is that the size of the MFI, and to a lesser extent its age, is associated with extent of outreach. The study shows that larger individual-based MFIs as well as larger group-based MFIs tend to extend larger

Table 1. Financial sustainability vs. depth of outreach

	Individual-Based		**Group-Based**		**Village Banks**	
	Mean	**St. Dev.**	**Mean**	**St. Dev.**	**Mean**	**St. Dev.**
Financial self-sufficiency	1.11	0.20	0.98	0.32	0.95	0.47
Operational self-sufficiency	1.23	0.28	1.12	0.35	1.09	0.75
Age	11.12	8.67	8.60	5.85	6.95	3.71
Size of MFI indicator	2.23	0.67	2.00	0.72	1.60	0.60
Average loan size/GNP per capital of the poorest 20%	4.80	4.92	1.63	1.97	0.63	0.39
Average loan Size (US$)	1220.23	1184.51	430.98	499.56	148.69	126.61
Women borrowers	0.46	0.16	0.75	0.21	0.88	0.21

Source: Adapted from Cull *et al* (2007)

loans and serve fewer numbers of female clients. Similarly, older individual-based MFIs perform worse on outreach measures than younger institutions. These findings need a note of caution however: as the research has not investigated MFIs over time, a dynamic element is missing. This is especially crucial when trying to determine whether there is "mission drift" or re-orientation from poorer to wealthier clients as MFIs strive to attain financial sustainability over time. In other words, the finding that as MFIs mature and grow they extend larger loan sizes, may demonstrate that existing clients' businesses have also matured and grown and hence larger loans are needed – rather than assuming that MFIs have abandoned the clients that were originally poor in favour of richer customers. Therefore, the findings of the study seem to show evidence in support of the existence of trade-offs but firm results are still lacking.

Among practical ways to overcome trade-offs, Matthaus-Maier and von Pischke (2009) explore new partnerships for innovation in microfinance and find that the potential of ICTs is significant. It is suggested that ICTs can contribute to the dual microfinance objectives simultaneously: they can help reduce transaction costs in the delivery of microfinance (which leads to contributing towards financial sustainability) while, at the same time, ICTs can help microfinance institutions to reach remote areas where most of the most marginalised and poorer populations in developing countries live (which contributes towards the goal of social inclusion and poverty reduction).

ICTS AND SOCIAL INCLUSION

Walsham and Sahay (2006) identified four categories of research that links ICTs with development in developing countries: broad issues linking ICTs and development, facilitating cross-cultural working, local adaptation and cultivation of Information Systems, and focusing on particular marginalised groups. These categories are not mutually exclusive and reflect major issues that have motivated research investigations to date. A wide range of ICTs have been examined in various sectors across locations. For instance, Sayed and Westrup (2003) looked into the role of enterprise resource planning systems in bringing together global networks in Egypt, while Silva and Figeura (2002) focused on improving use of ICTs in Chile; Aman and Nicholson (2003) explored the issue of remote cross-cultural communication in offshore development in Malaysia, while Okunoye and Karsten (2003) were concerned with the challenges of providing good access to knowledge sources via the Internet in sub-Saharan Africa.

With the variety of ICTs examined and the locations covered, most studies confirmed the idea that ICTs benefit development, although the correlation is unlikely to be a simple and linear one. Avgerou (2003), for instance, critiqued the discourse of a simple link between ICTs and development and highlighted the importance of intermediary. She argued that more investment in ICTs and other technologies may not directly impact on development without the establishment of standard market mechanisms, in which local government plays a crucial role. Gillard et al. (2008) used a fundamental human categorisation – gender – to investigate the uneven development brought about by ICTs. While international agencies such as IMF and the World Bank often measure development according to geography and economic zone, Gillard et al. explored the impact of ICTs on women's work and life through the 'feminization of employment' – the rise of women in the workplace. Avgerou and Madon (2005) argued that to follow the footsteps of developed countries in technology and social-organisational innovation has the potential of bringing out 'digital divide', where developing countries feel the threat of marginalisation in the free market and knowledge-intensive global system advocated by developed countries.

In a recent article, Zheng (2009) proposed a new way to analysing ICTs and social development – one that draws upon Amartya Sen's capability approach. Critiquing the narrow focus on economic development, Sen's capability approach is mainly concerned with 'the expansion of freedom...both as the primary end and as the principle means of development' (Sen, 1999: xii). Instead of focusing on income and consumption alone, the capability approach extends to much broader issues, such as poverty, sustainable development, gender equality, human rights, and democracy (Fukuda-Parr, 2003). The term 'capability' does not refer to trained potentials, including skills, abilities, and aptitudes, as in everyday usage, but the real opportunities (environmental opportunities and individual abilities) that a person has to lead a life that she values (Gasper, 2007).

At the heart of capabilities is one's 'freedom to achieve well-being' (Sen, 1987: 49), rather than the means to achieve. Commodities, or goods and services, are seen merely as some of the means to achieve (Sen, 1985). They are only important when they enable people to generate capabilities and to achieve freedom. Three sets of factors influence capability generation from goods and services: personal, social, and environmental (Sen, 1992). Personal characteristics, such as mental and physical conditions, literacy, and gender, influence the types of degrees of capabilities a person can generate from resources. Social factors include social norms, social institution, and power structure. Environmental factors, such as climate, infrastructure, institutions, and public goods, also play a role enabling or hindering individual freedom from available commodities (Zheng, 2009).

The capabilities approach is, in our view, informative to the study on ICTs in microfinance. The emphasis on 'freedom to achieve' rather than 'means to achieve' directs our attention to the recipients of microfinance, namely the users of ICTs instead of technology itself. By focusing on the end users and their aspiration to achieve 'freedom of well-being', our investigation may extend beyond the technologies that are currently available, to those that are not yet available, but may potentially bring social and economic achievements. Similarly, microfinance is no longer seen as mere commodity, the means to achieve, in our investigation, but an enabling factor to individuals' social freedom, influenced by the interplay of personal, social, and environmental factors.

ICTS USE IN MICROFINANCE

Although using ICTs in microfinance is a recent phenomenon, the technologies used and their economic benefits in microfinance have been well documented. ICTs play an important role in microfinance (Ivatury, 2009; Wieczorek-Zeul, 2009), particularly in the delivery of microfinance services. Ivatury (2009: 148), for instance, stated that ICTs contribute to microfinance in providing:

'...delivery channels that are inexpensive to set up, a wider range of financial services for poor customers, and the ability to handle transactions at low cost.'

In a survey by The Consultative Group to Assist the Poor (CGAP), it was found that 62 financial institutions in 32 countries reported using technology channels to handle transactions for poor people. The mostly commonly used technologies include:

Automated Teller Machines (ATMs) - There is a widespread use of ATMs in developing countries, mostly at urban and semi-urban locations. The reliance on electricity supply, constant telecommunication connections, and manually refilling and emptying cash increase the costs of using ATMs and to a certain extent limited the wider use in rural areas.

Point of Sale (POS) terminals – Particularly popular in Brazil, POS devices include card reader, mobile phone, personal computer, barcode scanner, or any hardware that can identify customers and receive instruction for the transfer of payments. Where transaction volume is expected to be high, or where wireless Internet access is available, computers may be used, but most POS devices are card-reading terminals. Each POS device uses a telephone line, mobile phone connection, or the Internet to send instructions for transferring money from one account to another. For example, after swiping a card through the POS device, the merchant presses a button on the terminal authorising payment from the customer's account. If the POS device is a mobile phone, the customer uses her phone to send a text message authorising payment from her bank account or from her account with the mobile phone company to the merchant's phone. Like ATMs, POS terminals have to be manned. Supermarkets, drugstores, postal office, and other retail outlets are ideal locations for POS terminals.

Mobile Banking – please note that mobile banking differs from using mobile phones as POS. Mobile banking is similar to Internet banking. The transaction is conducted unilaterally with the payer instructing the transaction via a mobile phone. Payment can be received at various kinds of accounts. Using mobile phones as POS requires multilateral interaction between the payer and the payee. The mobile phone is mainly used to identify the payer, and transactions are usually made between the payer's and payee's mobile phones only. Mobile banking has also been found to be particularly useful in rural areas where debit and credit cards, POS devices, ATMs, and even bank branches are virtually non-existent. In Zambia and the Democratic Republic of Congo (DRC), for instance, mobile users are able to make payments by using the special subscriber identity modules (SIM) cards issued by mobile phone companies (Ivatury, 2009).

In the absence of bank branches, the use of ICTs has the potential in extending the reach of microfinance to many people in poor and rural areas with limited infrastructure. How do ICTs contribute to social inclusion and poverty alleviation as a consequence of extended reach for microfinance, however, remains a question. Despite financial institutions' attempts to extend the reach of microfinance, including making unsecured loans, the reception in using ICTs from people in developing countries so far has been mixed. In a study in Brazil, Ivatury (2009: 160) argued:

'Still, one should not conclude that poor people will use formal financial services just because a technology channel is available. Witness the high proportion of inactive accounts opened at Banco Popular described earlier. More research should be done to understand why some poor people do not use these technology delivery channels.'

A FRAMEWORK FOR CAPACITY GENERATION

The literature review covered in this chapter highlights the fundamental fact that insufficient emphasis has been placed on what people need in terms of ICTs in order to facilitate capability generation, through microfinance, that would ultimately lead to social inclusion. In an effort to contribute to the existing body of work, we attempt to present here an original conceptual framework that puts people's capability building at the heart of the impact chain.

Figure 1 depicts graphically our conceptual framework. One important aspect of the framework is that it subscribes to a demand-led, bottom-up approach, meaning that ICT development needs to be responsive to the people microfinance aims to benefit. In this sense, the target group is composed of those marginalised poor women living in rural areas in developing countries – precisely the traditional recipients of group-based microfinance.

With this in mind, the assessment of existing ICTs and the development of more appropriate technology are based on the potential for generating capabilities in the target group that would allow

Figure 1. Conceptual framework

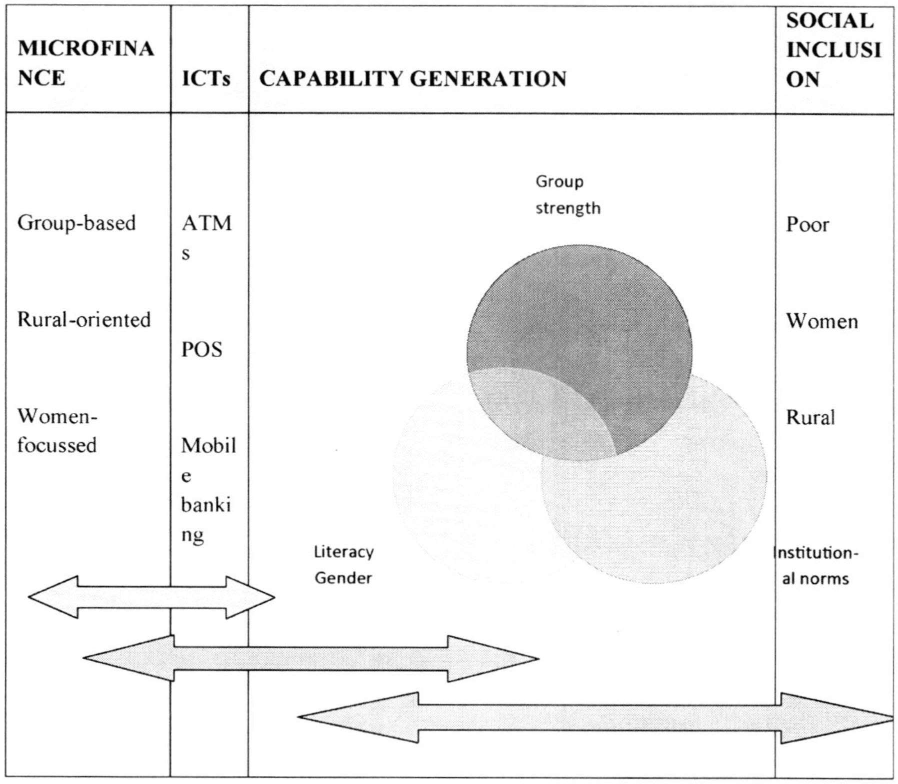

them to enhance their freedom to achieve social and economic inclusion. Three capability areas are key in the interplay of microfinance and ICTs, viz. people's literacy and gender, groups' strength and the norms of microfinance institutions.

It is conceivable that ICTs working with group-based microfinance are designed to be sensitive to the literacy levels and local languages of poor women in rural areas, while at the same time aim at elevating those literacy levels. Reaching larger numbers of poor women in rural communities may also be possible by use of ICTs that help strengthen group cohesion through schemes that allow members to work collectively. Institutional norms, meanwhile, can be influenced by the use and purpose of ICTs and, therefore, the impact on social inclusion could be enhanced if ICT development unlocks environmental opportunities for the poor.

In all this, our conceptual framework proposes an interaction between microfinance and ICTs that is deliberately aimed at helping build capabilities among the women living in poor, rural communities. As mentioned before, ICTs have the added benefit of helping reduce transaction costs in the delivery of microfinance services; and hence it is conceivable that microfinance dual goals – enhancing social inclusion while being financial sustainable – can be achieved with appropriately developed ICTs.

ICTS IN MICROFINANCE: FURTHER RESEARCH QUESTIONS

The above brings new and exciting challenges in terms of how best to investigate the interplay between ICTs and microfinance, which demands novel research questions. In the area of social inclusion in developing countries, the following are particularly relevant:

1. How can ICTs be modified or new ones developed in order to be accessible to low-literate women in rural communities?
2. To what extent would microfinance groups be strengthened and collective action promoted as a result of suitable ICTs?
3. What microfinance institutional norms and community norms are affected by ICTs, and how can ICTs be designed to improve overall environments that lead to greater social inclusion?

Above all, for ICTs to make a real impact on social inclusion, effective partnerships need to be forged between microfinance institutions and technology companies where common social goals are clearly stated. From the literature review presented in this chapter and the subsequent conceptual framework and research questions, there is every probability that there is great potential to be uncovered in future research.

REFERENCES

Aman, A., & Nicholson, B. (2003). The process of offshore development: Preliminary studies of UK companies in Malaysia. In Korpela, M., Montealegre, R., & Poulymenakou, A. (Eds.), *Organizational Information Systems in the context of globalization* (pp. 201–216). Boston: Kluwer Academic Publishers.

Avgerou, C. (2003). The link between ICT and economic growth in the discourse of development. In Korpela, M., Montealegre, R., & Poulymenakou, A. (Eds.), *Organizational Information Systems in the context of globalization* (pp. 373–386). Boston: Kluwer Academic Publishers.

Avgerou, C., & Madon, S. (2005). Information society and the digital divide problem in developing countries. In Berleur, J., & Avgerou, C. (Eds.), *Perspectives and policies on ICT in society* (pp. 205–218). New York: Springer. doi:10.1007/0-387-25588-5_15

Balkenhol, B. (Ed.). (2007). *Microfinance and public policy: Outreach, performance and efficiency*. Palgrave Macmillan.

Besley, T., & Coate, S. (1995). Group lending, repayment incentives and social collateral. *Journal of Development Economics*, *46*(1), 1–18. doi:10.1016/0304-3878(94)00045-E

Conning, J. (2000). *Group lending, moral hazard and the creation of social collateral.* IRIS Working Paper no. 165. University of Maryland.

Cull, R., Demirguç-Kunt, A., & Morduch, J. (2007). Financial performance and outreach: A global analysis of leading microbanks. *The Economic Journal*, *117*(517), F107–F133. doi:10.1111/j.1468-0297.2007.02017.x

Firpo, J. (2009). Banking the unbanked: Issues in designing technology to deliver financial services to the poor. In Matthäus-Maier, I., & von Pischke, J. D. (Eds.), *New partnerships for innovation in microfinance*. Springer Publishers. doi:10.1007/978-3-540-76641-4_11

Frederick, L. (2009). Information Technology innovations that extend rural microfinance outreach. In Matthäus-Maier, I., & von Pischke, J. D. (Eds.), *New partnerships for innovation in microfinance*. Springer Publishers. doi:10.1007/978-3-540-76641-4_10

Fukuda-Parr, S. (2003). The human development paradigm: Operationalizing Sen's ideas on capabilities. *Feminist Economics*, *9*(2–3), 301–317.

Ghatak, M., Gangopadhyay, S. & Lensink, R. (2005). Joint liability lending and the peer selection effect. *Economic Journal*.

Hermes, N., & Lensink, R. (2007). The empirics of microfinance: What do we know? *The Economic Journal, 117*(517), F1–F10. doi:10.1111/j.1468-0297.2007.02013.x

Ivatury, G. (2009). Using technology to build inclusive financial systems. In Matthäus-Maier, I., & von Pischke, J. D. (Eds.), *New partnerships for innovation in microfinance*. Springer Publishers. doi:10.1007/978-3-540-76641-4_9

Khandker, S. R. (2005). Microfinance and poverty: Evidence using panel data from Bangladesh. *The World Bank Economic Review, 19*(2), 263–286. doi:10.1093/wber/lhi008

Krishna, S., & Walsham, G. (2005). Implementing public Information Systems in developing countries: Learning from a success story. *Information Technology for Development, 11*(2), 123–140. doi:10.1002/itdj.20007

Lafourcade, A. C., & Isern, J. Mwangi. P. & Brown, M. (2005). *Overview of the outreach and financial performance of microfinance institutions in Africa*. Microfinance Information Exchange (MIX).

Marr, A. (2006). The limitations of group-based microfinance and ways to overcome them. *Small Enterprise Development, 17*(3), 28–40. doi:10.3362/0957-1329.2006.029

Matthäus-Maier, I., & von Pischke, J. D. (Eds.). (2009). *New partnerships for innovation in microfinance*. Springer Publishers.

Montgomery, R. (1996). Disciplining or protecting the poor? Avoiding the social costs of group pressure in micro-credit schemes. *Journal of International Development, 8*, 289–305. doi:10.1002/(SICI)1099-1328(199603)8:2<289::AID-JID368>3.0.CO;2-2

Morduch, J. (2000). The microfinance schism. *World Development, 28*, 617–629. doi:10.1016/S0305-750X(99)00151-5

Mosse, E., & Sahay, S. (2003). Counter networks, communication and health Information Systems: A case study from Mozambique. In Korpela, M., Montealegre, R., & Poulymenakou, A. (Eds.), *Organizational Information Systems in the context of globalization* (pp. 271–290). Boston: Kluwer Academic Publishers.

Navajas, S., Schreiner, M., Meyer, R. L., Gonzalez-Vega, C., & Rodriguez-Meza, J. (2000). Micro-credit and the poorest of the poor: Theory and evidence from Bolivia. *World Development, 28*, 333–346. doi:10.1016/S0305-750X(99)00121-7

Okunoye, A., & Karsten, H. (2003). Global access to knowledge: Findings from academic research organisations in Sub-Saharan Africa. *Information Technology & People, 16*(3), 353–373. doi:10.1108/09593840310489421

Robinson, M. (2001). *The microfinance revolution: Sustainable banking for the poor*. Washington, D.C.: The World Bank.

Sayed, E. H., & Westrup, C. (2003). Egypt and ICTs bring national initiatives. Global actors and local companies together. *Information Technology & People, 16*(1), 93–110. doi:10.1108/09593840310463041

Sen, A. (1985). *Commodities and capabilities*. Amsterdam: North-Holland.

Sen, A. (1987). *The standard of living*. Cambridge, UK: Cambridge University Press. doi:10.1017/CBO9780511570742

Sen, A. (1992). *Inequality re-examined*. Oxford: Oxford University Press.

Sen, A. (1999). *Development as freedom*. New York: Knopf.

Silva, L., & Figueroa, E. B. (2002). Institutional intervention and the expansion of ICTs in Latin America: The case of Chile. *Information Technology & People, 15*(1), 8–25. doi:10.1108/09593840210421499

Stiglitz, J. (1990). Peer monitoring and credit markets. *The World Bank Economic Review*, *4*(3), 351–366. doi:10.1093/wber/4.3.351

United Nations Population Fund. (2009) World population to exceed 50 billion by 2050. Retrieved on February 16, 2010, from http://www.un.org/esa/population/publications/wpp2008/pressrelease.pdf

von Pischke, J. (1996). Measuring the trade-off between outreach and sustainability of microenterprise lenders. *Journal of International Development*, *8*, 225–239. doi:10.1002/(SICI)1099-1328(199603)8:2<225::AID-JID370>3.0.CO;2-6

Walsham, G., Robey, D., & Sahay, S. (2007). Forward: Special issue on Information Systems in developing countries. *Management Information Systems Quarterly*, *31*(2), 317–326.

Walsham, G., & Sahay, S. (2006). Research on Information Systems in developing countries: Current landscape and future prospects. *Information Technology for Development*, *12*(1), 7–24. doi:10.1002/itdj.20020

Wieczorek-Zeul, H. (2009). The importance of new partnership. In Matthäus-Maier, I., & von Pischke, J. D. (Eds.), *New partnerships for innovation in microfinance*. Springer Publishers.

Zheng, Y. (2009). Different spaces for e-development: What can we learn from the capability approach? *Information Technology for Development*, *15*(2), 66–82. doi:10.1002/itdj.20115

ENDNOTES

1 ROSCAS stand for Rotating Savings and Credit Associations, by which people who are excluded from formal financial systems get together in groups and start saving money in a common pot. The saved money is then lent to each member in turn. ROSCAs are widely common the poorer areas of Asia, Africa and Latin America.

2 Financial Self-Sufficiency (FSS) ratio = Adjusted financial revenue divided by the sum of adjusted financial expenses, adjusted net loan loss provision expenses and adjusted operating expenses.

Chapter 8
Enterprise Risk Management: A Case Study in the Pharmaceutical Industry

Gary A. Stair
Pharmaceutical/Medical Device Executive, USA

ABSTRACT

How a company successfully implements an Enterprise Risk Management (ERM) program, to identify and manage potential risks, can mean the difference between financial freedom and financial despair. The Committee of Sponsoring Organizations (COSO) guidelines, a voluntary private-sector organization in the United States, has developed internal control guidelines to provide guidance to executive management and governance entities on critical aspects of organizational governance, business ethics, internal control, fraud, and financial reporting. This chapter will discuss an approach to build an ERM implementation plan within a pharmaceutical company by outlining the responsibilities and influences of industry participants, sales forces, middle-management and senior leadership and the ways in which they focus on monitoring and developing the risk mitigation process. The influences of technologies are integrated and new directions, such as e-media and e-detailing (Virtual Sales Representatives) are also explored.

INTRODUCTION

In today's business world, the accelerating pace of change and new product launches has increased at an alarming rate. Companies, which are first to market and launch their products to meet consumers' demand, have done so by effectively managing their internal and external information utilizing an environmental scanning process. Monitoring and evaluating economic indicators will help detect and aid in predicting future trends and outcomes. In strategic planning all organizations have the ability to benchmark economic conditions, which

DOI: 10.4018/978-1-60960-501-8.ch008

affect and contribute to improved profitability, market expansion and market penetration.

Predicting market trends, which may influence consumers' buying decisions, will help guide companies in making informed strategic decisions with the goal of achieving their short-term or long-term goals. By focusing on these forces and trends from each of their respective environments, the relevance of these forces or trends will determine the ability of these companies to strategically adapt to these forces or trends and possibly avoid potential threats facing their own companies. Several tools are available to assist a company in analyzing their environment, such as strategic thinking sessions where high-level goals and forecasting are developed that align with its mission, operational effectiveness measurements to ensure efficient use of resources, routine report reviews, and adherence to monitoring for compliance with applicable laws, processes and regulations.

By examining the strengths, weaknesses, opportunities, threats and managing the risks, companies will be better able to capitalize on emerging opportunities or minimize potential risks by leveraging the company's strengths and resources. The objectives of this chapter are to discuss the importance of ERM in the pharmaceutical Industry, relate ERM concepts to the pharmaceutical Industry's external environment, and training, managing, and motivating employees involved with the ERM process.

BACKGROUND

In the midst of several high-profile corporate scandals and subsequent regulatory legislations and compliance policy violations, "TAP Pharmaceuticals settles with Dept of Justice for $875 million" (Healthcare Financial Management, 2001) or more recently "Pfizer's $2.3 Billion-Dollar Settlement" (Forbes, 2009), the development and implementation of strict internal control systems has become an organizational requirement. These requirements have led many organizations to review their risk management areas and deem them as vitally important. Although the Sarbanes Oxley Act of 2002, a major regulatory policy in the United States, covered a part of an organization's total risk management policy, many organizations felt compelled to develop comprehensive systems, which covered all aspects of risk management. The development and implementation of these comprehensive systems are termed Enterprise Risk Management programs.

Enterprise Risk Management programs are simply a framework of preventable, detective and corrective measures in which the effective application of them to an overarching strategy across the entire organization, designed to identify potential events or risks within a specified comfort level and to provide reasonable assurance on the achievement of company's objectives. Several risk management best practices are available by reviewing the Committee of Sponsoring Organizations of the Tread way's (COSO) Internal Control Integrated Framework system (COSO, 2008) and considered industry standards in building solid internal controls systems. COSO formed in 1985 to sponsor the National Commission on Fraudulent Financial Reporting; an independent private sector initiative aimed at studying the causal factors, which could lead to fraudulent financial reporting in the USA. This committee developed recommendations for public companies and their independent auditors, the US Security Exchange Committee and other regulators, and educational institutions. Their outcome provided an internal control process, which enacted "by an entity's board of directors, management and other personnel was designed to provide reasonable assurance regarding the achievement of objectives in the following categories:

- Effectiveness and efficiency of operations
- Reliability of financial reporting

- Compliance with applicable laws and regulations " (COSO, 2008, p3)

These guidelines improved "the quality of financial reporting through business ethics, effective internal controls, and corporate governance" (COSO, 2008, p1). Furthermore, the COSO asked PricewaterhouseCoopers to develop a framework which management could use to understand and improve their organizations' enterprise risk management systems. Several high-profile business scandals (e.g. Enron, Tyco International and WorldCom) led to enhanced corporate governance and risk management programs. This project led to the development of the Sarbanes-Oxley act. This act required public companies to maintain systems of internal control, requiring management to certify with the usage of independent auditors to test to the effectiveness of their monitoring and measurement systems. COSO believes by adding this internal frameworks of control, companies have a more robust understanding on the subject of enterprise risk management.

The level of dedication the senior leadership team places on creating an ERM framework will define how important ERM is for the organization and how the organization uses this framework will determine how it will fit the organization's needs. Building an ERM and an internal control system "is a *process*. It is a means to an end, not an end in itself" (COSO, 2008, p1). The organization should assemble a broad team of motivated, well-experienced, diverse individuals from different areas, where each team member represents each area or discipline. These team members should be committed and motivated on developing and implementing a strategic, overarching risk management program.

A team leader will oversee the project and internal developments. This team leader should be an independent contractor and an expert in risk management. Collaborating with the organization's internal and external advisors, the team leader should be ultimately responsible for reporting all discrepancies in accounting or financial reporting directly to the board of directors. The individual team representatives will address any identifiable risks to their own areas or disciplines and the entire team should develop preventative, detective or corrective strategies and solutions to mitigate future risks.

Communication is an integral part in developing a successful outcome; all team members need communications training as well as a clear understanding of the final deliverables. In some cases, it may be wise to develop a separate committee of key stakeholders to oversee the entire project. Before implementing the ERM process, understanding the needs, wants, and concerns and communicating with the stakeholders prior to the development of the ERM plan is an important part (preventative measure). By engaging this group of stakeholders, the organization can determine stakeholder needs and alignment to future organizational goals and objectives (detective measure). Next, developing a clear understanding on the level of commitment with senior leadership is also an integral part due to the final implementation and follow up actions needed to maintain on-going support of the ERM process (corrective measure). Therefore, senior leadership, such as the CEO and CFO, should be included in this oversight committee since they are liable for the "appropriateness of the financial statements and disclosures contained in their periodic report" (Martin, 2004, p34).

Once the assembly of the initial internal research and task forces are complete, the creation of a project vision or charter will clearly articulate how to monitor achievement of goals, objectives, and metrics as well as create preventative, detective and corrective control measures. Each teammate will also understand their responsibilities and have their roles clearly outlined should a conflict arise or if the organization falls behind on its implementation plan. The outcome of the project vision and mission should be to identify and mitigate any risks, which may threaten the shareholder value.

Upon completion of the strategic framework, the process of identifying and prioritizing the main risks will be critical. Acquiring risk identification issues through several formats, such as questionnaires, interviewing, and comparisons to competitors or by benchmarking can help the team gather anonymous wide-ranging problematic execution issues within the organization. Furthermore, by developing a grading criterion for each risk and prioritizing those issues according to the importance can help categorize different risks and the application of specific mitigation techniques. Risks, such as financial risks, operational risks, and hazardous risks prioritization are according to their risk level and timing importance. It would be important to categorize risks, which can threaten shareholder value with a higher criterion.

Even though the pharmaceutical industry is "not currently bound by law to comply with the provisions of Sarbanes-Oxley, health care organizations and their boards may wish to use the Act's guidelines to conduct an internal audit of ethics policies and conflict-of-interest procedures that will protect and better position the organization in this era of reform" (Maltz, 2003, p32). Pharmaceutical companies are, in particular, are at risk of revenue loss due to increased financial reporting pressures and governance by other members of the Pharmaceutical Research and Manufacturers of America (PhRMA), who represents research-based pharmaceutical and biotechnology companies. These members have developed a code on how to establish ethical relationships with healthcare professionals, while developing new medicines for patients. Their mission is to ensure healthcare professionals have the latest, most accurate information regarding prescription medicines (PhRMA 2010), the US Office of the Inspector General Guidelines as well as domestic and foreign competitors.

CASE STUDY: PHARMACEUTICAL COMPANY A

In this case study, Pharmaceutical A is not a market leader in any specific disease state category, however, they have performed well over the past decade with their improved sales, pricing and marketing techniques. However, to maintain sales leadership in a highly competitive environment, it may be important for senior leadership to review and understand their internal and external environments.

Pharmaceutical A has faced the risk of noncompliance with the Sarbanes Oxley Act of 2002. Due to US Securities and Exchange Commission (SEC) allegations, Pharmaceutical A conducted an internal investigation of their accounting practices. The investigation identified several possible areas of accounting errors and showed evidence of misconduct, such as possible false reporting to make the company more attractive to investors. It was clear the internal control systems of the company were weak and needed improvement. This issue led the company to revise some of their financial statements from prior years. Therefore, if Pharmaceutical A had implemented a detective measurement objective of an ERM program, the system would have alerted their internal team to this important risk, which may have threatened shareholder value, and the team could have taken appropriate actions to avoid financial mishaps and despair in a timely manner.

Risk Analysis

After identification and prioritization of potential areas of exposure based on internal controls and monitoring systems, each organization will determine how to respond to these identifiable risks. The concept of risk is one which may denote a pending negative impact or influence to an asset or downgrading value characteristic arising from a future event or situation. However, risk is the possibility of an event occurring which will have impact on not attaining or achieving a desired result of goal. Therefore, the evaluation of risk, its communication, likelihood and perception should be weighed and evaluated before concluding a decision making process. When evaluating risk, one will always have certain options which have higher potential risk than others. Organizations will have to make sure to evaluate all options (internally and externally) carefully to help guarantee risk mitigation success.

By implementing an internal control system, the risk of non-compliance behavior occurring is significantly reduced. The organization will monitor their accounting and financial reporting practices and if they identify problems and make the necessary steps to counter them, the organization can reduce the non-compliance risks to an acceptable level. If Pharmaceutical A had implemented a preventative measurement, such as routine education with all internal employees regarding the risks of lost business opportunities, penalties associated with non-compliance and the identification of problem areas of concern before they become a problem, they could have saved the organization significant time, effort, energy and expenses. A risk management committee should also have reviewed the organization's financial and accounting monitoring practices. By independently monitoring each of the organization's transactions as they occur and validating procedures, the organization is practicing fair and standard reporting, which will further reduce the risk of non-compliance.

Pharmaceutical A could also diminish competitor risk by training existing sales forces in all aspects of decision-making and sales ethics, which can mitigate non-compliance with the PhRMA Code and US Officer of Inspector General (OIG) guidelines. Apart from conducting risk management training for the organization as a whole, each business unit will need to develop strategies to mitigate risks in each of their individual disciplines. A corrective measurement process delivered in a way of giving responsibilities, training on how to use the tools, teaching the overall concepts and educating on the expected benefits, will help gain a commitment from employees performing the risk management as a part of their daily functions. The training should also serve as a measure to ensure a successful risk management culture within the organization.

Awareness of the risks will provide the employee a better understanding of how to avoid and mitigate each risk. This is a very important step for the successful implementation of an effective ERM process. By implementing a successful top-down communication process between senior management and the employees, organizations should avoid a lack of communication dilemma, which may lead to senior management being unaware of what is going on within the middle layers of the organization. In the Pharmaceutical A scenario, a lack of communication situation existed, resulting in middle management making decisions, which negatively affected employees and the overall organization.

Developing an ERM program is an ongoing process. Successful implementation involves integration at all levels of the organization and is a part of daily business activities and function. As stated by Sobel (2004, p29), "melding enterprise risk management with governance means directors, senior management, internal and external auditors and risk owners must work interdependently." If senior management continuously remains engaged, measuring and monitoring the ERM program, they can better evaluate the success of the implementation process and the overall success of the program. The committee team members will also have the responsibility to determine if the new ERM program is an integral part of employees' daily activities. The results of these findings would serve as a new benchmark for future comparisons. In addition, if an identified area of concern arises within these findings, the organization can proactively take the necessary steps to avoid non-compliance situations. Lastly, as the monitoring process becomes more efficient and effective, the organization may wish to make additional investments in technologies, which are capable of continuously monitoring the ERM system to aid in uncovering risks and mitigating issues prior to falling into non-compliance areas.

Investment strategies inherently contain a set of processes, rules, beliefs, behaviors or procedures, ultimately designed to guide an investor's selection on an investment or project. Usually their strategy will be designed around the evaluation of risk and reward. Some investors wish to maximize their returns by investing in risky projects, while others will prefer minimal risk. Nevertheless, most will rely on a strategy somewhere in between. Corporations depend on thorough planning and observation and the use of measurement tool such as calculating net present values. Typically, the higher net present value, although it does not guarantee the success of a project, it does however lessen the risk associated with its pursuit.

Ultimately, corporations must evaluate external and internal investment strategies to help them analyze the risks associated with investment decisions of expanding into existing market or launching into the new markets. Measuring the cash flow projects or investments generate will determine the success the fine balance between the amount of capital investment expended versus the return on investment may bring. Therefore, Pharmaceutical A must build the best plan which demonstrates the best investment of time, effort, energy and resources to launch products into new market spaces, thereby balancing the usage of existing resources with little financial (capital investment) outlay in order to maximize the highest rate of return on their investment.

Communication and Management Structure

The Pharmaceutical industry has had its share of difficulties over the past decade. Its continuing effort to bring quality pharmaceuticals to the market has helped ward off, comfort and cure many diseases for those patients suffering pain or misery. The process in which those products are brought to market depends largely on a clear, transparent and consistent internal communications structure. By comparing and contrasting two different pharmaceutical companies, Pharmaceutical A – a multi-billion dollar organization and Pharmaceutical B - a small start-up biotech, both have a similar goal in mind: partnering with health care professionals, providing quality medicines for patients and to develop a consistent revenue flow. These organizations primarily generate their revenue through a sales force which delivers promotional sales messages to health care professionals in an effort to gain brand loyalty and have their products become the agents of choice for patients.

Pharmaceutical A's leadership structure has the elements of a "Top – Down" communication structure. This structure does not give employees a great deal of exposure to the senior leadership team. Pharmaceutical A's communication structure from the outside appears to work very well; however, as seen in many large organizations, the senior leadership team primarily makes decisions from their level or vantage point, without the involvement of the management team tasked with its execution. Furthermore, implementation of these strategies is dependent on the clarity and understanding of these strategies by their front line management teams.

Many of the Pharmaceutical A front line managers implementing these strategies may not have had "hands on" field sales experience, due to being hired straight from Ivy League MBA schools and/or lacking skills necessary to have a thorough understanding how to communicate or elicit feedback on how to positively execute strategies. Due to these deficiencies, the information is usually presented in a more rigid, structured format such as a mandate or consequence. Therefore, information is not positively shared or unilaterally expressed across the organization in a positive manner which ultimately causes confusion and misunderstandings. As explained by Mindszenthy (2000, p29), "Cones of Confusion emerge when there is inconsistent sharing of information, in our experience resulting from either team leaders not understanding the implication of the corporate message, or not knowing how to share it or sometimes both."

This dysfunctional communication (lack of clarity, transparency, not consistent or positively delivered) to a sales force by the senior leadership team builds mistrust, misunderstanding and a sense of protecting oneself. This behavior may indicate a withdrawal from the project for fear of not wanting to be involved or result from being assigned to a project that will fail. If this is the case, the project most likely will fail and may ultimately lead Pharmaceutical A to not meet their objective. To create an open transparent communicative environment as described by the COSO, Pharmaceutical B, on the other hand, ensures that open lines of communications are preserved at every level of the organization regardless of title or position. They understand the importance of frequent communication, eliciting feedback and stating what finish or the end state goal should look like. When employees at Pharmaceutical B had their team meetings, they gathered the necessary information and had a better understanding of "what the finish should look like" from their management. Thereby the employees are better equipped to communicate the vision and empower themselves to action. This management approach helps build a better sense of involvement, alignment and expands upon the TEAM (Together Everyone Achieves More) concept. (HeartQuotes, 2010)

Some of the vehicles used by Pharmaceutical B to build upon the TEAM concept created an aligned workforce and empowered employees to action and commitment. The vehicles most often used were individual group meetings, town hall meetings with senior leadership, bulletin email broadcasts and shareholder meetings in which everyone was encouraged to attend and share their ideas. This is a measure Pharmaceutical A should incorporate in future meetings to gain company wide buy in and support. The openness to discuss needs, wants and concerns in a positive transparent, open environment avoids the "Cones of Confusion", as stated previously by Mindszenthy, (2000), and clearly set up Pharmaceutical B with an effective communications structure which was well ahead of its time.

Pharmaceutical B was able to integrate internal and external communication between all levels of the organization. This ability allowed them to focus on what was important, build a positive common vision and not spend time trying to sell employees on management directives or sort through clutter. They were able to stay ahead of the competition and shape the way their organization functioned. It will be vitally important to leverage communication resources, including new and emerging Web 2.0 tools, which create an environment within the organization where everyone has the appropriate tools to do their jobs effectively that will ensure that relationships are built on trust or and to avoid decentralization. This communications structure, which was implemented and enhanced by Pharmaceutical B, and could serve as a unique benchmark for Pharmaceutical A to review, emulate and build upon as they pursue an internal and external environment built on achieving successful outcomes.

External Environment

Pharmaceutical A's external environment consists of three components: its remote environment (macroeconomic), industry environment, and operating environment. In these three environments, there is a chance that one, two or all three of its external environments may change. Failure to frequently monitor or scan all of these environmental factors may result in disaster for Pharmaceutical A. All of these environmental components affect the firm's operations at a domestic and foreign level and it is very possible for changes to occur at any time. Change is a constant factor in business and the pharmaceutical industry is no different. There are numerous factors that will impact the internal and external environment of Pharmaceutical A. These environmental factors are inter-related and can not be excluded from one another. Upon reviewing these three environmental factors, associated risks, and possible solutions, failure to respond to these issues may negatively impact Pharmaceutical A.

The remote environment consists of changes which occur beyond the normal functioning operations of Pharmaceutical A. Five remote environment factors exist: economic, technical, political, social and ecological (Pearce & Robinson, 2005). Many of these factors cannot solely be influenced by Pharmaceutical A; they do, however, have a rippling effect on Pharmaceutical A's operations. One of the major issues Pharmaceutical A will face over the next decade is the partnering/developing of products as a result of the advancements in the human genome. DNA mapping, stem cell research and manufacturing will all take place in the remote environment. Burgers et al (1993) stressed the importance of collaboration among companies with complimentary resources is important for growth and survival. Strategic alliances will allow companies to share risks and explore emerging uses of technology to manage their resources and capabilities.

As technology changes, Pharmaceutical A must determine how they will upgrade, modify or partner with other companies which have invested in the latest technology to a level needed to stay competitive in the marketplace. Pressures on pharmaceutical and biotechnology organizations to expedite the research and development of new drug products have increased over the past few decades. Not only are drug developers challenged with developing drugs faster, but they are equally challenged to do so while incurring less expense. Technological advancements over the past decade have greatly increased the data management efficiencies for drug development companies. Today, pre-clinical and clinical study trials can be managed globally with integrated data management systems. Technological improvements have not only expedited the capture and regression of study data, but such improvements have reduced the amount of labor needed to process such tasks hence decreasing incurred labor expense.

As technological advancements increase the ability to manage study data on a global scale with relatively few resources, Pharmaceutical A has the opportunity to build upon its service lines for drug development organizations. Pharmaceutical A prides itself on the ability to offer comprehensive drug development services to fulfill its client's drug development needs. Data management systems are becoming more advanced each year and are trending toward web-based applications. Pharmaceutical A's investment upgrades in data management system, while significant, must be continually improved to keep pace with global standards.

Pharmaceutical A adapts to technology changes in their remote environment through configuration and customization of their data management system for each client's study needs. Providing tailored data management services serves a multitude of technological functionality for their clients. Not all studies are designed in a similar manner. Therefore, customization allows Pharmaceutical A clients to select what features of data management are most effective as it pertains to each study. Pharmaceutical A's strategic adaptability to technological advancements help its clients process data more effectively and efficiently ultimately resulting in the marketing of drugs in a more timely and cost effective manner.

If USA political influences impact laws, policy changes and regulations, such as Medicare, Medicaid and Managed Health Care reform, many pharmaceutical companies will have to change the way they conduct business. Social factors such as the aging Baby Boomers, Generation Y and Generation X'ers face an increased prevalence of diabetes, hypertension, cholesterol and several different forms of cancers, all of which will create a need for additional research and development in drugs to delay or cure these devastating disease states. Global warming, population and waste decay will continue to challenge many businesses, including Pharmaceutical A, which will have to invest in technological solutions to manufacture products or modify facilities to meet federal ecology standards or requirements.

The industry environment focuses primarily on entries to market barriers, suppliers influence, buyer influence, and competition. Competition among firms within the same industry determines the profitability of the industry and these competitive forces become very important in strategy development. For example, if a leading pharmaceutical company has the number one selling drug internationally to treat high cholesterol and its patent is soon to expire, low-costs substitutes from rival generic companies will try to enter the market early and drastically cut into this company's revenues and market share. In this situation, trying to negotiate market entry with the substitute product promoter may become a top strategic priority for the company.

The entry of new companies or competitors into the pharmaceutical industry brings new products, new science, expanded research and a strong desire to gain market share. "The seriousness of the threat of entry depends on the barriers present and the reaction from existing competitors that the entrant can expect" (Pearce & Robinson, 2005, p92). Product differentiation is one major source of barriers to entry. The ability of a company to successful differentiate its product will determine if the company can gain entry into a new market.

Pharmaceutical A hoped to use a product differentiation method to enter the Migraine market. Their product, a serotonin receptor agonists, is used to treat migraine headaches. This drugs narrows (constrict) blood vessels in the brain and relieve swelling, reducing the symptoms of migraine headaches (WebMD, 2010). A similar top selling product's patent expired in February 2009 allowing Pharmaceutical A to use its proprietary technology to differentiate its own serotonin receptor agonist's product, enabling needle-free subcutaneous delivery of product for the treatment of acute migraine.

Pharmaceutical A's strategy to enter the pharmaceutical industry was to use technology to create product differentiation for its product. This technology was the most advanced pre-filled needle-free injection system in development. Their patented design facilitated comfortable, precise injections of liquid formulations to the subcutaneous (below the skin) space. Their simple, three step dosing process was deemed to work well for patient self-administration. The needle-based forms of serotonin receptor agonists have several use-limiting characteristics such as patient fear of needles, needle stick risk, and complexity of use. Product differentiation through technological advances gave Pharmaceutical A the competitive edge it needed to enter the highly competitive pharmaceutical migraine prevention industry.

These influences helped maintain Pharmaceutical A dominance in certain therapeutic categories. As new product class competitors entered the marketplace, more similarities than differences have existed. Another issue Pharmaceutical A will face is their ability to continually differentiate products in order to maintain their competitive edge. Differentiation is critically important if a company is to maintain its superiority edge and generate revenue for the organization. In an excerpt from Charles King III (2000, p1) on product differentiation:

"Three aspects of the pharmaceutical industry make it ideally suited for studying the role of marketing in product differentiation. First, pharmaceutical firms spend an unusually large percentage of their revenues promoting their products.... Second, pharmaceutical companies mostly concentrate their marketing efforts in a single form of marketing, visits to doctors' office, which typically represent some 70 to 80 percent of industry marketing expenditures, and simplifies the analysis. Finally, drugs characteristics are essentially fixed, determined by the chemical properties of the particular molecule that constitutes the drug....The FDA also strictly regulates the advertising and promotion of prescription drugs"

Therefore, if Pharmaceutical A is unable to market a feature or benefit which the marketplace will bear as a unique advantage (i.e., flexible dosing, delay of action, fewer side effects), a price war will ensue and many drugs will be evaluated strictly on cost. If the price is lower, the buyers, managed care plans, wholesalers, pharmacies and consumers, will all benefit due to this low price or when slight differences in benefits do not outweigh the costs. On the other hand, if this lower price can increase volume, then Pharmaceutical A has to weigh how much it has to sell at the lower price to make a profit. Many generic companies sell solely on price in order to sell larger quantities of products. These manufacturers, however, do not have to follow the same strict guidelines as the branded pharmaceutical manufacturer in terms of quality controls on their products. As more product manufacturers enter the marketplace, suppliers of bulk materials are able to sell more for less or if insufficient quantities are available, they may charge a higher price for the same bulk or raw material manufacturers bought for less only a few months earlier. Pharmaceutical A must evaluate the market at all times to determine how they will price and compete in the pharmaceutical market.

The operating environment is the core of Pharmaceutical A's business. How they relate to their competitors, creditors, customers, suppliers, and especially employees will determine their enduring success. Pharmaceutical A has a very effective sales force, budgeting structure, distribution system and marketing presence. The company relies heavily on its suppliers and co-promotion partners for materials and resources for product manufacture and sales promotion. While Pharmaceutical A has been recognized as one of the better financially managed pharmaceutical companies, this recognition allows the company the ability to negotiate with several creditors on the financial resources it may need to fund research and development projects. However, if Pharmaceutical A is unable to maintain this reputation, they will not be able to invest in emerging disease state technologies advancements.

Therefore suppliers, creditors and co-promotion partner relationships may change if the company cannot maintain its financial reputation. Maintaining a strong financial position and a clear strategic vision will help Pharmaceutical A and its employees remain successful over the next decade. As the workforce ages and the reduction in the number of company-owned sales representatives throughout the industry declines, maintaining and recruiting the best marketing and sales talent throughout the organization will be necessary as contract sales forces have surpassed the tipping point and are now the rule - not the exception in today's marketplace.

As a result of constant change, companies must continually evaluate their benefit plans, incentive compensation structures and upward mobility for their newly recruited employees. The relationships with their employees, suppliers, and creditors must be well-established if Pharmaceutical A will be recognized as a leading pharmaceutical company in the USA. Expressing their value routinely will help reinforce and secure a feel of content and well being. Once this type of environment can be established, training and managing the needs, wants and concerns of employees should be a relatively easy task.

Training and Managing Employees

Implementing Enterprise Risk Management (ERM) is not just for senior management, as "employees at all levels within the organization are responsible for the success of the risk management initiative. Existing risk managers should be enlisted in this effort to help train and educate all employees about risks and risk management" (Burnaby, Hass 2009). Employees are an important part of any organization. Therefore, it is important for management and human resources to determine methods and practices to keep employees satisfied in their position and to motivate employees to embrace ERM processes and procedures. Anything less can have a negative effect on individual, team, and company performance and ultimately hurt the company's success and profits. For employees to succeed, they need ability, motivation, and opportunity (Dreher & Dougherty, 2001). An understanding of these issues can influence human resource practices allowing management to diagnose problems effecting employee behavior and focus on necessary solutions to modify the behavior. Ability is knowledge, skills, and aptitudes of employees.

Human resources need to identify employees or candidates with certain abilities and then shape/nurture these abilities further. Morals, values and motivation are key drivers to successful working relationships'. Human resources must enhance employee motivation through expectancies (i.e., mentoring, performance feedback, job-related training), instrumentalities (i.e., bonuses, incentives, recognition programs), and valence (i.e., salary increase, promotions). Lastly, opportunity is the constraint that affects work outcomes (Dreher & Dougherty, 2001). These include access to the necessary information, appropriate tools, software, equipment, training and development, challenging tasks, and budgetary/staff support.

Understanding one's moral and ethical values will give an insight into an employee's actions and decision-making ability. Similarly, if the leadership team of Pharmaceutical A were to demonstrate and provide real-life examples of ethical behavior and ERM decision-making, their employees, vendors and suppliers should be more willing to follow. Maxwell (2002) provides the following leadership levels for increasing influences and potential:

1. Position (Rights: People follow because they have to)
2. Permission (Relationships: People follow because they want to)
3. Production (Results: People follow because of what you have done for the organization)
4. People Development (Reproduction: People follow because of what you have done for them)
5. Personhood (Respect: People follow because of who you are and what you represent." (p75)

Therefore, the leadership team has an ethical duty to be honest with their employees, vendors and suppliers. In addition, they have a responsibility to establish a code of conduct, a set of performance standards and moral values for the company. It is equally imperative that the company have open lines of communication with their employees, vendors and suppliers at all times. All internal and external members of the team desire equal and fair treatment at work. Additionally, Maxwell states "people need a standard, which is the golden rule" and further explains, "If I treat you as well as I desire to be treated, you win, if you treat me likewise, I win" (2002, page 90). This should be the standard for every business to follow when evaluating and treating their employees, especially when developing an ERM system..

Another approach many companies use is a combination of strict compliance guidelines and rules when issues or violations occur within their organizations. Having these guidelines and rules will help protect the employee, as well as the company. and levels the playing filed in regards to legal situations and human resources policy violations. Therefore, continuous dialogue and developing ways to improve and monitor these policies will maintain and encourage open discussion as well as build motivation with all members of the team.

Finally, some of the challenges that a company faces will be in aligning their commercial and ethical agendas as they manage their day-to-day transactions with outside vendors, suppliers and financial institutions. By examining and establishing ERM policies on sales force conduct, repaying financial obligations, purchasing practices and investment decisions, it will be very crucial for companies such as Pharmaceutical A to maintain a high ethical standard and not get into the habit of poor financial planning or venturing into unethical behavior in order to increase sales or marketability. "Because when it is all said and done, your ability as a leader will not be judged by what you achieved personally or even by what your team accomplished during your tenure. You will be judged by how well your people and your organization did after you were gone. Your lasting value will be measured by succession" (Maxwell, 2002, page 102).

Motivating Employees

Pharmaceutical A's motivational strategies and tactics are a major focus for the future of this organization. Pharmaceutical A has implemented a new team-based philosophy. In order to execute this change, Pharmaceutical A will need to reassure their employees that the compensation plan will not be affected and must focus on what motivates employees. How they execute their plan to inspire motivation will encompass many facets. Senior leadership may believe that they are providing a great environment and addressing the needs of their employees. If these needs are met, then a person can feel fulfilled and progress to the next level of satisfaction. So, if Pharmaceutical A is planning to motivate employees with a goal-setting program, it may fail if employees feel unappreciated or misunderstood. These employees may not be able to let go of previous company wrong-doings. If this is the case, these previous wrong-doings or "baggage" will hinder the employee's ability to move forward and release this past excess baggage. By denying the employee's basic needs, this could cause further failure within Pharmaceutical A's organization. Companies must first understand what motivates employees and then once their basic inner needs are met, companies, like Pharmaceutical A, can establish an effective ERM foundation for success.

Rewards and recognition are an important part of managing people. Motivation is a "willingness to exert effort toward a goal" (Dreher & Dougherty, 2001, page 31). When a leader develops a system for recognition and communicates it clearly to their employees, they will begin to align their employees to the company's strategic goals. Communication is an important part when implementing any recognition system. The employee must understand the expectations and the recognition or return for their efforts. Jensen (Chew, 2005) mentions in his article "Self-Interest, Altruism, Incentives, and Agency Theory" that people do care immensely about failure and success; they do have feelings, emotions and do care about their honor and self-esteem; and they feel shame and pride. Therefore, holding employees accountable for their careers and praising their successes can only help build open lines of communication, avoid significant risks of underachievement and avoid unmet organizational goals due to lack of enthusiasm from employees. The important link between establishing an effective Enterprise Risk Management (ERM) system and training/developing employees at all levels should not be overlooked. Embedding ERM decision-making methods, practices and strategies into ongoing staff development programs will enable employees to understand and embrace ERM concepts.

Therefore, Pharmaceutical A can maintain its competitive edge and meet its financial goals and objectives, if they recognize, review and evaluate the factors impacting their remote environment, industry environment and operating environment, which, in turn, may impact their business positively or negatively. Technology will be the lifeblood of the organization. By buying, partnering or modifying existing facilities to accommodate new advancements, they will ensure successful product availability in the future. Pharmaceutical A must stay current and aware of changes in the political, social, and ecological factors to be considered a viable player in the pharmaceutical industry. This competitive superiority will ensure success over the next several decades. Several companies will challenge Pharmaceutical A by bringing products to market which may be similar, priced lower or better positioned on managed health care plans. Pharmaceutical A must continue its innovation and pipeline development. By treating their employees, suppliers and creditors as a valued part of their organization, Pharmaceutical A should be able, in the short term, continue on a path of success, However, by establishing long term or future research goals the organization will continue on its mission to provide a healthy lifestyle to those customers and patients it serves. Management has a stake in career development programs. Dreher and Dougherty (2001, pages 233-249.) state that "of the top 100 best companies listed in January 10, 2000 issue of *Fortune*, many provide employees with development opportunities". These companies have to compete with each other in the marketplace and providing extensive programs, such as benefits, training and career development programs. These can increase employee performance and also impact human resources' ability to attract top talent to the company and retain excelling employees within the company. In addition, career development programs teach the necessary skills and tools to employees to eventually replace outgoing managers and executives.

FUTURE RESEARCH DIRECTIONS

As companies have to develop long term enterprise risk management programs and procedures in light of changing economic indicators, many pharmaceutical companies will have to heavily rely on emerging technologies and sales forces reductions to manage their balance sheets to remain profitable. Technologies like e-Media and Video Sales Representatives will likely be the trend versus the customary "feet on the street' mentality. Research by Fugh-Berman and Ahari (2007) indicates that "pharmaceutical sales organizations spent about $150,000 per primary care rep and $330,000 per specialty drug rep," Justifying the return on investment for these representatives is very difficult without the advent of launching new blockbuster drugs.

The trend of reducing sales forces (to save money in an economically depressed time) and using contract sales forces has reached its "tipping point". As stated in Malcolm Gladwell's book "The Tipping Point", a simple idea and little things can make a big difference causing a "tip" in a circumstance. *The Tipping Point* is one dramatic moment in an epidemic when everything can change all at once. If an average cost per call or detail to a physician by a sales representative "ranges between $200 (low side) and $400 (high side)" (Pharma Marketing 2008), while e-detailing costs closer to $100 per e-detail with a Video Sales Representative. The trend to move to lower costs forms of promotions seems evident. As large sales force reductions continue and territories expand, the need to cover areas which are not as easily accessible by "feet on the street" representatives, e-detailing appears to be a very viable solution. Partnering with vendors which staff call floors or have relationship marketing departments designed to promote products online will help ease the burden of significant investment or lost revenue for start up companies or existing companies with small or large sales forces. (Alkhateeba et al 2009).

When managing the risk associated with emerging technologies, using outside vendors can be easily and accurately managed as expectations are set upfront and finalized in the negotiated contract. Thereby making e-Media a very viable option for the external environment, as hiring companies will still have an added measure of control on target lists, messaging and fulfillment of resources. The outside vendor will have an obligation to meet the expectations of the contract and be able to make the number of calls on the targets via the web or conference call.

Managing the risk of training employees will be in the same manner as current pharmaceutical representatives due to the fact that the same rules apply for HR, messaging, compliance and field training. Employee career progression within the vendor will be managed outside the company unless the hiring company wishes to have the vendor reps and teammates "rolled into" the company at the end of the contracted period. Lastly, working with a vendor allows the hiring company to turn off or turn on a sales force or inside sales team consisting of video sales reps at any time, depended on contract obligations.

This shift in sale force ownership and the financial viability of the vendor could allow video sales reps with the same career progression they would have experienced with a pharmaceutical company depending on the length of contract and inflow of new contracts. Sales representatives could be promoted to District manager or Regional Manager or even further dependant on the flexibility of the technology vendor. All of this responsibility which originally would have been absorbed by the pharmaceutical companies can now be subcontracted out to other vendors, reducing the overall risk expenditures to the organization.

CONCLUSION

Understanding the environment and the value of an enterprise risk management process including a clear and concise ten step process, as described by Barnes (2009):

1. Mandating from the top.
2. Establishing an ERM department and buy-in.
3. Deciding on a control framework.
4. Determining all risks.
5. Assessing risks.
6. Identifying business unit objectives and performance measures.
7. Reviewing objectives and control summaries.
8. Monitoring monthly ERM reporting system.
9. Conducting analysis by ERM department.
10. Continuously monitoring the process.

These steps will aid in possible corrections and are critically important for companies such as Pharmaceutical A to build effective sales and marketing strategies. By routinely reviewing, updating and evaluating the internal and external factors (i.e. training, career development, managing employees, future directions and trends) Pharmaceutical A would be better prepared to build and meet long-term objectives and goals. Although economic challenges are in constant flux, Pharmaceutical A has been able to reach their goals with strict adherence to their core mission and values. If Pharmaceutical A were to have implemented an ERM system, they would have had a better understanding on how to control and manage their risks and mitigation techniques and avoid losses in shareholder and company value.

Integrating an ERM system into an organization's culture requires an open, honest, transparent, and clear commitment from employees all the way to the top senior management level. This system will help an organization have a better understanding on how to identify, prioritize and mitigate their main risks as well as help them implement actions proactively before entering areas of non-compliant behavior.

Organizations which use all the tools associated with Enterprise Risk Management policies and procedures will be better able to adapt and survive in continually challenging economic downturns or reach new revenue goals in prosperous economic times. The most important factor will be hiring the right forecasters and being religious in testing the values and outcomes of the proposed processes. Periods of challenges and economies change quickly and missing even the little things that Malcolm Gladwell (2000) describes in his book "The Tipping Point" can be detrimental or the difference to being ahead of the market and being viewed as an industry leader.

Enterprise Risk Management starts with the idea of what the finish should look like and that a foundation needs to be built and engrained with each employee from the ground up all the way top. With this level of commitment, organizations will be able to endure any storm and continue to surpass each and every financial milestone placed in front of them.

REFERENCES

Alkhateeba, F. M., Khanfarb, N. M., Doucettec, W. R., & Loudond, D. (2009). Characteristics of physicians targeted by the pharmaceutical industry to participate in e-detailing. *Health Marketing Quarterly*, *26*(2), 98–116. doi:10.1080/07359680802619792

Burgers, W. P., Hill, C. W. L., & Kim, W. C. (1993). A theory of global strategic alliances: The case of the global auto industry. *Strategic Management Journal*, *14*(6), 419–432. doi:10.1002/smj.4250140603

Burnaby, P., & Hass, S. (2009). Ten steps to enterprise-wide risk management. *Corporate Governance*, *9*(5), 539–550. doi:10.1108/14720700910998111

Chew, D. H., & Gillan, S. L. (2005). *Corporate governance at the crossroads, a book of readings*. New York: McGraw-Hill.

COSO. (2008). *The Committee of Sponsoring Organizations*. Retrieved December 25, 2009, from http://www.coso.org/default.htm

D'Aprix, R., Clemons, J. G., Windsor-Lewis, S., Likely, F., Baron, A. L., & Moorcroft, D. (2006). *Fast forward: Future trends in corporate communication*. UK: Strategic Communication Management. (Original work published 2005)

Dreher & Dougherty. (2001). *Human resource strategy: A behavioral perspective for the general manager*. New York: McGraw-Hill.

Garrette, B., & Mitchell, W. (1998). *Acquiring partners' capabilities: Outcomes of scale and link alliances between competitors. Managing strategically in an interconnected world.* New York: John Wiley and Sons.

Gladwell, M. (2000). *The tipping point–how little things can make a big difference*. New York: Little, Brown, and Company.

HeartQuotes. (2009). TEAM–Together Everyone Achieves More. Retrieved December 24, 2009, from http://www.heartquotes.net/teamwork-quotes.html

King, C. III. (2000). *Marketing, product differentiation, and competition in the market for antiulcer drugs*. Harvard Business School Review.

Lachance, J. R. (2000). International Symposium of the International Personnel Management Association. *Public Personnel Management*, *29*(3), 305.

Maltz, A. (2003). Codes of ethics in the wake of Sarbanes-Oxley. *Trustee*, *56*(10), 31.

Marketing, P. (2008). *Fewer sales reps lead to higher costs*. Retrieved June 5, 2010, from http://pharmamkting.blogspot.com/2008/11/less-sales-reps-lead-to-higher-costs.html

Martin, A. G. (2004). Making SOX work for your company is just good business. *Fort Worth Business Press*, *17*(24), 34.

Maxwell, J. C. (2002). [What every leader needs to know. Nashville, TN: Thomas Nelson Publishing.]. *Leadership*, 101.

Milkovich, G., & Newman, J. (2004). *Compensation*. New York: The McGraw-Hill Companies.

Mindszenthy, B., & Roberts, G. (2000/2001). *Team leaders and the communication loop*. Strategic Communication Management.

Pearce, J., & Robinson, R. (2004). *Strategic management* (9th ed.). New York: The McGraw-Hill Companies.

Sobel, P. J., & Reding, K. F. (2004). Aligning corporate governance with enterprise risk management. *Management Accounting Quarterly, 5*(2), 29.

Web, M. D. (2008). *Migranes and headaches.* Retrieved January 24, 2010, from http://www.webmd.com/migraines-headaches/triptans-serotonin-receptor-agonists-for-migraine-headaches

ADDITIONAL READING

Brealey, R. Myers, S. C., & Allen, F, (2005). Principles of Corporate Finance, Eighth Edition. New York: McGraw–Hill.

Burrill, C. W., & Ledolter, J. (1999). *Achieving quality through continual improvement.* New York, NY: Wiley.

Chase, R. B., Jacobs, F. R., & Aquilano, N. J. (2005). *Operations Management for Competitive Advantage 11ed.* New York: McGraw-Hill.

eJournal USA. (2005). The business perspective on corporate governance. [Retrieved January 24, 2010 from http://usinfo.state.gov/journals/ites/0205/ijee/kenney.htm]

Eskenazi, J. Workforce Planning: The Key to True Staff Forecasting and Strategic Staffing. SHRM White Paper. Retrieved December 18, 2009, from http://www.countrywatch.com.ezproxy.apollolibrary.com/

Eye for Pharma (2009) Fixing the physician-rep relationship [Retrieved June 5, 2010 from http://social.eyeforpharma.com/blogs/yanis-saradjian/fixing-rep-physician-relationship]

Forbes. (2009). Pfizer's $2.3 Billion-Dollar Settlement. [Retrieved January 24, 2010 from http://www.forbes.com/2009/09/08/pfizer-bilion-dollar-settlement-fda-opinions-contributors-david-r-henderson-charles-l-hooper.html

Fugh-Berman, A., & Ahari, S. (2007) Following the Script: How Drug Reps Make Friends and Influence Doctors. PLoS Med 4(4): Retrieved January 24, 2010 from http://www.plosmedicine.org/article/info:doi/10.1371/journal.pmed.0040150

Gray, C. F., & Larson, E. W. (2006). *Project Management: The Managerial Process, 3e.* New York: The McGraw-Hill Companies.

Healthcare Financial Management. (2001). TAP Pharmaceuticals settles with DOJ for $875 million - Brief Article [Retrieved January 24, 2010 from http://findarticles.com/p/articles/mi_m3257/is_12_55/ai_82481865/?tag=content;col1

Katsanis, L. P., & Pitta, D. (2006). The Limits of Strategic rationality: Ethics, Enterprise Risk Management and Governance. *Journal of Product and Brand Management, 15*(4), 250–254. doi:10.1108/10610420610679610

Koch, C. (Jan. 2006). The ABCs of ERP. Retrieved February 26, 2010 from: http://www.cio.com/research/erp/edit/erpbasics.html

McConnell, C., & Brue, S. (2005). *Economics: Principles, Problems and Policies* (16th ed.). New York: McGraw-Hill.

McShane, S., & Von Glinow, M. (2000). *Organizational culture. Organizational Behavior.* New York: McGraw-Hill.

Noe, (2004). *Fundamentals of human resource management.* New York: McGraw-Hill.

Parker, G. (2003, February). Leading a team of strangers: most teams now are diverse or virtual - Fundamentals Formerly Training 101 - cross-functional teams. Retrieved December 21, 2009, from LookSmart Web site: http://findarticles.com/p/articles/mi_m0MNT/is_2_57/ai_97650312

Paul, R. (2005). Repeal Sarbanes-Oxley! *Hope for America*. Retrieved April 18, 2010 from http://www.ronpaul2008.com/articles/289/repeal-sarbanes-oxley

PhRMA. (2010). Code on interactions with Healthcare Professionals. [Retrieved January 24, 2010 from http://www.phrma.org/code_on_interactions_ with_healthcare_professionals

Pharma Reform (2010) Healthcare Market considerations for Eliminating Pharmaceutical Sales Representatives [Retrieved June 24, 2010 from http://www.pharmareform.com/2010/06/03/healthcare-market-considerations-for-eliminating-pharmaceutical-sales-representatives/]

Stowers, J. (2004). *Yes, You Can Achieve Financial Independence* (4th ed.). Kansas City: Stowers Innovations.

The Alcoa Story. (2007) Retrieved February 11, 2010 from www.corportearchitects@aia.org.

The Committee of Sponsoring Organizations of the Treadway Commission" [Retrieved January 19, 2010 from the http://www.coso.org/]

Thompson, J. (2005). Wachovia: Merging businesses and cultures. *Chief Learning Officer*, *4*(12), 22.

Weber, R. A., & Camerer, C. F. (2003). Cultural conflict and merger failure. *Management Science*, *49*(3), 400. doi:10.1287/mnsc.49.4.400.14430

Weitzner, D., & Darroch, J. (2009). The Limits of Strategic Rationality: Ethics, Enterprise Risk Management and Governance. [Emerald Group Publishing LTD. Toronto, Canada]. *Journal of Business Ethics*, *92*, 361–372. doi:10.1007/s10551-009-0159-0

White, M. (2005). A Cure for the Upset Company. Retrieved February 3, 2010, from www.pepsifacts.com.

Young, C. March 13, 2007. *Management: Employee Retention Efforts Drive Turnaround at Sears*. Retrieved February 11, 2010 from http://getmyarticles.com/article_display.php?catID=33&artID=101894.

KEY TERMS AND DEFINITIONS

Committee of Sponsoring Organizations (COSO): A committee created by the National Commission on Fraudulent Financial Reporting. COSO published a document which outlines and provides a guide on internal control and framework. Members of this committee comprise the American Accounting Association, the Financial Executives Institute and the Institute of Management Accountants.

E-Detailing or Video Sales Representative Messaging: A process in which representatives use computer technology to enhance, deliver or bypass the traditional sales call to healthcare providers. (A traditional method of delivering sales message is known as "detailing.")

Enterprise Risk Management: The methods and processes used by organizations to manage risks and seize opportunities related to the achievement of their companies overall objective.

PhRMA Code: A voluntary code of conduct by medical healthcare companies on how to manage relationships with U.S. healthcare professionals. This Code addresses interactions with respect to marketed products and related pre-launch activities.

Risk Mitigation Process: The process of identification, assessment, and prioritization of risks, whether positive or negative followed by a concerted effort of resources to minimize, monitor, and control the probability and maximize opportunities.

Sarbanes-Oxley Act: Developed due to continuous accounting scandals in 2002. This act was developed by two Senators, Paul Sarbanes and

Michael Oxley. The act mandates the regulation of public Corporations within the United State, it consist of 11 requirements for financial reporting.

U.S. Officer of Inspector General (OIG): Is an agency within the United States federal government which is charged with identifying, auditing, and investigating fraud, waste, abuse, and mismanagement.

Chapter 9
The End of the Job Title: The Prospects of Analytics in the Staffing Industry and How to Deliver Them

Georg Juelke
Capgemini, The Netherlands

ABSTRACT

The staffing industry, despite being a global, multi-billion dollar business, has not yet widely exploited the use of business intelligence to make companies more competitive. Staffing companies are far removed from developing enterprise wide analytics and their analytical capabilities are either impaired or localized in their approach. Business intelligence commonly used in many other industries to optimize processes, reduce costs, or develop new services is dormant in staffing. This chapter analyses some of the root causes that impair the industry's ability to develop analytics. While some originate in specific market conditions that are reflected in the design of IT systems, it is the absence of a common nomenclature to classify job categories that prevents consistent data management and the ability to integrate data across divisions and geographies. The chapter introduces the application of information extraction and expert system to generate artificial job classifications that could replace existing ones, which are largely based on conventional semantic notions. Under the assumption that companies in the staffing industry can deploy shared and common job classifications across their IT systems this chapter presents a range of service improvements, new services and data driven insights that are presently unrealized.

INTRODUCTION

When Yahoo appointed a senior executive in the role of "Head of the Hive for Cranium" (translation: director of public relations), the note did not create much of a stir. The reader easily comprehends that the job title was meant to be humorous and convey an air of irreverence and creative wordplay than indicating what the person actually *did*. The seemingly incongruous invention of a job

DOI: 10.4018/978-1-60960-501-8.ch009

titles is symptomatic of a larger trend that sees the meaning of job titles rapidly deteriorating to the point that it becomes the subject of ridicule or irony.

Anyone working in the Information Technology (IT) sector or in an IT related industry will know the exasperated looks of relatives who inquire about one's profession and given an answer, inevitably reply: "But what is it that you do?"

The diversifying of the professional landscape creates all sorts of new, overlapping and hybrid functions and thus again breeds a multitude of job titles, which are increasingly nonsensical and considerably detached from general comprehension.

As the descriptive function of a job title becomes less important, other interests and purposes impose themselves. Inspired by a sense of political correctness, job titles are invented to try to dignify menial tasks. Education Centre Nourishment Production Assistant (dinner lady) came only second place in a poll about silly job names (Coles, 2007).

Instead of a pay rise, an employer might choose to give a staff member a more gratifying job title that leads to a burgeoning number of pompous sounding names that eventually arouses not respect but suspicion. Frequently, employees see their job titles changing while the nature of their work or the degree of responsibility did not change at all.

This very recent phenomenon contrasts with the way a professional title used to be the subject of fierce protection going back to medieval times when craftsmen and artisans formed guilds, which rewarded their members with titles that were given out only after a strictly defined process of apprenticeships, exams and accreditations. These vouched for its bearer to confirm that he possessed all the skills and the knowledge that were associated with a title. To usurp such a title, without proper accreditation, was considered an offence and subject to punishment.

The primary aspect and significance of a job title rests therefore in its semantic compactness, the ability to compress a wealth of information into a short string of words. In an ideal world of unadulterated job titles, sender and receiver would exchange within a single term, a vast range of professional skills and experience that otherwise required a lengthy description. The clarity and preciseness of a job title in the past was the result of rigorous codification, aggregating a defined set of data (skills) into a universally understood category.

The job title however is now a relic from a time when the professional landscape was less diverse and careers used to follow a clearly defined and predictable path, when the choice of an academic field for example correlated with a husbanded profession.

Practical Application of Job Titles

The dizzying number of job titles and the process of inflation that comes along with it has real, tangible and costly economic consequences. The job title used to serve a vital function as the prime matching criteria that brought together a suitable applicant and a potential employer in need of specific skills. As this function has been irreversibly eroded, both sides, applicant and employer, are now forced to descend to a level below: A lengthy narrative of qualities and skills to advertise demand and offer respectively. A useful aggregate has been irretrievably dissolved into its components.

Recruitment is concerned with finding and selecting the best candidates in a talent market to ensure the competitive fitness of a company. It is a time consuming and complicated task that consists in both drafting and advertising a job profile and then best matching the incoming flood of applications (CVs) against it. It is costly due to the sheer elaborateness of the process itself and the countless man hours that are invested in it, as well as the expenses that it might incur when one gets it wrong and employs a person who turns out to be not suitable for the position. Industry experts estimate the cost of replacing an exempt employee ("exempt" from wage and hourly regulations) at

1X - 1.5X the employee's salary plus benefits. For an executive, the cost increases to 3X - 5X the executive's salary plus benefits. Recruitment is a vital, continuous and daunting business activity; not only for a multi-billion recruitment industry and HR departments, but for the majority of business managers who have to make hiring decisions. Despite all advancement in machine intelligence, it remains quintessentially a human task unaided by computerised process automation. It still consists of - as it always did - sifting through pages and pages of CVs scanning for the right set of skills and experience in a candidate who might be best matched to fit the requirements.

How to Make Career Data Accessible to Data Mining?

The professional aptitude of a person is the sum of skills that she or he acquired both through education as well as professional experience. To solve the data problem behind the concept of a professional title, the entirety of a person's skills needs be aggregated into a uniquely identified and recognised label. The taxonomic and semantic anarchy that invaded the invention of professional titles renders the existing system unsuitable. Semantics itself are overburdened to compress the complexity of skills and experience into a sequence of maybe two or three words.

The solution - of course - is to restore the precise correlation between a job category and the set of skills, accreditations, education and professional experience that are required to meet the demands of a profession. It must be based on a system that is both open and flexible enough to accommodate the constant learning and changes that characterize modern career development with the rigors of a standardized, transparent process that guarantees that a professional category convenes with the qualifications of its bearer.

The absence of universally understood professional categories and the need to rapidly indentify available skills in a pool of candidates, especially in an internet based recruitment process led to the emergence of a new search practise, which narrowly focuses on skills which are needed by a potential employer and can be unmistakably identified. They very often relate to skills or knowledge that are subject to some sort of standardized test or accreditation process and therefore retained a degree of control and significance. But this comes at a cost; the entire range of a professional's knowledge and job experience is stripped down to a set of key words. For a few tokens of reliable information about a person's skills, any other kind of professional qualification that does not fall into a narrow array of key words is being sacrificed. Reliability of information falls into two categories catering to different requirements. In the first instance, the accreditation that an applicant claims to possess needs to be easily certifiable. Secondly, it should be semantically unambiguous. Information cannot be encoded in the syntactic structure of sentences but needs to match a set of key words that a search programme can detect without the help of human judgement.

This chapter will discuss anecdotal evidence of the speed, efficiency and ultimately the limitations of such a process. Having posted the author's curriculum vitae on a well known job board, it was a disappointing realisation of how little market interest was elicited by ten years of experience in a senior business intelligence function. The update of the exact same résumé with a certification acquired in conjunction with a widely known software product triggered - within minutes - a flurry of recruitment activity. Somehow, the words describing this certification functioned as a trip wire, triggering search agents which in return were sending notifications to recruitment consultants and HR departments who then could spring into action: A string of three words draws a reaction in a way that a diligently drafted CV of three pages could not.

On closer inspection the professional certification holds the secret on how the original compactness of a job title can be restored. What has so powerfully helped the advertising of, or search for a professional talent? Three things: Semantic compactness, implicit richness and reliability of information. Three words, universally understood, described a certified professional qualification. What these words indeed stand for is a curriculum of knowledge its bearer had to study and subsequently give proof of his knowledge in a standardized test.

If this principle could be widened into a general practise that captures any type of professional skill and condenses it into a professional category according to defined rule, novel opportunities of data processing would crop up. A system is required where a precisely defined set of skills, acquired through study, courses, vocational training or professional experience is registered and qualified. It subsequently rolls up (i.e., aggregates)

into a professional category that reflects exactly the sum of skills and experiences that were entered. These categories will be represented by relatively fine grained codes far removed from the sweeping and blunt descriptions of job titles. They need to possess the right degree of granularity to incorporate the acquisition of new skills or consolidation of certain skills through experience in time within the same profession. The rules that aggregate a set of skills into a category need to be pre-defined and dynamically adjust the additions or updates of professional expertise into a different, usually higher ranking class. Skills can be weighed with regards to their overall significance and qualified respective to degree of competence and duration of practical experience. This translates into a degree of transparency and standardization, which will reliably correlate the entire range of a person's professional competence to a code. Codes that describe a professional group can eventually be accrued into a broader semantic description as it is provided by the ILO (International labour organisation), a comprehensive and structured compilation of professions published in "Resolution Concerning Updating the International Standard Classification of Occupations" (Geneva 2007) which rolls up into a nested hierarchy of more general professional categories and that ultimately condenses into 10 major groups:

1 **Managers**
2 **Professionals**
25 **Information and communications technology professionals**
251 **Software and applications developers and analysts**
2511 **Systems analysts**
2512 **Software developers**
2513 **Web and multimedia developers**
2514 **Applications programmers**
2519 **Software and applications developers and analysts not elsewhere classified**
3 **Technicians and associate professionals**
4 **Clerical support workers**
5 **Service and sales workers**
6 **Skilled agricultural, forestry and fishery workers**
7 **Craft and related trades workers**
8 **Plant and machine operators, and assemblers**
9 **Elementary occupations**
10 **Armed forces occupations**

The Standard Classification of Occupations at its most granular level (i.e. 2519 Software and applications developers and analysts not elsewhere classified) would not be suitable to describe a person's actual profession. It also leaves the question unanswered what criteria exactly apply defining that level. And it is here where we can find the information that holds the key to create consistency in the universal definition of job classifications. Finding the rules that underpin those criteria will ultimately unlock the analytical potential of career data.

What Would Change?

Career related data, despite its tremendous relevance for individuals, companies or governments alike, has so far largely escaped a penetrating analysis as it is common in many other fields where operational or statistical data is accessible to data mining. The relative anarchy that currently rules in naming professions detached from any standardized and transparent system requires additional efforts in matching candidates to requirements in recruitment or reporting on recruitment related services. What would change if a standardized, skill based system of professional categories became commonplace?

The recruitment and temporary staffing business developed into a global, multi-billion industry with a broad range of services delivered to local and international customers. Its core competency is the recruitment of candidates and the matching of a candidate's skills with the requirements of a customer. This process still requires the review of countless resumes, interviews and tests. It also depends on the judgement of a recruiter to interpret the requirements of customers and apply it to the selection of potential candidates. Apart from being a time consuming and therefore costly process, it is fraught with the danger of misinterpretation. The way a request for talent is defined might differ from the manner the information about the available talent is stored in a candidate database. If both customer and the recruitment agent would base their request and the related recruitment service on the same categorical system, communication between the two parties would be notably simplified. The request of a client for particular professional competence could translate directly into a computerized match finding the best possible candidates in a database. In areas where certain skill sets are rare and a direct match difficult, this system could apply algorithms that find the best possible match against available resources.

The trend in the industry for the larger staffing companies is towards high volume national and international accounts. Staffing companies deliver multimillion services to large multinational corporations in many of the developed countries. These companies in return demand consistency of services, quality controls and reporting that ties into their purchasing and financial systems. Financial reporting on staffing services inevitably refers to aggregated job categories defined by the client which bear no relationship with job categories in invoicing and pay-rolling systems that are the source of the reporting. It requires the national operations of staffing providers to shoehorn their data into the reporting requirements of their clients. Language becomes an additional obstacle in this endeavour. As job titles are stored in the respective national languages it requires the orchestrated effort of multilingual data administrators in a laborious mapping exercise. What from the outset appears to be a sensible and clear-cut client request to have visibility of their spending on staffing services; subsequently turns into a time consuming and ultimately useless task. The client has to accept the financial reporting in good faith insofar as the aggregation of job titles into the categories he has imposed on his suppliers does not follow any rule based system. If both customer and supplier had used a standardized system, in which financial data for supplied services made reference to codified job categories and not just the effort required to produce the reporting, it would be reduced but the quality of data vastly improved.

Standardized, skill based job categories would not only produce consistent data in national candidate databases and matching tools. It would effortlessly support multiple languages, relating without ambiguity, the various semantic descriptions to a given codified job category. Defining job categories in a universal, language independent fashion will leverage the geographical coverage of large multi-national staffing provider or enable collaboration between smaller companies across different countries. Migration of skilled labour is a century old reality. Excess of labour in one country and the demand for that labour in another will inevitably provide a migratory incentive in which individuals and entire economies benefit. The European Union specifically encourages labour mobility by allowing its members to work anywhere within its geographic boundaries. Staffing companies could provide a lucrative service by maintaining integrated candidate databases that would facilitate the recruitment and deployment of skilled labour across national borders. International recruitment is already a service offered by some staffing companies but it requires an extraordinary administrative effort. A standardized system of job categories would simplify the task tremendously. The demand of skills in one country could be precisely matched against the availability in other countries without the interference of language bound specifics. Staffing providers could extend the geographical reach of their recruitment efforts and leveraging their international presence. Smaller companies lacking that reach could easily engage in collaborations with local counterparts in other countries to capitalize on differences between demand and availability of skills between regions or countries.

From the Author's Perspective

Frequently, career development does no longer follow a straight path. Academia might not translate into a chosen profession. One profession might be given up for another one unrelated to the first. Careers even within the same company might take a course so that a current role bears hardly any relation with the function an employee was once hired for. Knowledge and experience wherever they were acquired from turn into the content of a resume, painstakingly drafted and updated. They are often re-written to accentuate recently attained skills and omitting others that might have no relation to the position a candidate applies for. A person's diverse skills and experience might qualify her or him for different positions which belong to seemingly unrelated professions, career paths that might be pursuit equally if opportunity knocks. However, the economy of resume maintenance demands a focus from the outset, the deliberate sacrifice of viable options. If hybrid skills should leave open the chance for a varied pursuit of job opportunities it necessitates multiple resumes with different foci that prune any job experience or skill that are not relevant to a specific job application.

Many highly skilled professionals consider careers in foreign countries where their talents are sought after and better remunerated. It requires them to write and maintain their CVs in two or more languages. Someone might not consider applying for a position in a country whose language he or she does not speak, but it might be that in the eyes of a prospective employer, his skills override any linguistic shortcomings. Not being able to advertise his skills and availability in a given language turns into an opportunity not realised.

Job boards, recruitment agencies or head hunters maintain their own forms or templates which an applicant is asked to populate with his data. It demands considerable time and patience to register with even the major players in the market. A standard, skill based form would require a professional to maintain only a single, personal data set. It could be stored in an electronic file format that will be exchanged between multiple parties without further adjustments. It would allow the whole depth and range of someone's professional experience to enter dynamically into a job category. Additional training, newly acquired skills, promotions and new managerial responsibilities could dynamically be added to the existing profile possibly resulting in a higher classification. Existing job skills – if practised – increase in seniority over time. They all aggregate according to a weighed and pre-defined manner into a specific job category. By simply updating all relevant data into a structured input template, a programme will calculate all values into one or many job categories. Individual job categories are tied into broader professional clusters. Job categories within clusters are graded relative to each other indicating scaling degrees of seniority. This relation of the individual category to neighbouring ones will make sure that the totality of skills renders the highest category within that cluster. It leaves open the possibility that a different set of skills translate into a separate category of a different cluster thus enabling a candidate to advertise his suitability in different professional areas.

New Services and Insights

A staffing provider or recruitment firm that accumulates data of such quality and consistency would put itself in a position to offer completely new services and set the table for a vast area of business intelligence that allows for unprecedented insights into the hidden dynamics of career developments. It could turn its operational data itself into a marketable or billable service for their customers and candidates alike. The world of employment is of vital interest for all parties involved in the economic process: Candidates or employees, companies, governments and academic institutes. No other institution generates more data on employment and employment services than staffing companies. Especially larger companies find themselves in the unique position to produce labour data across all developed economies. However, none of them seem to have a prominent voice in employment matters. They do not advise governments or companies, they do not shape education or counsel job seekers. Terabytes of data that remain information unrealised.

If a staffing company has information about the exact skill set of their candidates. it can data mine it for a quantitative and qualitative analysis that might single out specific skills and/or accreditations which can produce an inordinate effect on career and salary development:

- What does it mean for a career if you add skill x or y, learned a foreign language, worked abroad, changed careers, quit work to raise a child?
- What specific skill combination is the most powerful for career success?
- How well do certain skills or academic backgrounds transfer into different career paths?
- Is there are correlation between how well a candidate is matched against a job and the distance he is prepared to commute?
- What is the relationship between skills, gender, age and labour mobility?
- How does labour mobility between two countries differ for people of comparable skill and demographics?

The analysis of career and salary development, assessment of employability based on such data will enable a staffing provider to effectively function as a consultancy. It could advise, for example, a candidate in the IT market that - based on historical data and statistical analysis – the addition of programming language X or certification in module Y in addition to his current qualifications would increase his salary by 16,5% within 2 years. It could issue recommendations to job centres or government bodies concerned with unemployment that people with job profile x would increase their average employability by a given percentage if they added training y to their current portfolio.

A staffing company that can competently answer these questions based on analytics will position itself as a provider of employment competence and not simply what is casually described in industry lingo as "bums on seats". A recruitment firm that could decipher the fine grained relationship between skill acquisition and career development or employability would instantaneously add value to the staffing and workforce management process. It would provide tools that could open up a new field of human resource management. The development and deployment of human talent could be put on an empirical footing. Project teams, corporations, entire economies could scrutinize the balance of skills they have assembled and how it translates into competitive fitness. Recruitment, training, large scale national education and immigration policies could be guided by a data driven approach.

To accomplish such a feat it requires a common data dictionary first. Skills represent an atomic level below which no further differentiation occurs. A starting point would be an inventory of skills which need to be defined with a degree of granularity that has to find the balance between detail and manageability. If skills were too narrowly defined their number would mushroom beyond what is manageable from a data administrative perspective. If defined too broadly the specific competence set and level of a worker cannot be compiled accurately.

These skills need to be consolidated into job categories, something that rests on two conditions. It requires the definition of job categories and the exact description of rules that correlate to a collection of skills into a job category. These rules will be algorithms defined per job category that check whether a given skill set meets its criteria. This type of standardization will invite discussions across many professional organizations that watch over the accreditations of its affiliated members. It would require the governance and endorsement of an international body like the ILO (International Labour Organisation) that could support the creation of a widely accepted, international standard.

Ultimately, the creation and maintenance of skills based job categories would rest on two pillars. Data services that would:

- Set up a global data dictionary identifying the entirety of skills and job categories
- Provide the ongoing mapping between skills and job categories.

The opportunities that open up in terms of information potential would be realised in a new generation of business application software: matching software, CV databases, HR systems and data warehouses that convert the enormous quantity of data into business intelligence. Data services and application design need to be developed hand in hand. Standardization and the commercial use on standardized data are mutually dependent and it requires heavy lifting to get both off the ground. But whoever pulls it off will re-define how we recruit, supply recruitment services and manage careers on the strength of data driven insights.

Solutions and Recommendations: Bottom Up or Top Down?

As earlier stated, it is preferable to solve the taxonomic problem of arriving at unique and reliable job categories with a bottom up approach. While both approaches are conceptually on equal footing, it was always the relative ease of the top down approach that defined existing categorizations, exemplified by the nested hierarchies of the ILO or the Standard Occupational Classification of the US Bureau of Labor Statistics. The inherent inflexibility of such a method renders it only marginally useful for the requirements of career and labour analytics that are our focus.

How to Create a Common Taxonomyof Job Categories

In the codifying process, every job category will get its own unique identifier while extending the granularity of established taxonomies like the ones provided by the ILO or the US Bureau of Labour Statistics. From the lowest level of existing taxonomies it will probably have to extend one or two levels down to incorporate a fine grained description of professional categories required for effective recruitment. One level to indicate a profile that correlates with actual professions found on CVs and respective requirements, and a marker for degrees of competence or seniority.

In current candidate management systems, skill and job categories are mostly maintained in a text based, descriptive format, less so in uniquely identified codes. In cases where job categories are codified a direct and dynamic relationship between a skill set and a job category is not being kept.

The comprehension of that relationship however is specific to the subject knowledge of a candidate or a recruiter, who in countless instances define how professional competencies are correlated to given job categories.

The purpose of a universal professional taxonomy is to build system that translates the collection of defined skills, knowledge and experience can be dynamically into the most likely and accurate job category - guided by machine intelligence alone. It would result in an expert system that applies human expertise in that subject area and combines it with the speed and efficiency of computer based processing. An expert system has to transfer specific human competence or knowledge into a rule based system that can be executed by a program. It acquires the ability to distil accurate and useful generalizations out of ambiguity and fuzziness typical of natural language and turn them into reasonably precise rules that are the domain of computer based logic. Self-evidently, that knowledge exists. It is contained in hundred millions of resumes that circulate either in hardcopy or electronically. A standardized taxonomy is not understood as a naming exercise but the definition of rules that renders rigorous and language independent codification.

In a first step, the skills and job categories are codified independently without necessarily establishing any correlation between the two. The term "skill" is here used comprehensively and loosely in a way that includes the entire range of human competencies, job skills per se, professional experience, vocational or academic training. A standardized taxonomy in this sense is not so much a naming exercise but rather the application of rigorous definition of classes that define an individual skill sufficiently different from any other to warrant its own code. Unaided by a computerized system, such a task would be beyond the capacity of purely human effort. It would require human agents to scan hundred thousands of documents and making an inventory of professional and academic competencies. It would require further effort to specify the statistical relationship between competencies (skills) and job categories. While this task is performed in countless individual cases (and every CV gives a testimony of that) it has been a challenge to apply it to the enormous corpora of text that all CVs in circulation represent.

However, new systems are being built and optimized that *"aim to automate the tedious process of extracting large collections of facts from the Web in an unsupervised, domain-independent and scalable manner"* (Etzioni, et al, 2004, p1). The identification and definition of skills can be done by a process called information extraction, the task of automatically extracting knowledge from text. In this case, it will be in the shape of a system that extracts a large list of skills from suitable documents (CVs) based on grammatical or linguistic features. This could be done in conjunction with hand labelled training examples that would provide such a system with suitable samples as input for rule forming. Information extraction mines information from a large body of text either guided or unsupervised. It could utilize a number of domain specific extraction patterns to generate a list of skills which could then serve as extraction rules and validation patterns to evaluate the accuracy of instances extracted by these rules. A categorization and coding of skills would consider the following when assigning unique identifiers.

- Semantic description of skills
- Degree of competence
- Duration a skill was practiced or applied

The process of identifying distinct classes of professions and their subsequent categorization follows a similar method. The taxonomy of professional categories is partly established by a number of national and international bodies. These do not necessarily reflect the subtle distinction that applies to actual job titles which can be found in resumes.

The most granular level of the taxonomy provided by the ILO (International Labour Organisation) specifies, in one instance, a category named "Software and Application Developer". In recruitment such a category would inevitably demand further specification. It probably mentions certain types of software development tools or applications preferably with some indication of the level of competence. Text based information extraction could compile classes and subclasses that reflect the necessary balance between granularity and aggregation that originates from the exact documents that communicate a candidate's competences. A random example taken from an IT job board illustrates the case: "Senior Business Analyst / BA, Software Development, Agile, .NET" solves the problem by awkwardly concatenating professional category, seniority, subject area and specific software into a single line that represents a job category in direct relation to the demands of the vacancy. Information extraction has to assemble a comprehensive list of job categories based on extensive text search that aggregates bottom up.

Human intelligence possesses a remarkable capacity to recognize patterns and relations with which to synthesize from a resume a reasonable generalization (description) of a person's professional aptitude. A recruiter quickly scanning a CV will browse through three to five pages of a CV and arrive at a general idea of a person's profile including competence, experience and seniority. Human intelligence is well equipped to handle the fuzziness and complexity of the presented information by making sweeping generalizations. Transferring this ability to an expert system has proven to be frustratingly elusive.

To illustrate the breadth of this capacity, consider the following hypothetical case of an IT project manager who used to work for many years as a developer of finance software. In this function, he acquired extensive knowledge of various programming languages and industry specific business processes. His professional focus shifted a few years ago to the managerial side of software development. A simple mapping of skills against job categories would categorize him equally as a developer and project manager without applying a nuanced weighing that considers context and relations. A human agent would factor in that skills, regardless of previous proficiency will erode over time due to the lack of practice. This will be even more accentuated in a field that undergoes rapid technological changes. A professional shift from developer to project manager would indicate that the career focus of a person has irrevocably taken a new direction. But a recruiter would also recognise that the experience as a software developer would enhance and complement the profile of an IT project manager. When matching a person's skill set against job openings, an experienced recruiter would discard a software developer role for our candidate even though his qualifications fit the requirements.

Contextual knowledge and understanding the implications of career decisions are high level capacities that an expert system has to incorporate. With large enough corpora of textual information to mine and gradually refined extraction rules, it is possible to build a knowledge base that will simulate the human reasoning process.

Skills and Job Categories: How to Combine the Two

This chapter has described how the process of both guided and unsupervised information extraction (UIE) could be employed to generate a comprehensive list of skills and job categories from a large number of CVs. These are preliminary steps to understand the correlation between the two. The information could be used to build electronic candidate management systems in which resumes exist as individual patterns of skill codes that reflect professional competence and career history referring to specific job categories. From such a system, one could easily infer the statistical relevance of a particular skill by counting the frequency with which it appears in relation to a particular job category. General abilities could identified that might be unspecific for a particular job which would be weighed less in determining the relevance of a skill to fit the skill pattern of a job category. Eventually, one could identify the characteristic skill set of individual job categories or entire job classes. A job category that has been deducted from a particular skill set would restore a former descriptive authority. It would provide a bottom up approach to create a universal categorization of job categories that is non-arbitrary and entirely driven by probabilistic and statistical approaches. The knowledge base amassed in a system would enable the design of candidate management applications that only require the definition and listing of skills as input and from which the system would derive matching job categories.

FUTURE RESEARCH DIRECTIONS

Staffing Industry and Analytics

Business intelligence and analytics are nowadays ubiquitously found in all major industries. Logistics, sales and distribution, pharmaceutical, electronic, utility and service companies are routinely using business intelligence to gain insights into their business processes and clients. Business Intelligence has long proven to be a key driver for greater efficiency and profitability. However, the staffing industry presents us with a paradox that warrants deeper analysis. There is hardly an industry that generates such a copious amount of data holding potential insights which could not only help making businesses more competitive but which is also exceedingly interesting for individuals, businesses, academia and governments. In no other multinational and IT supported industry we will find a wider gap between the promise of profound and varied data driven insights and how little of that data has been analyzed so far. It would be difficult to find more interesting data that has been subject to such humble scrutiny. Employment related research in the academic field inevitably runs into the challenge of generating sufficient data would hold the key for the novel understanding of career and migratory phenomena. Compiling large enough samples of career data would require prohibitively expensive surveys even at a regional or national scale and are usually outside the reach of academic research in the social sciences. While governments and public administrations study and publish employment data it is without exception of low granularity and significantly time delayed. It is hard to overestimate the universal relevance of employment and career data and the interest invested into it.

Manpower, one of the world's biggest staffing companies publishes an annual "Employment Outlook Service" which is derived from a survey of employers in various countries. The publication solicits considerable interest in the business world and is frequently quoted. However, it represents an investigative effort outside of the companies' normal service offering and is un-related to the abundant data that the company generates as part of its operational efforts.

Why is it then that employment agencies and staffing companies do not subject their operational data to comprehensive analysis? The reasons are varied and in their sum have proven to be insurmountable to this day.

- The staffing industry was traditionally a local business that in relatively short time grew into an international business. The design of many IT systems in the recruitment are still reflecting that business approach. IT systems are designed to help managing local operational processes with no thoughts given to an integrated or enterprise wide management of data.
- Candidate data is very much seen as proprietary of a given branch or recruiter

with little incentive to be shared across an organization

- The business processes within the recruitment industry are so different from one another that IT system are frequently moulded around them. A pay rolling and invoicing system that supports temporary staffing or recruitment cannot be extended to outsourcing or outplacement. A large multinational staffing company is most likely to operate a portfolio of independent subsidiaries that manage their own IT systems. The data generated within them has hardly any structural or semantic communality that would permit a meaningful analysis.

Master Data Management (MDM) in the Staffing Industry

Most software systems have lists of data that are shared and used across applications or even enterprise wide critical for reporting and analytics. Typical master data items are customers or products representing the key assets of a company. The key assets of the staffing industry are their candidates present in all of their business processes and subsequently their transactional data.

The staffing industry is unfortunately beset with all evils of inconsistent data and master data management. In almost canonical fashion, any incident that works against the creation of consistent, enterprise wide distribution of master data is somehow present in larger, multinational staffing companies, such as:

- Larger companies in the industry have frequently undergone re-organizations, mergers and acquisitions that impact data integrity.
- IT landscapes are complex
- The data entry process itself often allows for inconsistency, especially since the two most frequently used master data objects (candidate and client data) can often be created at an operational level without a central data integrity control process.
- A large organisation is likely to run multiple databases that are not standardized.
- Data Ownership is scattered.

Many of these problems will prove to be persistent for a long time. The cost of migrating to common, integrated IT systems is prohibitive. However, the emergence of comprehensive, codified skill and job categories would enable companies to maintain consistent candidate master data across the entire enterprise as a shared service. Different divisions in large multinational staffing companies provide different services that translate into different business processes. What they share in common is that their services deal with human resources, job seekers, candidates or people being out-placed. In that one area a company and its subsidiaries or even an entire industry could consistent, corporate wide master data by setting up their candidate management systems with universally consistent, comprehensive and shared taxonomies of skills and job categories. Creating a common data structure at the heart of their service delivery would enable the type of analytics driven development which currently lay dormant.

CONCLUSION

Predictive Analyticsin the Staffing Industry

For all the reasons given above, business intelligence in the staffing industry is still in an infant state. It can barely report on the business processes that have happened in its past. But it is possible to provide an outlook into the fascinating and transcendent potential of the staffing industry's data into its fully realized state and imagine a company that has mastered the following:

- Enterprise wide analytics across all divisions, services and national boundaries
- A standardized taxonomy of job categories across all operational IT systems (Candidate data bases, invoicing and payroll systems)
- A fine grained relationship between skills and job categories as the foundation of a standardized, enterprise wide taxonomy.
- Integrated enterprise wide reporting with high granularity data into a centralized data warehouse within 24 hours.

These conditions represent ambitious goals for an industry that has not yet been achieved anywhere but are comfortably within reach of what is technically feasible. Once accomplished, it would set the table for predictive analytics, the ability to make reasonable accurate forecasts based on past or current events. These have, as we will show, special significance and varied application in relation to staffing data. The correlation of employment data with macro-economic trends is self evident. Temporary staffing is highly sensitive to changes in the economic climate. Temporary workers are usually the first who will be laid off in the event of an economic downturn followed by job losses of permanent employees.

At the outset of an economic recovery, and the start of increased economic output, companies usually hire temporary workers to meet the additional workload before they consider hiring on a permanent basis. This makes temporary staffing a reliable indicator for large scale economic trends. With current reporting capabilities, staffing companies would register this trend long after it has manifested itself at a larger scale and became common knowledge. It reports on what is already visible to the general public.

However, if data is sufficiently fine grained, consolidated and analyzed with little time delay it will reveal subtle but tangible patterns, those first markers of larger economic movements. It would require data of the described quality to be correlated over a certain time period with macro economic data to spot reliable trend indicators. These trend indicators could be something so subtle and previously unrecognised like the hiring or letting go of people in specific profession before overall employment numbers go visibly up or down. We could envision investors turning to staffing companies with powerful analytics to learn first of economic developments at large corporations, industries or entire sectors that enable them to make knowledge based investment decisions ahead of markets. Local hiring decisions based on local knowledge would quickly aggregate to a characteristic pattern that is indicative of an emerging trend. This is equally applicable to understand individual companies, industries or macro-economic trends. The use of a complementary workforce (temporary staffing) follows seasonal trends that are highly predictable. Deviations of the norm would provide indications of market or corporate developments that have predictive power the expected performance of these markets and corporations. The quick consolidation of local trends into an emerging overall picture ahead of common market knowledge is one powerful example of the unrealised promise that analytics hold in relation to the staffing industry.

REFERENCES

Coles, J. (2007). *New job titles are loopy labels.* Retrieved January 15, 2010, from http://www.thesun.co.uk/sol/homepage/news/article24172.ece

Davenport, T. H., & Harris, J. G. (2007). *Competing on analytics.* Cambridge, MA: Harvard Business School Press.

Downey, D., Etzioni, O., & Soderland, S. (2005). *A probabilistic model of redundancy in information extraction.* Seattle: University of Washington.

Etzioni, O., Cafarella, M., Downey, D., Popescu, A., Shaked, T., & Soderland, S. (2004). *Unsupervised named-entity extraction from the Web: An experimental study*. Seattle: University of Washington.

Frasconi, P., Gori, M., & Sperduti, A. (1997). *A general frame work for adaptive processing of data structures*. Italy: Universita degli Studi di Firenze.

ILO (International labour organisation). (2007). *Resolution concerning updating the international standard classification of occupations*. Geneva, Switzerland.

Schumacher, S. (2007). Probabilistic versus deterministic data matching: making an accurate decision. *Information Management*.

Wolter, R., & Haselden, K. (2008). *The what, why, and how of master Data Management*. Microsoft Corporation. Retrieved February 23, 2010, from http://msdn2.microsoft.com/en-us/library/bb190163.aspx

ADDITIONAL READING

Agichtein, E., & Gravano, L. (2003) Querying Text Databases for Efficient Information Extraction. In Proceedings of the 19th IEEE International Conference on Data Engineering (ICDE 2003), Bangalore, India

Agichtein, E., Gravano, L., Pavel, J., Sokolova, V., & Voskoboynik, A. (2001) Snowball: A Prototype System for Extracting Relations from Large Text Collections. In Proceedings of the 2001 ACM SIGMOD International Conference on Management of Data, Santa Barbara, California.

Akerkar, R. And Sajja, P., (2010) Knowledge-Based Systems, Sudbury, MA: Jones & Bartlett Publishers.

Ayres, I. (2007). *Super Crunchers: Why Thinking-By-Numbers is the New Way To Be Smart*. New York: Random House.

Babcock, C. (2006), Data, Data, Everywhere, InformationWeek. Retrieved February 23, 2010, from http://www.informationweek.com/news/global-cio/showArticle.jhtml? articleID=175801775

Blum, A., & Mitchell, T. (1998) Combining Labeled and Unlabeled Data with Co-Training. In Proceedings of the 11th Annual Conference on Computational Learning Theory, Madison, Wisconsin.

Bonde, A., & Kuckuk, M. (2004) Real-World Business Intelligence: The Implementation Perspective, Information Management Magazine, Retrieved March 13, 2010, from http://www.information-management.com/issues/20040401/1000839-1.html

Brachman, R., & Levesque, H. (2004). *Knowledge Representation and Reasoning*. San Mateo, CA: Morgan Kaufmann.

Califf, M. E., & Mooney, R. J. (1998) Relational Learning of Pattern-Match Rules for Information Extraction In Working Notes of AAAI Spring Symposium on Applying Machine Learning to Discourse Processing, Menlo Park, CA: AAAI Press.

Craven, M., DiPasquo, D., Freitag, D., McCallum, A., Mitchell, T., Nigam, K., & Slattery, S. (2000). Learning to Construct Knowledge Bases from the World Wide Web. *Artificial Intelligence*, *118*(1-2), 69–113. doi:10.1016/S0004-3702(00)00004-7

Downey. D, Etzioni, O., Soderland, S. and Weld, D.S., (2001) Learning Text Patterns for Web Information Extraction and Assessment, Technical Report University of Washington- Department of Computer Science and Engineering-04-05-01.

Freitag, D. (1997) Using grammatical inference to improve precision in information extraction. In ICML-97 Workshop on Automation Induction, Grammatical Inference, and Language Acquisition, Nashville, TN, Mateo, CA: Morgan Kaufmann.

Kaplan, R. S., & Norton, D. P. (2004). *Strategy Maps: Converting Intangible Assets into Tangible Outcomes*. Cambridge, MA: Harvard Business School Publishing.

Liebowitz, J., & Wilcox, L. C. (1997). *Knowledge Management and Its Integrative Elements*. New York: CRC Press.

Lin, W., Yangarber, R., & Grishman, R. (2003) Bootstrapped Learning of Semantic Classes from Positive and Negative Examples. In Proceedings of ICML-2003 Workshop on the Continuum from Labeled to Unlabeled Data, pages 103–111, Washington, D.C.

Marchand, D. A., & Kettinger, W. J. (2001). *Making the Invisible Visible: How Companies Win with the Right Information, People and IT*. New Jersey: Wiley.

Soderland, S. (1999). Learning information extraction rules for semi-structured and free text. *Machine Learning*, *34*(1), 233–272. doi:10.1023/A:1007562322031

KEY TERMS AND DEFINITIONS

Candidate Management Systems (CMS): The staffing industry create structured, electronic records of candidates' demographic and address data, their capabilities, skills and remuneration. Systems provide the ability to interpret curricula and enter relevant data to applicable database fields. They serve a key function in recruitment, assignment and evaluation and provide the basis for matching available candidates against the requirement of clients.

Data Integrity: Accuracy and consistency of stored data, indicated by an absence of any alteration in data between two updates of a data record. Data integrity is imposed within a database at its design stage through the use of standard rules and procedures, and is maintained through the use of error checking and validation routines.

Data Mining: Sifting through very large amounts of data for useful information. Data mining uses artificial intelligence techniques, neural networks, and advanced statistical tools (such as cluster analysis) to reveal trends, patterns, and relationships, which might otherwise have remained undetected.

Enterprise Wide Analytics: Is a strategic approach to business intelligence in which key data and analytics are managed enterprise-wide. This represents a departure of data management in a multitude of silos that is carried out at a departmental, divisional or national level. Frequently it involves the central management and control of key data that is then distributed across an organization.

Expert System: Artificial intelligence based system that converts the knowledge of an expert in a specific subject into a software code. This code can be merged with other such codes (based on the knowledge of other experts) and used for answering questions (queries) submitted through a computer. Expert systems typically consist of three parts: (1) a knowledge base which contains the information acquired by interviewing experts, and logic rules that govern how that information is applied; (2) an Inference engine that interprets the submitted problem against the rules and logic of information stored in the knowledge base; and an (3) Interface that allows the user to express the problem in a human language such as English. Despite its earlier high hopes, expert systems technology has found application only in areas where information can be reduced to a set of computational rules, such as insurance underwriting or some aspects of securities trading. Also called rule based system.

Information Extraction (Unsupervised Information Extraction (UIE)): Automated process to extract large collection of facts from a corpora of text usually the web.

Master Data Management (MDM): Most software systems have lists of data that are shared and used by several of the applications that make up the system. Commonly agreed on master-data items are "customer" and "product."

Matching Software: A module and function of candidate management systems that electronically matches the requirement for an open position against the pool of suitable candidates.

Predictive Analytics: Is a recent trend in analytics which involves studying past historical data to predict future events like predicting the behavior of customers or stocks.

Probabilistic (vs. Deterministic): Situation or model where there are multiple possible outcomes, each having varying degrees of certainty or uncertainty of its occurrence. Probabilistic matching uses likelihood ratio theory to assign comparison outcomes to the correct, or more likely decision. This method leverages statistical theory and data analysis and, thus, can establish more accurate links between records with more complex typographical errors and error patterns than deterministic systems.

Taxonomy: The orderly classification of items (related to one another in an actual or presumed manner) into distinct groups.

Chapter 10
The Promises and Challenges of Health Information Technology

Brian Gugerty
Gugerty Consulting, LLC, USA

Michael J. Maranda
Evaluation Researcher, Belgium

ABSTRACT

This chapter explores the application of Information Technology to healthcare in the United States. Recent developments and trends in healthcare information technology (HIT) are presented and discussed. Widespread adoption of HIT promises to save lives, save money, and improve health. Definitions, descriptions, and examples of electronic health records (EHRs) and personal health records (PHRs) are provided. The significant efforts to broadly and meaningfully adopt HIT over the next several years are discussed. The significant challenges in implementing EHRs are discussed, including transformation of clinical processes. Finally, the impact of HIT on the concept of ownership of the healthcare record and how it may change the relationship between the patient and healthcare provider are explored. Implementing effective HIT on a nationwide scale will require considerable effort.

INTRODUCTION

The application of information technology to healthcare offers the promise of the comprehensive management of health information, including the exchange of information between patient and healthcare providers, among providers, and between providers and payers/government agencies. Healthcare information technology (HIT), if implemented extensively, has the potential to improve the quality of healthcare, to increase efficiency of providing care, and to reduce medical errors. This result translates into saving lives and saving money. The systematic application of

DOI: 10.4018/978-1-60960-501-8.ch010

HIT through a country can result in the general improvement of the population's health.

Despite the fact that much of HIT was pioneered and developed in the US and this country remains an innovator in HIT, the US lags behind some other countries and regions in the implementation of HIT. Currently, there is a federal initiative in the US to promote adoption of HIT by healthcare providers. It is hoped that HIT will be implemented in the US broadly enough so that—in the words of David Blumenthal, MD, U.S. National Coordinator for Health Information Technology—a "nationwide electronic health information system" will develop (Blumenthal, 2010).

The Alliance for Health Reform (2009) describes HIT as encompassing "a broad array of technologies involved in managing and sharing patient information electronically, rather than through paper records." The central component of this broad array of technology is the electronic health record (EHR). An EHR is a record of an individual's health and healthcare-related information that is created, managed, and consulted by authorized clinicians and staff within one or more healthcare settings. Computer hardware and software technologies and standards underpin the EHR, and networking technologies and standards extend the EHR into the area of health information exchange (HIE). HIE enables the important information contained in the EHR to be shared across healthcare delivery organizations, regions, and nations.

Despite several decades of vigorous efforts to encourage the adoption of EHRs, most physician office practices and many hospitals have not done so in the United States. Even hospitals in the US that have begun to adopt HIT have a long way to go to achieve the desired end points, as will be discussed in detail later in this chapter. In fact, several countries or regions, including the Netherlands, the UK, France, Hong Kong, Singapore, and Australia, are significantly more advanced in the basic penetration and use of the EHR than the US. The US federal government through the Office of the National Coordinator of Health Information Technology and other agencies are trying to change that. The Bush administration set a goal of a fully functional EHR being used by every American by 2014. The Obama administration concurs with that goal and has appropriated tens of billions of dollars to make it a reality. This high-level focus on EHR adoption and the significant money that is available for such adoption over the next five years has generated great interest and activity in the US and beyond. Although the US has far to go in implementing HIT on a national scale, it remains a major innovator of HIT. The fragmented healthcare system with diverse types of providers allows for much innovation and experimentation in HIT.

There is the promise that HIT can decrease costs and improve healthcare quality, the highest levels of government and society support the implementation of HIT, and there is money available for the "meaningful use" of EHRs. Therefore, what, if anything, is holding the US back from quickly achieving the goal of implementing HIT? The answer is severalfold: engrained processes in clinical practice, fear of change, and a history of focusing on the technology rather than on how to make best use of technological innovation. HIT per se is indeed necessary, but it is not sufficient in itself to change healthcare for the better. Clinical processes and business processes need to be designed/redesigned *in conjunction with HIT* to advance toward the goals now set. For decades, too much of the focus has been on the technology and not on how to use the technology in ways that transform work, clinical practice, and outcomes. Already, much progress has been made using HIT to automate processes that easily lend themselves to automation, such as clinical laboratory functions and healthcare organization business processes such as accounting and billing. The challenge now is to use HIT in areas that are more difficult to automate. Using HIT as part of transforming fundamental clinical processes, such as planning and deciding about what patient care to provide, ordering care, and documenting (aka, charting) patient care, is now on the agenda for HIT. Furthermore, many of these processes are not standardized to begin with and quite a few are substantially suboptimized presently. Adoption of HIT and the EHR must therefore be intimately involved with clinical practice and the improvement of clinical processes. Moreover, people generally do not like change. These factors all need to be addressed to make the *implementation of HIT*, not merely the installation of HIT, a success.

If the aforementioned issues are not enough of a challenge to those striving for an effective EHR for every citizen and healthcare provider, another challenge is the trend toward empowering patients and other consumers. The Internet and other technological and social factors have combined to make more medical information directly available to the consumer than ever before. This has led to the development of the personal health record (PHR). Patients can self-schedule an office visit and even surgery, check the results of a lab or radiology exam, pose questions to a doctor or nurse, and much more with a PHR. A PHR may be distinct from an EHR or it may be integrated with an EHR. Although clinicians have not fully made the transition from the paper medical record to the EHR, principles about who the healthcare/health record fundamentally serves, who "owns" the record, and related issues are being rethought.

This chapter will explore, and hopefully shed light on, these developments and issues. We will define and describe examples of EHRs and PHRs. We will elaborate on the significant efforts to broadly and meaningfully adopt HIT in the next five years and beyond. We will describe and explore the significant implementation challenges of EHRs, including clinical process transformation. Finally, we will explore the exciting changes unfolding in the very "ownership" of the health and healthcare record that HIT promises to improve.

Chapter Objectives

1. Define HIT, EHRs, and related terms.
2. Describe examples of HIT and how they are used currently.
3. Discuss the current major US initiative to spur the adoption of EHRs.
4. Describe the relationship between EHRs and clinical processes.
5. Discuss the challenges of HIT implementation.
6. Explore issues in the use and "ownership" of PHRs.

BACKGROUND

As stated in the Introduction, HIT can be described as "a broad array of technologies involved in managing and sharing patient information electronically, rather than through paper records." This definition importantly highlights the primary purpose of HIT as managing and sharing patient information electronically. The US Government Agency for Healthcare Research and Quality's (AHRQ) informative National Resource Center for Health Information Technology website (healthit.ahrq.gov) lists many examples of HIT. As is generally the case, better understanding of a new concept can be gained from concrete examples of the concept or components thereof. We will take several of these examples of HIT—authentication, bar coding, biosurveillance, clinical decision support, computerized provider order entry, continuity of care document, and remote monitoring—and discuss them further.

Authentication

Authentication is defined as a system of "knowing who you are" and authorizing the proper level of access and control over said system. Authentication for HIT is not different than authentication for other types of systems in other industries. However, perhaps except for matters of national security, authentication for HIT is equal to or greater in importance than authentication for any *purpose*, such as retail or banking. Given the recent explosive growth of HIT, the extension of HIT into direct consumer use via PHRs, and the sensitive nature of the data, authentication is currently a hot issue in HIT. Authentication may be based on what you have (a special card or token), what you know (password or PIN), who you are (a fingerprint or other biometric scan), or some combination of these methods. Many HIT applications, such as EHRs, frequently require quick access, so biometric authentication mechanisms are starting to grow in popularity in healthcare. Furthermore, single sign-on solutions are increasing in use because many users of HIT need to access several systems at one time, and being authenticated by multiple systems via single sign-on offers distinct advantages.

Bar Coding

A common technology in many industries, especially the retail industry, bar coding is used in many ways to enhance efficacy and patient safety in healthcare. From more effective wrist identification bands to material management processes to barcode medication administration applications, this mature technology in other industries is still making productive inroads into healthcare. A good example of the application of this technology in healthcare is barcode medication administration applications. One study found that barcode medication administration was instrumental in reducing medication errors at Veterans Administration hospitals by 64.5% (Malcolm, Carlson, Tucker, 2000). Here is how barcode medication administration works in a typical hospital: The nurse wheels the medication cart into the patient's room. Her wireless laptop is on top of the cart. She scans the patient's wristband barcode with a scanner and the patient's medication list pops up on the screen. She confirms the patient's identity and then she scans each medication barcode that the pharmacy dispensed in individually wrapped unit doses that the patient is due to receive in that timeframe. If she scans the wrong medication or the wrong amount of medication, an audible and visual alert prominently displays, potentially preventing a medication error (Greenly & Gugerty, 2002).

Biosurveillance

While there is no generally accepted definition of biosurveillance, "...it generally refers to the automated monitoring of information sources of potential value in detecting an emerging epidemic, whether naturally occurring or the result of bioterrorism" (United States Government Accountability Office, 2005). The University of Pittsburgh Medical Center, Center for Biosecurity, describes a four-step process for biosurveillance consisting of the following:

- **Surveillance:** Systematic monitoring for a case (or cases) of an unusual disease and/or an unusual cluster of disease—the stimuli for the initial disease outbreak report
- **Reporting:** An account of the initial disease outbreak that reflects an assessment of credibility and of potential public health significance
- **Verification:** Inquiry to affected country and the initial investigation to verify the disease outbreak
- **Response:** Intervention and containment of the disease outbreak (University of Pittsburgh Medical Center, 2009)

Information sources for biosurveillance can include data from environmental monitoring systems, prescriptions, purchases of over-the-counter medication, and medical symptoms obtained electronically from ambulatory care records.

Clinical Decision Support (CDS)

Jerome Osheroff, MD, a leading authority on clinical decision support takes a broad view of CDS, defining it as "refers(ing) broadly to providing clinicians or patients with clinical knowledge and patient-related information, intelligently filtered, or presented at appropriate times, to enhance patient care. Clinical knowledge of interest could include simple facts and relationships, established best practices for managing patients with specific disease states, new medical knowledge from clinical research, and many other types of information" (Osheroff, 2009). The "old school' notion of CDS was rule-based alerts and reminders, which is still certainly included in a broad-based definition of clinical decision support. However, clinical decision support can also include context-sensitive data presentation such as flowsheets, medical order creation facilitators such as evidence-based order sets, reference information such as links to relevant journal articles, and documentation templates such as history-and-physical and encounter note templates. This is all intended to improve clinical care by providing the right information to the right person (clinician and or patient) in the right format (alert, order set, etc.) through the right channel (EHR, PDA [personal digital assistant], etc.) at the right time (to influence important decisions or actions).

Computerized Provider Order Entry (CPOE)

CPOE is the process by which physicians and qualified non-physician providers place patient care orders—for medications, tests, procedures, and other services—directly into the computer. CPOE systems can be a stand-alone system or a module of an integrated multi-module EHR.

The US Federal Government's Agency for Healthcare Research and Quality's PSNet Primer on CPOE (2009) describes the general capability of CPOE as follows:

"A CPOE system, at a minimum, ensures standardized, legible, and complete orders and thus has the potential to greatly reduce errors at the ordering and transcribing stages. CPOE systems are generally paired with some type of Clinical Decision Support (CDS). A typical CDS suggests default values for drug doses, routes of administration, or frequency and may offer more sophisticated drug safety features such as checking for drug allergies or drug–drug or even drug–laboratory (e.g., warning a clinician before ordering a nephrotoxic medication in a patient with elevated creatinine) interactions."

Continuity of Care Document

HL7 is a standard setting body for HIT standards. HL7's Continuity of Care Document (CCD) specification is an XML-based markup standard that specifies the encoding, structure, and semantics of a patient summary clinical document for exchange. CCD specifies a standardized core dataset of the most relevant administrative, demographic, and clinical information regarding a patient's healthcare services received during one or more healthcare encounter. CCD is important because patients routinely transition from care setting to care setting in the healthcare system, and clinicians can greatly benefit from current and relevant summary information from a provider from the predecessor location in the patient's medical journey. Adoption of the CCD standard and practice is seen as a good first step in healthcare organizations use of HIE (HL7, 2007).

Remote Monitoring

An important trend in consumer health is home care. People are living longer and desiring to receive care in their homes and not in hospitals or extended care facilities. HIT is facilitating home care by enabling remote monitoring of readings from scales, blood pressure cuffs, electrocardiograms, glucometers, and many other medical devices in the home. For example, patients with congestive heart failure can easily decompensate and require hospitalization to stabilize their cardiopulmonary status. Some patients with congestive heart failure are now being remotely monitored in the home from physician offices or hospital-based programs. Patients weigh themselves, obtain blood pressure and pulse readings from automatic blood pressure machines, and noninvasively measure their oxygen level with a finger clip probe daily. The data from these devices are electronically transferred via Bluetooth technology to a computer in the home and then via the Internet to a physician office or hospital-based program. Nurse practitioners or other medical staff monitor the readings and contact the patient if the readings fall out of desired parameters and make adjustments to the patient's diet, exercise, and/or medication regimen to keep the patient stable and out of the hospital (Butterfield, 2010; Health Information Technology Standards Panel, 2008).

The main example of HIT is the electronic health—or healthcare—record (EHR). The EHR is formally defined by the National Alliance for Health Information Technology as an electronic record of health-related information on an individual that conforms to nationally recognized interoperability standards and that can be created, managed, and consulted by authorized clinicians and staff *across more than one healthcare organization*. The National Alliance defines an electronic medical record (EMR) as an electronic record of health-related information on an individual that can be created, gathered, managed, and consulted by authorized clinicians and staff *within one healthcare organization*. Finally, a personal health record (PHR) is defined by the National Alliance as an electronic record of health-related information on an individual that conforms to nationally recognized interoperability standards and that can be drawn from multiple sources while being *managed, shared, and controlled by the individual*.

The digital patient record is an umbrella concept under which EHR, EMR, and PHR similar terms can be grouped. Simpler definitions of the EHR, EMR, and PHR terms can be useful in some situations and are as follows:

EMR: A digital patient record used within a single care setting or environment.

EHR: A digital patient record that can be used across more than one healthcare organization.

PHR: A digital patient record either maintained and controlled by the consumer or jointly controlled by the consumer and his or her healthcare provider.

Health information exchange (HIE) is defined be the National Alliance as the electronic movement of health-related information among organizations according to nationally recognized standards. A Regional Health Information Organization (RHIO) is HIE operationalized in a geographic region, such as a large city, state, or province. The National Alliance for Health Information Technology defines an RHIO as an organization that oversees and governs the exchange of health-related information among organizations according to nationally recognized standards (National Alliance for Health Information Technology, 2007).

OPPORTUNITIES AND CHALLENGES FOR THE MEANINGFUL USE OF HEALTHCARE INFORMATION TECHNOLOGY

The American Reinvestment and Recovery Act of 2009 legislation and the regulations that guide its implementation call for the meaningful use of certified EHRs by 2015 (H.R. 1–111th Congress, 2009; Department of Health and Human Services, 2010). The legislation includes monetary incentive for hospitals and physician office practices to adopt EHRs for a period of five years beginning in 2011 and enforces penalties for those who have not adopted EHRs beginning in 2015. The incentive money is significant—approximately $5,000,000 per hospital initially and then increased reimbursement through Medicare or Medicaid thereafter for five years, and up to $44,000 per physician or licensed independent provider for medical office practices. This incentive has led to a great deal of interest in the adoption of EHRs. Key aspects of this initiative are the terms "certified EHRs" and "meaningful use."

Much work has gone into developing the criteria by which EHRs are and will be certified, and although there is still policy being set to define what and how many certifying bodies will exist, there is solid information available for vendors to build certified EHRs and for healthcare organizations to purchase and use certified EHRs today. By contrast, the "meaningful use" criteria have yet to be finally defined and specified by the federal government, but most of the interim work leading to the final definition and regulations for "meaningful use" has been done as of this writing to describe what will likely be finalized.

The Certification Commission for Health Information Technology (CCHIT), founded in 2004, is a not-for-profit entity that establishes comprehensive criteria for the capabilities of EHRs. Hundreds of diverse volunteers representing clinicians, healthcare operations personnel, and informaticians engage in a consensus-based process to create and validate criteria. CCHIT then certifies commercial and noncommercial products as satisfying the criteria. The Commission is recognized by the federal government as a certifying body and has certified over 200 EHR products to date. The classes of products the commission certifies are as follows: Ambulatory EHR, Inpatient EHR, Emergency Department EHR, and Enterprise EHR. Table 1 presents a sampling of Inpatient EHR certification criteria. It is hoped that these specific criteria give the reader a good understanding of the scope, type, and level of granularity the criteria address.

There has been a 40-year history of HIT in the healthcare industry, but much of that history involved automating back office functions such as billing and departmental functions like laboratory, pharmacy, and radiology processing. With the current focus on the EHR, clinicians need to use HIT—via EHRs—to plan for care, make decisions about care, order care, and document care. This is a profound culture change in medicine and healthcare. Up until now, use of HIT and EHRs for these types of function—integral to how physicians and a broad array of clinicians practice medicine and healthcare—have been unavailable or nonexistent. For example, less than 18% of providers prescribe medication via CPOE in hospitals (Jha, Desroches, & Campbell, 2009), and even less do so using e-prescribing in medical office practice. The meaningful use criteria is an effort by the federal government to make sure hospitals and medical office practices do not merely install certified EHRs, but they use them meaningfully.

To that end, the Office of the National Coordinator for Healthcare Information Technology has issued preliminary criteria for meaningful use of certified EHRs. These criteria will be amended somewhat in 2010, but it is likely they will be adopted more or less intact. It any case, the criteria are helpful to understand what type of practice the US government is trying to foster with these criteria (see Table 2).

Many clinical informaticians have known for years that implementation matters. The best clinical system, if implemented suboptimally, is likely to yield poor results. This heuristic is even more important in the high stakes era of the broad-based meaningful use of certified EHRs. The Healthcare Information Management Systems Society (HIMSS) seven-stage HIMSS Analytics U.S. EMR Adoption Model is helpful when trying to understand the types of implementation of HIT in which healthcare organizations are engaged presently and will be engaged for the next several years (see Figure 1) (HIMSS, 2010)

Approximately 85% of healthcare organizations in the US are at an EMR adoption of stage 0–3. Most of these organizations have had significant penetration of EMR throughout the organization and at least some successful imple-

Table 1. Sample of select inpatient EHR certification criteria as of February 10, 2010

Category	Criteria
Patient demographics and administrative information	The system shall provide the ability to access demographic information such as name, date of birth, and gender needed for patient care functions.
Provider information	The system shall provide the ability to uniquely identify clinicians for the provision of care.
Problem lists	The system shall provide the ability to display different views of the problem/diagnosis list based on the status of the problem (for example, active, all, or resolved or charted in error).
Allergy information	The system shall provide the ability to capture the severity of an allergic or adverse reaction.
Medication list	The system shall provide the ability to display the name of the ordering clinician, medication order (name, dose, route, and frequency), a start date and time, and an end date and time or duration for entries on the medication list.
General ordering requirements	The system shall provide the ability for clinicians to write all patient care orders electronically, including, but not limited to, nursing care, medications/immunizations, diagnostic testing, nutrition and food service, consultation, and blood products.
Order sets	The system shall provide the ability to incorporate multiple choices of medications or other types of orders within an order set for clinician selection.
Ordering: medication orders	The system shall provide the ability to allow the clinician to order medication doses in mg/kg/min, microgram/kg, and microgram/kg/min. The system shall provide the ability to compute drug doses, based on appropriate dosage ranges, using the patient's body surface area and ideal body weight.
Decision support for medication and immunization orders	The system shall provide the ability to check for potential interactions between medications to be prescribed/ordered and current medications and alert the user at the time of medication prescribing/ordering if potential interactions exist. The system shall provide the ability to detect a drug dose that falls outside the minimum-maximum range for a single medication dose and to inform the clinician during ordering.
Medication administration	The system shall provide the ability to capture medication administration details as discrete data, including: (1) the medication name, strength, and dose; (2) date and time of administration; (3) route and site; (4) user name and credentials. When using barcode scanners, the system shall provide the ability to alert the end user that the medication being administered has triggered one or more of the following errors: wrong patient, wrong medication, wrong time, wrong route, or wrong dose.
Clinical task management	The system shall provide the ability to establish time periods for designating medication administration tasks overdue.
Generate and record patient- specific instructions	The system shall provide the ability to provide access to test and procedure instructions that can be modified by the end user.
Health record management	The system shall provide the ability to include patient identifying information as well as time and date report printed, on each page of individual patient-specific reports generated.
Clinical documentation	The system shall provide the ability to display ASTM Continuity of Care Record documents and file them as intact documents in the EHR. Summary patient record content information will include the following: patient demographics, medication list, immunizations, medication allergy list, problem list, procedures, discharge summary, and diagnostic test results.
Access control	The system shall enforce the most restrictive set of rights/privileges or accesses needed by users/groups (e.g., system administration, clerical, nurse, doctor, etc.), or processes acting on behalf of users, for the performance of specified tasks.

Table 1. Continued

Category	Criteria
Audit	The system shall record within each audit record the following information when it is available: (1) date and time of the event; (2) the component of the system (e.g., software component, hardware component) where the event occurred; (3) type of event (including data description and patient identifier when relevant); (4) subject identity (e.g., user identity); and (5) the outcome (success or failure) of the event.
Authentication	The system shall authenticate the user before any access to protected resources (e.g., protected health information) is allowed, including when not connected to a network, e.g., mobile devices.

Table 2. Select initial proposed criteria for "Eligible Providers" (Medical Office Practices) for EHRs

Objective	Measure
Implement drug–drug, drug–allergy, and drug–formulary checks.	The eligible provider (EP) has enabled this functionality.
Maintain an up-to-date problem list of current and active diagnoses based on ICD-9-CM or SNOMED CT®.	At least 80% of all unique patients seen by the EP have at least one entry or an indication of none recorded as structured data.
Generate and transmit permissible prescriptions electronically (eRx).	At least 75% of all permissible prescriptions written by the EP are transmitted electronically using certified EHR technology.
Maintain an active medication list.	At least 80% of all unique patients seen by the EP have at least one entry (or an indication of "none" if the patient is not currently prescribed any medication) recorded as structured data.
Record and chart changes in vital signs.	For at least 80% of all unique patients age 2 years or older seen by the EP, record blood pressure and BMI; additionally, plot growth chart for children age 2 to 20.
Incorporate clinical lab test results into EHR as structured data.	At least 50% of all clinical lab tests results ordered by the EP or by an authorized provider of the eligible hospital during the EHR reporting period whose results are in either a positive/negative or numerical format are incorporated in certified EHR technology as structured data.
Generate lists of patients by specific conditions to use for quality improvement, reduction of disparities, research, and outreach.	Generate at least one report listing patients of the EP with a specific condition.
Report ambulatory quality measures to Center for Medicare and Medicaid Services or the states	For 2011, an EP would provide the aggregate numerator and denominator through attestation. For 2012, an EP would electronically submit the measures.
Send reminders to patients per patient preference for preventive/follow-up care.	Reminder is sent to at least 50% of all unique patients seen by the EP who are age 50 and over.
Implement five clinical decision support rules relevant to specialty or high clinical priority, including for diagnostic test ordering, along with the ability to track compliance with those rules.	Implement five clinical decision support rules relevant to the clinical quality metrics the EP is responsible for.
Submit claims electronically to public and private payers.	At least 80% of all claims filed electronically by the EP.
Provide patients with an electronic copy of their health information (including diagnostic test results, problem list, medication lists, and allergies) upon request.	At least 80% of all patients who request an electronic copy of their health information are provided it within 48 hours.

Table 2. Continued

Objective	Measure
Provide clinical summaries to patients for each office visit.	Clinical summaries are provided to patients for at least 80% of all office visits.
Provides the capability to exchange key clinical information (for example, problem list, medication list, allergies, and diagnostic test results), among providers of care and patient authorized entities electronically.	Perform at least one test of certified EHR technology's capacity to electronically exchange key clinical information.
Provide summary care record for each transition of care and referral.	Provide summary of care record for at least 80% of transitions of care and referrals.
Provides the capability to provide electronic syndromic surveillance data to public health agencies and actual transmission according to applicable law and practice.	Perform at least one test of certified EHR technology's capacity to provide electronic syndromic surveillance data to public health agencies (unless none of the public health agencies to which an EP or eligible hospital submits such information have the capacity to receive the information electronically).
Protect electronic health information maintained using certified EHR technology through the implementation of appropriate technical capabilities.	Conduct or review a security risk analysis and implement security updates as necessary

Figure 1. The HIMSS analytics U.S. EMR adoption model

US EMR Adoption Model[SM]

Stage	Cumulative Capabilities	2009 Final	2010 Q1
Stage 7	Complete EMR; CCD transactions to share data; Data warehousing; Data continuity with ED, ambulatory, OP	0.7%	0.7%
Stage 6	Physician documentation (structured templates), full CDSS (variance & compliance), full R-PACS	1.6%	1.8%
Stage 5	Closed loop medication administration	3.8%	5.0%
Stage 4	CPOE, Clinical Decision Support (clinical protocols)	7.4%	7.7%
Stage 3	Nursing/clinical documentation (flow sheets), CDSS (error checking), PACS available outside Radiology	50.9%	50.0%
Stage 2	CDR, Controlled Medical Vocabulary, CDS, may have Document Imaging; HIE capable	16.9%	16.5%
Stage 1	Ancillaries – Lab, Rad, Pharmacy - All Installed	7.2%	6.9%
Stage 0	All Three Ancillaries Not Installed	11.5%	11.4%
Data from HIMSS Analytics[TM] Database © 2010		N = 5235	N = 5223

mentations of EMR modules over the past 10–15 years. A full 50% of healthcare organizations are now knocking at the door of stage 4 (CPOE, Clinical Decision Support). Levels 4 and above fundamentally alter clinicians' (especially physicians') ways to plan for, decide about, order, and document the care of patients. Because of the intensity of the changes required and the new specialized user populations involved, not all the healthcare information technology implementation approaches that led to success at lower levels of adoption will lead to success at higher levels of adoption.

The significant difference between stage 3 and 4 is the adoption and use of CPOE. Several lessons-learned reviews of CPOE implementation have noted underestimation of both the complexity of the implementations and depth and breadth of CPOE's impact on most clinical processes in a hospital (Aarts, Ash, & Berg, 2007; Ash, Stavri, & Kuperman, 2003; Campbell, Guappone, Sittig, Dykstra, & Ash, 2008). While the reports of benefits of CPOE have largely been attributed to the CPOE technology per se, it is important to note that the medical ordering process is a core fundamental clinical process in a hospital. This core ordering process affects many processes; along with the workflow of physicians and other licensed independent providers, a host of other clinical and operational staff are affected. "There is a profound impact (of CPOE) on workflow beyond that of the provider" (Aarts, Ash, & Berg, 2007, p. S4) Drucker (2002) characterized a modern hospital as "altogether the most complex human organization ever devised," so information technology that fundamentally changes a core process in a hospital, which has tentacles into many other processes, by definition is a significant, complex change. Since processes in hospitals are not yet widely standardized across hospitals, CPOE must be customized and adapted to fit the needs of the new improved local processes that occur after CPOE implementation. In this context, there is a growing realization that the implementation of EHRs is critically important—perhaps more important than the technology per se.

Provider and staff satisfaction with the implementation of EHR, which will significantly affect their workflow and practice, is important to measure (Sitting, Campbell, Guappone, Dystra, & Ash, 2007; Kuperman & Gibson, 2003). Gugerty, Maranda, and Rook (2006) found that a measure of staff satisfaction with implementation of a critical care clinical information system, used formatively in the post-implementation period, allowed for adjustment of the system and its use, enabling increased staff satisfaction with the implementation.

Several large healthcare organizations are now adapting techniques from other industries and applying them to the *implementation* of HIT and EHR, focusing on the specific technique of lean product development. The Toyota Way, otherwise known as the Lean Product Development System (LPDS), is a common framework these organizations use to achieve a successful combined clinical transformation/HIT implementation at stage 4 and above in the HIMSS Adoption Model. Lean Six Sigma techniques are commonly borrowed as well. LPDS has its theoretical roots in sociotechnical systems theory (STS), which Morgan and Liker (2006) describe as follows: "When reduced to the simplest terms, STS says that, to be successful, an organization must find the appropriate fit between the social and technical system that fits the organizational purpose and external environment." The technical system includes technology used in the organization *and* policies and standard operating procedures in the organization. The social system involves the selection, development, and characteristics of the organization's people as well as the as the culture that emerges from their interaction. According to STS, multiple independent components of the system interact to create a complex whole, which changes over time in response to changes in the environment" (Liker, 2004).

This section began with a description of the broad topic of HIT. There are many examples of HIT being used to improve effectiveness and efficiency of care but none more comprehensively than EHRs. EHRs are being connected by regional health information organizations, which are a manifestation on HIE. HIE is enabling better informed care to be planned and delivered by providers and received by consumers who routinely access a variety of healthcare providers at different locations over time. The US federal government desires to increase adoption of EHRs and make HIE the norm across the United States. To that end, the government is promoting and incentivizing the "meaningful use" of certified EHRs by hospitals and medical practices. Some of the meaningful use criteria appear to be stretch goals to quite a few healthcare organizations, but some consumer groups are urging the government maintain the goals and timelines for adoption of EHRs because substantial benefits to lowering healthcare costs and increasing healthcare quality are expected. However, the overall goals of achieving widespread meaningful use of certified EHRs and a nationwide electronic health information system powered by HIE standards and practices is a daunting one. To achieve these goals, a substantial amount of complex process change has to occur throughout the US healthcare system. Acquiring EHRs and HIE platforms will be insufficient. Implementing them effectively while changing hundreds of processes in thousands of organizations will be the much bigger challenge. Luckily, theory and practices that can mainly be traced back to sociotechnical systems theory are available and increasingly used to enable the effective implementation of EHRs and HIE that is needed. At the same time that all of this intense activity with EHRs and HIE is going on, the PHR is rapidly evolving and being adopted by millions of health consumers worldwide. PHRs are in some ways distinct from EHRs, especially in terms of "ownership" and use, and in other ways closely related to EHRs, especially in terms of data and standards. Both are facilitated by HIE. Many PHR issues and several examples of PHRs are discussed in the following section.

Personal Health Record

The PHR, according to advocates, promises to revolutionize healthcare. Furthermore, it will change the power relationship between physician and patient. What is a PHR? To put it simply, a PHR is a record of an individual's basic health metrics (e.g., blood type, allergies, and current and past health problems), physical and laboratory test results, heath treatment, and other information relevant to health that is usually initiated and maintained by the individual or by a designee (e.g., parent, guardian, or other relevant individual). The basic idea is common sense. Having accurate and useable information about one's health can save time and effort, avoid confusion, and prevent errors when one visits a healthcare provider for routine treatment or is receiving emergency care.

The idea of the PHR predates the widespread use of computers and information technology in healthcare. As early as 1956, Carl Dragstedt in a guest editorial for the *Journal of the American Medical Association* wrote the following:

I would like to attract a little attention to the serious statement that what everyone in this country needs is a good personal health log. By that I mean a suitable booklet in the permanent possession of everyone, in which would be recorded some of the important aspects of his health record, encompassing items from his family history, and data on such things as his vaccinations and inoculations, his diseases and operations, his blood pressure, blood cell counts, and similar laboratory findings. It would be for him, and all of his contacts with hospitals and physicians, somewhat comparable to a permanent passport for travelers. Upon consulting a physician or entering a hospital, he would submit his health log.

The idea of the personal health log or personal medical log is essentially the same as the PHR. The term "personal health record" can be applied to any health record created and maintained by consumers. Sadly, Dragstedt's suggestion did not result in widespread use of the PHR.

Concern about the cost of healthcare as well as medical errors combined with advances in HIT has resulted in a movement to promote the PHR. Now with the advances in HIT, electronic versions of PHR are now available, and today, the PHR is usually conceived of as an electronic record. Raisinghani and Young (2008) observed the following: "The paper-based and fragmented healthcare system is no longer appropriate for the digital economy of the 21st century. Integrated health information technology system is the solution to transform clinical practice to consumer centric and information driven. Tools such as PHRs are means to an end that provide better, safer, and more affordable healthcare for consumers."

There are clearly many advantages to the PHR. When visiting a new provider, patients will no longer have to fill in forms with the same basic health information that they have given before. Furthermore, patients will not have to rely on their memory for when they had an operation or for the name of a prescription medication they are taking and will not need to carry a copy of their latest test results. The PHR, if connected to an EHR, can become a comprehensive longitudinal health record that would not only serve the individual, but also the physician, insurers, employers, and other stakeholders (Lang, 2009). Also, when integrated into an EHR, the PHR can be used as a communication channel between the consumer and the provider, as illustrated below. The PHR can be used to schedule appointments, cancel appoints, and review test results.

The advantages of the PHR are illustrated by an email received by a fictional consumer from his healthcare provider.

`<<begin example letter>>`

Date: Sat, 23 Jan 20XX 00:16:48-0500

From: outgoingmail@anycityhealthcareprovider.com

To: John_Ferraro@jmail.com

Subject: Doctor's appointment in 3 days

Dear John,

This is an automated email reminder for your upcoming appointment at Any City Healthcare on Wednesday, 02-25-2010. To view the time and description of your appointment, click on the Appointments section of your eChart at www.anycityhealthcareprovidercom.

Your physician would like you to know:

Any City Healthcare Provider is now providing you with very valuable healthcare innovation. You can now access your medical records available online. If you go to another healthcare provider anywhere in the world that has access to the Internet, with your permission, they will be able to access your medical record.

To begin this process, we need your assistance. On the www.anycityhealthcareprovider. com homepage, there is a link to start the process. You will be asked a series of questions about your medical history and other relevant questions. We will verify this information with you when you visit us. Although we will update this information based on your medical visits and tests, we ask that you access your medical record to update your record with information to which we do not have access. If you do this, your medical record will always be accurate and complete.

A password will be given to you. You may share this password with other physicians or anyone you believe should have access to your health record.

This innovative service needs your cooperation to be effective and useful. Thank you for your efforts in beginning this process, which will ultimately help to serve you better in your health care needs.

Sincerely,

Jane Smith
Any City Healthcare PHR Manager

<<end letter>>

Providing secure emails between the consumer and providers is only one example of how the PHR can be useful. Before the functions of the PHR are discussed, it is necessary to describe the types of data that the PHR should ideally contain. The following is a list of data the PHR should contain:

1. Basic information about the consumer—name, address, phone number, email address, other contact information, next of kin, marital status, gender, age, height and weight, blood type, relevant information regarding parents' health, and insurance provider.
2. Record of the individual's health history—known health conditions, known allergies, previous ailments and surgeries, current treatments and prognosis. These medical topics deserve special consideration:
   a. Medication list, i.e., the list of medications the patient is currently taking including details of administration, and history of the consumer's previous medication and the response to the medication.
   b. Laboratory, radiological, and other test results.
   c. Procedure/operative reports.
   d. List of current medical issues or problems (symptoms).

Some of the uses of the PHR have already been briefly discussed. Basically, it is a repository of information and, if web-based, is integrated with provider websites and databases. As mentioned above, the PHR can be used to schedule appointments with providers and as a secure means of communication between providers and the consumer. The excerpt above illustrated a secure email from a provider reminding the consumer that the he has an appointment. The consumer can use the PHR to cancel or change the appointment. Laboratory or other test results can be posted to the PHR for the consumer to view. Depending on the nature of the test or the policy of the physician ordering the test, the results may have to be first reviewed by the physician before they are posted to the PHR. If the results of the test may frighten the consumer (e.g., a biopsy to determine if a tumor is malignant) or if the results may be confusing or difficult for the consumer to understand (e.g., results from an MRI), it is appropriate for the physician to review the results first. The physician may first discuss the results with the consumer before they are posted to the PHR or, if appropriate, the physician may simply append an explanation to the test results so that the consumer can better understand them.

Providers can use the PHR to indirectly communicate with each other about the consumer. Before a physician orders a treatment, test, or medication, he or she can check the PHR to see what other treatments, tests, or medications the patient has received or is scheduled to receive. Having a source of information like the PHR that physicians can check reduces the likelihood of medical errors. PHRs may also save time and money; if the relevant test has been recently done, then the physician may simply review the results of the test. If a test is already scheduled but not done, the physician may simply modify the test if it does not include the parameters needed for that physician's purposes. For example, a patient is schedule for a blood draw to measure cholesterol levels. Another physician may request that the patient's blood also be tested for glucose level.

Consumers can use the PHR to check the verbal instructions the physician has given regarding treatment or medication. Additionally, the PHR can be linked to health information and educational material. For example, information about a procedure that the consumer needs to undergo may be available for downloading or viewing. The consumer may learn that MRIs are painless, but that he or she will be exposed to loud noise and may be confined to a small space. Consumers may also be able to access information to help them choose between treatment options.

PHRs that are currently available can be classified by the provider and the medium used to store the information. If we classify the PHR based on how the data are stored, there are two basic types: portable storage device PHRs and Internet PHRs.

Portable storage devices that can be used as PHRs include smart cards and USB storage devices. These PHRs include files for generating the PHR on any personal computer. Data can be uploaded or downloaded to any compatible computer. Usually, the data have to be input manually into this type of PHR. Consumers can carry the PHR with them at all times.

Among Internet PHRs, there are two subtypes: those that are part of an EHR and those that are not. As stated above, PHRs that are integrated into an EHR have many advantages besides being accessed via the Internet. Based on who provides the PHR, there are three types: vendor, provider, and third-party payer. PHR may be sold by a vendor to the consumer, who then owns and controls the PHR. Alternatively, a healthcare provider may provide the PHR. In this approach, the provider owns the PHR. The parameters of use by the consumer are determined by the provider. A third class of provider of PHRs is third-party payers, i.e., primarily insurers. The insurer or other third-party payer owns the PHR, which is designed to serve the needs of the owner. These PHRs are usually concerned with administrative and billing data. Lang (2009) suggests the approach that holds the most promise would be a hybrid where the consumer, provider, and third-party provider would be equal contributors and stakeholders. However, he does not suggest who would initiate this approach. Unless the government at the national or individual state level or large-scale professional associations or consumer groups promote such a plan, it is not likely to be implemented. More likely, healthcare providers' or third-party payers' PHR will dominate. Large-scale health providers or consortiums of providers are more likely to have EHRs than the individual physician or small practices.

Although PHRs can provide many benefits and save money for healthcare providers, there are a number of obstacles that make the widespread adoption of PHR difficult in the United States. Problems with the widespread implementation of PHR can be analytically divided into technical, legal/administrative, and social.

Technical Problems with PHR Implementation

In the United States, a majority of healthcare providers do not have an EHR, and many providers who do have an EHR do not participate in HIE, which limits the usefulness of the PHR. Furthermore, an EHR is needed for data to electronically transfer between the providers' records and the PHR. There are improving and evolving HIE technical standards, but they have not matured enough to allow PHRs to easily connect to and share information with EHRs and each other.

Legal/Administrative Problems with PHR Implementation

A major concern for PHR administrators is member authentication. If the PHR is not part of an HIE, then the PHR's right to access data needs to be authenticated. Also, online PHR administrators need to provide a means to verify the user's identity and right to access the PHR information before providing access. Although the technology exists to authenticate a user, as evidence by online banking, to implement PHRs on a massive scale is difficult logistically. There are also administrative issues related to this—do all providers in a practice have access to all consumers' PHRs who use the practice without explicit permission from the consumer? Are consumers notified if there is a change to their PHR? If the PHR is part of HIE, are there additional administrative issues? The rationale of the HIE is that you share information between organizations or among organizations. Raisinghani and Young (2008) argue that HIE implies a radical change in the thinking of healthcare organizations that typically have a "silo" mentality and operate on hierarchical manner. With the widespread adoption of HIEs, healthcare organizations will have to change their way of thinking and share information with other organizations. Possibly the most important administrative issue is the consumer establishing a PHR in the first place. Does the consumer simply enter the data as best they can, as they would with a paper form? Do they enter the data in consultation with a healthcare professional? Because the PHR is only as good as its content, Kern (2009) suggests that a physician phone the consumer and discuss and verify the content of the PHR. He notes, however, that a physical examination is necessary to establish the baseline health data.

PHRs also pose legal issues in the United States because of federal and state legislation about the confidentiality of healthcare information. The Health Insurance Portability and Accountability Act (HIPAA) is the most important piece of legislation regarding the privacy and security of healthcare information. HIPAA sets the minimum standard for privacy and security practices of healthcare providers, health insurers, and clearinghouses. A report by the World Privacy Forum (2008, p. 3) observes that HIPAA does not cover a PHR "...unless the PHR vendor is itself a covered entity under HIPAA." Thus, individuals would lose the privacy protection afforded by HIPAA. The message for consumers is clear: choose a PHR provided by an entity covered by HIPAA (i.e., healthcare providers and insurers). The problem for administrators of covered entities providing PHRs is how to maximize the usefulness and ease of use of the PHR and still comply with the relevant federal and state privacy and security regulations.

Social Problems with PHR Implementation

There are a number of interrelated social factors that make the widespread adoption of the PHR difficult.

Functional illiteracy and medical illiteracy. Functional and medical illiteracy are interrelated. Depending on how you define each of them, they can be for many purposes equivalent. Although the rate of illiteracy (those who cannot read or write a simple sentence) in the United States is quite low, the rate of functional illiteracy is sufficiently high to pose challenges to healthcare providers. According to the 2003 National Assessment of Adult Literacy, about 15% of the adult population of the United States functions below basic prose literacy, which means that they cannot perform everyday, simple literacy activities in English. The results are based on a test of literacy rather than respondents' self-reports. The 2003 survey was the first national survey to measure the health literacy of American adults. Only 12% of adults had proficient health literacy, and 14% were below the basic English health literacy. A majority of adults (53%) were rated as having intermediate health literacy. A higher proportion of older adults (65 years of age and older) were rated as both below basic prose literacy and below basic health literacy.

Based on the survey, there is clearly a link between prose literacy and health literacy. Individuals who have low prose literacy will also have low health literacy. Moreover, computer literacy depend on basic literacy. People who do not have basic prose literacy are unlikely to be able to use a computer. The results of the 2003 survey suggest that, with our present technology, many consumers will need help using their PHR. However, once these individuals with low literacy establish a PHR with accurate information, it will reduce the likelihood of medical errors that would result from these individuals misinterpreting medical forms and supplying the wrong background data. However, they will most likely still need help using the PHR.

Digital divide. According to the Pew Research Center in December 2009, 74% of US adults used the Internet and 60% had broadband at home. Internet use drops to 60% for individuals with household incomes of less than $30,000 a year, and for those with less than a high school education, the usage is 39%. Older adults also have a lower rate of Internet use—of those 65 years and older, only 38% go online. These statistics indicate that there is a small but significant part of the US population that does not have access to the Internet. Of particular concern is the low rate of Internet use by the elderly, whose use of medical care is disproportionately higher than their proportion of the population.

Consumer indifference. There are people who believe that they are too busy or who are simply not interested in investing much time or effort on their health. They put as little time as possible into their healthcare. To put it bluntly, some people will need incentives or to be pushed to establish a PHR. The ease of use, once a PHR is established, should encourage its use.

Despite these difficulties, there have been two notable national implementations of the PHR in the US—the Department of Defense (DoD) and the Veterans Administration (VA). In 2008, the DoD launched MiCARE, the Military Health System's prototype PHR in Tacoma, Washington, at Madigan Army Medical Center. The pilot project uses both Microsoft® HealthVault™ and Google Health™ repositories for the data. MiCARE is designed for military personnel and their families and is a voluntary program. Health information stored through MiCARE is protected under the "Connecting for Health" guidelines, which were developed by the Markle Foundation in collaboration with PHR developers. These guidelines assist in keeping sensitive information private. The VA launched its online PHR known as My HealtheV*e*t nationwide. Registering and using My Heathe*e*Vet is voluntary and free for veterans and their dependents. In 2008, My Health*e*Vet had more than 630,000 registered users (Nazi & Woods, 2008). My Health*e*Vet is an evolving system. Communicating with VA health providers through secure messaging is planned for 2009 and 2010. Also scheduled for 2010 are VA appointment reminders, viewing VA appointments, and viewing VA laboratory results.

Solutions and Recommendations for Implementing a PHR

The PHR shows great promise. There are a number of serious issues regarding its implementation and its use by the general public. Because of low literacy, lack of financial resources, or physical or mental disability, there will be a significant minority of the American population who will have difficulty establishing and using the PHR. There are also technical, administrative, and legal issues that need to be addressed. Kaelber, Jha, Johnston, Middleton, and Bates (2008) note that, "Studies of adoption generally show relatively low adoption rates among patients who have been offered access to a PHR, which have typically been free to them." A national strategy is needed for widespread adoption of PHRs by the American public. First the problem needs to be properly researched and then a concrete strategy needs to be developed that takes into account all stakeholder concerns. Such a strategy needs to ensure that the next generation of Americans is health literate and that those Americans who are not health literate are provided with the resources needed to make effective use of the PHR.

CONCLUSION

Although there has been a 40-year history of HIT in the healthcare industry, most of that history has involved automating functions such as billing, laboratory, pharmacy, and radiology processing while ignoring basic clinical processes. The current emphasis of HIT is applying this technology to basic clinical functions, literally affecting the way medicine and healthcare is practiced. The lynch pin to a compressive approach to HIT is the EHR. It integrates health information from all sources. The interface of the EHR in the future is likely to be the PHR. The PHR is both a source of information for the patient and means for the patient to input relevant information. Clinicians now need to use HIT—via EHRs—to plan for care, make decisions about care, order care, and document care using HIT/EHRs. This is a profound culture change in medicine and healthcare.

The overall goal of achieving widespread meaningful use of certified EHRs within a nationwide electronic health information system is a challenge. To achieve this goal, a substantial number of complex processes need to change throughout the US healthcare system. Acquiring EHRs and HIE platforms are insufficient by themselves. Implementing them effectively while changing hundreds of processes in thousands of organizations is the task ahead for the US. This will change the culture of healthcare and the relationship of the patient to the healthcare provider. The promise of improved healthcare is real, but to fully achieve it on a nationwide scale will require considerable effort.

The implications of applying HIT to basic clinical processes have implications for the consumer of healthcare beyond *receiving* better healthcare. The full benefits of HIT for a population can only be realized if the patient becomes an active, participating consumer. Full and effective use of a PHR requires a health literate (or the provision of services to ameliorate health illiteracy) and actively involved consumer.

Although this chapter has focused on the United States, the issues discussed here are applicable to most countries. Applying HIT to a whole nation is not simply a matter of money. It requires changing many processes in many organizations. Despite lagging behind some countries in fully implementing EHRs on a nationwide scale, the US still remains a leader in the development of HIT. The fragmented healthcare system of the United States may allow for more experimentation and innovation in HIT. Considerable progress has been made in the US in HIT developments such as the PHR.

Fully implementing HIT in any country will change the culture of healthcare. And, the full implementation of PHR will change the power relationships in healthcare. Increased patient (or consumer) power comes with the price of increased patient responsibility and effort.

REFERENCES

Aarts, J., Ash, J., & Berg, M. (2007). Extending the understanding of computerized physician order entry: Implications for professional collaboration, workflow and quality of care. *International Journal of Medical Informatics*, *76*, S4–S13. doi:10.1016/j.ijmedinf.2006.05.009

Agency for Healthcare Research and Quality. (2009). *National Resource Center for Health Information Technology*. Retrieved November 13, 2009, from http://healthit.ahrq.gov

Agency for Healthcare Research and Quality. (2009). *PSNet Primer on CPOE*. Retrieved November 13, 2009, from http://healthit.ahrq.gov

Alliance for Healthcare Reform. (2009). *Definition of healthcare information technology*. Retrieved November 13, 2009, from http://www.allhealth.org

Ash, J. S., Stavri, Z., & Kuperman, G. J. (2003). A consensus statement on considerations for a successful CPOE implementation. *Journal of the American Medical Informatics Association*, *10*(3), 229–234. doi:10.1197/jamia.M1204

Blumenthal, D. (2010). *Homepage information*. Retrieved March 28, 2010, from http://healthit.hhs.gov

Butterfield, S. (2010). Remote monitoring: Out of sight, right in line. *ACP Internist*, 1.

Campbell, E. M., Guappone, K. P., Sittig, D. F., Dykstra, R. H., & Ash, J. S. (2008). Computerized provider order entry adoption: Implications for clinical workflow. *Journal of General Internal Medicine*, *24*(1), 21–26. doi:10.1007/s11606-008-0857-9

Department of Health and Human Services, Office of the Secretary. (2010). *Health Information Technology: Initial set of standards, implementation specifications, and certification criteria for Electronic Health Record technology*. Federal register: Rules and regulations.

Dragstedt, C. A. (1956). Personal health log: Guest editorial. *Journal of the American Medical Association*, *160*(15), 1320.

Drucker, P. F. (2002). *Managing in the next society* (p. 118). New York: St. Martin Press.

Greenly, M., & Gugerty, B. (2002). How bar coding reduces medication errors. *Nursing*, *32*(5), 70.

Gugerty, B., Maranda, M., & Rook, D. (2006). The clinical Information System implementation evaluation scale. *Studies in Health Technology and Informatics*, *122*, 621–625.

HL7. (2007). *HL7 continuity of care document, a healthcare IT interoperability standard, is approved by balloting process and endorsed by healthcare IT standards panel*. Retrieved March 17, 2010, from http://www.hl7.org/documentcenter/public/pressreleases/20070212.pdf

Health Information Technology Standards Panel (HITSP). (2008). *IS 77: Remote monitoring interoperability specification*. Retrieved March 22, 2010, from http://www.hitsp.org

HIMSS. (2010). *HIMSS analytics U.S. EMR adoption model*. Retrieved March 30, 2010, from http://www.himssanalytics.org

H.R. 1–111th Congress. (2009). American Recovery and Reinvestment Act of 2009. Retrieved January 12, 2010, from http://www.govtrack.us/congress/bill.xpd?bill=h111-1

Institute for Educational Sciences, National Center for Educational Statistics. (2006). *The health literacy of America's adults. Results from the 2003 National Assessment of Adult Literacy.*

Jha, A. K., Desroches, C. M., Campbell, E. G., Donelan, K., Rao, S. R., & Ferris, T. G. (2009). Use of electronic health records in US hospitals. *The New England Journal of Medicine*, *360*(16), 1628–1638. doi:10.1056/NEJMsa0900592

Kaelber, D. C., Jha, A. K., Johnston, D., Middleton, B., & Bates, D. W. (2008). A research agenda for personal health records (PHRs). *Journal of the American Medical Association*, *15*, 729–736.

Kern, S. I. (2009). PHRs can boost efficiency, but also hold risk. *Medical Economics*, *86*(15).

Kuperman, G. J., & Gibson, R. F. (2003). Computer physician order entry: Benefits, costs, and issues. *Annals of Internal Medicine*, *139*, 31–39.

Lang, R. D. (2009). Personal health records ready for prime time? *Journal of Healthcare Information Management*, *23*(3).

Liker, J. K. (2004). *The Toyota way*. New York: McGraw-Hill.

Malcolm, B., Carlson, R. A., Tucker, C. L., et al. (2000). *Veterans Affairs: Eliminating medication errors through point-of-care devices*. Paper presented at Healthcare Information and Management Systems Society Conference, Dallas, TX.

Morgan, J. M., & Liker, J. K. (2006). *The Toyota product development system: Integrating people, process and technology*. New York: Productivity Press.

My HealtheVet website. (2010). *About my HealtheVet.*

Nazi, K. M., & Woods, S. S. (2008). *MyHealtheVet PHR: A description of users and patient portal use.* AMIA Annual Symposium.

Osheroff, J. A. (2009). *Improving medication use and outcomes with clinical decision support: Step-by-step guide*. Chicago: HIMSS Press.

Pew Internet & American Life Project. (2010). *Internet, broadband, and cell phone statistics.*

Raisinghani, M. S., & Young, E. (2008). Personal health records: Key adoption issues and implications for management. *International Journal of Electronic Healthcare*, *4*(1), 67–77. doi:10.1504/IJEH.2008.018921

Sitting, D. F., Campbell, E., Guappone, K., Dystra, R., & Ash, J. S. (2007). Recommendations for monitoring and evaluation of in-patient computer-based provider order entry systems: Results of a Delphi survey. In *proceedings of the AMIA Annual Symposium*. (pp. 671–675).

United States Department of Defense. (2008). *DoD Launches Personal Health Record Pilot.* Military health systems press release.

United States Government Accountability Office. (2005). *Federal agencies face challenges in implementing initiatives to improve public health infrastructure*. Retrieved October 13, 2009, from http://www.gao.gov/new.items/d05308.pdf

University of Pittsburgh Medical Center. (2009). *Bio-security.* Retrieved from www.upmc-biosecurity.org

World Privacy Forum. (2008). *A legal and policy analysis of Personal Health Records: Why many PHRs threaten privacy.*

KEY TERMS AND DEFINITIONS

Electronic Health Records (EHR): Provides an individual's health and healthcare-related data that is created, managed, and consulted by authorized personnel within one or more healthcare settings.

Healthcare Information Technology (HIT): The use of technologies that allows the effective and efficient management of patient records so that the information can be shared between physicians, hospitals and healthcare companies.

Personal Health Records (PHR): An electronic record that has been developed and maintained by the patient. It may or may not be in collaboration with the healthcare provider.

Chapter 11
Evaluation and Benefits Realization of E-Business Investments in Healthcare Organizations

Chad Lin
Curtin University of Technology, Australia

ABSTRACT

Despite high expectations for the value of e-business in healthcare, its adoption remains poorly understood and under-researched. The adoption and effective utilization of e-business in the healthcare industry can potentially lead to many benefits, such as: an increased accessibility to healthcare providers, improved work-flow efficiency, a higher quality of healthcare services, decreased scheduling conflicts, and reduction in healthcare costs. Healthcare organizations have started to fully make use of e-business technologies to overcome many of the challenges they face. However, healthcare executives have found it increasingly difficult to justify rising IT expenditures and are often under pressure to find a way to evaluate the contribution of their e-business investments to business performance, as well as to find reliable ways to ensure that the business benefits from the investments are actually realized. Hence, the case study approach was conducted to examine the evaluation practices and processes of their e-business investment. The key objectives of the study are to: (1) examine current e-business investment evaluation and benefits realization practices and processes among different types of healthcare organizations; and (2) identify issues and factors that are critical in evaluating and managing e-business investments in healthcare organizations. One contribution of the study is the examination of the relationship between e-business investments and evaluation practices and processes of healthcare organizations.

DOI: 10.4018/978-1-60960-501-8.ch011

INTRODUCTION

The adoption of electronic business (e-business) offers a significant opportunity for organizations (including healthcare) to access potential global customers and suppliers via the Internet. It has been defined as the utilization of information and communication technologies (e.g. the Web, the Internet, intranets, extranets) in support of all business activities (Lin et al., 2007; Standing and Lin, 2007). It has played a critical role in the growth of world economy (Timmers, 1999). For example, Forrester Research has estimated that US e-business sales will grow from US$155billion in 2009 to US$250 billion in 2014, with a compound annual growth rate of 10% (Forrester Research, 2010a). In Western Europe, the e-business sales has been forecasted to increase from US$93 billion in 2009 to US$156 billion in 2014, with an annual growth rate of 11% (Forrester Research, 2010b).

Not too long ago, Australian healthcare procurement system has been still a mainly paper based system with manual processing (GS1, 2004). Effective utilization of e-business technologies in the healthcare industry has the potential to lead to an increased accessibility to healthcare providers, improved work-flow efficiency, a higher quality of healthcare services, improved inventory management, and improved supply chain of manufacturing and distribution of goods and services, reduction in healthcare costs and medical errors (Bhakoo & Sohal, 2008; Heeks, 2006; Lin et al., 2008a, 2008b). Driven by concerns about escalating procurement costs in the industry, several e-business related initiatives (e.g. the Monash, PECC and PEG projects) have been launched by Australian Federal and State governments during the last decade to develop better ways of ordering and procuring medical supplies electronically within the Australian healthcare industry (More and McGrath, 2002).

Despite this, the adoption of IT in healthcare is often difficult, time consuming, and expensive exercises (Sallas et al., 2007). They may proceed slowly, and difficulties in successful adoption of IT by healthcare organizations are well-documented (Heeks, 2006; Lorenzi & Riley, 2004). Factors contributing to low adoption include high investment costs, productivity loss, and lack of financial incentives (Heeks, 2006; Lorenzi & Riley, 2004; Reardon & Davidson, 2007). Another important factor is the inability by healthcare organizations to evaluate the business value of their IT investments (e.g. e-business) and in ensuring the expected benefits are eventually realized or delivered (Devaraj & Kohi, 2003). In particular, the problems and difficulties in measuring benefits and costs are often the main reason for uncertainty about the expected benefits of e-business investments and hence are the major constraints to investments in e-business (Lin et al., 2007).

Indeed, the evaluation of investments in e-business is a complex tangle of financial, organizational, social, procedural and technical threads, many of which are currently either avoided or dealt with ineffectively (Lin et al., 2008; Standing & Lin, 2007). Often e-business projects fail to deliver what is expected of them because most organizations focus on implementing the technology without the adoption of the tools or processes necessary to help to track and measure the e-business projects (Lin et al., 2007). For example, a survey by PriceWaterhouseCoopers (2003) found that organizations achieved expected pay-offs only 25-75% of the time. In addition, the major benefits organizations can gain from IT investments in e-business are inherently qualitative (i.e. customer satisfaction and systems efficiency) and cannot be easily assessed beforehand and calculated in monetary terms (Soliman & Janz, 2004). The less precisely bounded environment of e-business technology adds more complexity to the traditional IT measurement problem as this type of investment is physically distributed between suppliers and customers (Torkzadeh & Dhillon, 2002; Straub et al., 2002). Some of the problems and challenges associated with evaluation of e-business investments are listed in Table 1.

Table 1. Difficulties associated with e-business investment evaluation

Issues	Authors
The cross organizational nature of e-business technology makes it much for difficult for organizations (e.g. healthcare organizations) to allocate and assign costs and benefits to projects.	Lin and Huang (2009); Subramani (2004)
Failure to measure their e-business investments and identify relevant e-business costs and benefits.	Lin et al. (2007)
Insufficient time and effort devoted to measuring the long term e-business benefits.	Standing and Lin (2007); Stamoulis et al. (2002)
Traditional financially oriented evaluation methods (e.g. ROI, NPV) can be problematic as they have difficulty in quantifying relevant e-business benefits and costs.	Tsao et al. (2004); Sugumaran and Arogyaswamy (2004)
It is very difficult to evaluate intangibles and make relationship between e-business and profitability explicit.	Lin et al. (2008); Torkzadeh and Dhillon (2002)

While an IT investment evaluation methodology (IEM) focuses on selecting and monitoring e-business investments that offer the greatest benefits, an IT benefits realization methodology (BRM) or process needs to be used in order to extend investment evaluation further into the project life cycle by ensuring expected benefits are eventually realized after a decision to invest has been taken (Standing & Lin, 2007). The adoption of an evaluation methodology on e-business investments alone is insufficient to ensure that the benefits identified by healthcare organizations are realized (Lin et al., 2008; Ward & Daniel, 2006). The use of a benefits realization methodology would enable healthcare organizations to ensure that desired business changes have been clearly defined, and they are measurable (Ward et al., 1996). Furthermore, it has been also suggested that there might be a connection between the adoption of IEM/BRM and IT maturity (e.g. Hackbarth & Kettinger, 2004; Lin et al., 2007). Organizations which have place more emphasis on increasing their IT maturity are more likely to adopt IEM and BRP (Lin et al., 2005).

Hence, the key objectives of this study are to: (1) examine current e-business investment evaluation and benefits realization practices and processes among different types of healthcare organizations; and (2) identify issues and factors that are critical in evaluating and managing e-business investments in healthcare organizations. One contribution of this study is the examination of the relationship between e-business investments and evaluation and benefits realization practices among healthcare organizations.

BACKGROUND

IT Investments in E-Business

Past research has focused on a wide range of aspects specific to organizations and their adoption of e-business (Korchak & Rodman, 2001). For example, several studies had been conducted to examine the determinants and inhibitors for IT adoption in organizations in several countries (e.g. Enterprise Ireland (2004) in Ireland, Buhalis and Deimezi (2003) in Greece, Levy and Powell (2003) in UK, and Locke and Cave (2002) in New Zealand). The many levels of e-business adoption have been examined and strategies for increasing adoption developed (Levy and Powell, 2003). In order to minimize the problems of investing in e-business, a number of adoption issues and factors have been identified which include a lack of business policy and strategies, technical expertise and experience, user involvement, top management commitment, level of IT maturity, e-business evaluation and benefits realization processes (Daniel et al., 2002; Korchak & Rodman, 2001; Lawson et al., 2003; Lin et al., 2007; Rao et al., 2003).

IT Investment Evaluation (IEM) in E-Business

Research on IT investments in e-business initiatives by Australian organizations indicate that many organizations had no measures of success and most did not carry out post-implementation reviews for their investments (Marshall & McKay, 2002; Lin et al., 2008). To-date the research has delivered contradictory findings on the effect of the e-business expenditures on organizational productivity (Thatcher & Pingry, 2004). Therefore, it is not difficult to see that the measurement of the business value of IT investment in e-commerce has been the subject of considerable debate by many academics and practitioners (Sugumaran & Arogyaswamy, 2004). Although some IT productivity studies have produced inconclusive and negative results, or the interpretation of results may depend on many factors (e.g. mismeasurement of outputs and inputs, the difficulty of establishing the overall value of IT, the choice of inappropriate methods of evaluation, and lags in learning and adjustments) (e.g. Stratopoulos & Dehning, 2000), a number of researchers have indicated that IT spending in e-business is directly related to organizational performance (e.g. Brynjolfsson & Hitt, 2003) with effective leverage and evaluation of IT investments in e-business resulting in improved organizational performance (Melville et al., 2004). According to Lee and Runge (2001), organizations that evaluate their IT adoption and investments are better able to exploit the Internet's potential for their organization, and thus create short-term competitive advantages.

The major benefits organizations can gain from e-business investments are inherently qualitative and cannot be easily assessed beforehand and calculated in monetary terms (Giaglis et al., 1999). The problem becomes more evident as IT is used to link the supply chain or to change the structure of industries, and costs and benefits have to be tracked across functional and organizational boundaries (McKay & Marshall, 2004). The less precisely bounded environment of e-business technology adds more complexity to the measurement problem as this type of investment is physically distributed between suppliers and customers (Straub et al., 2002; Torkzadeh & Dhillon, 2002). Furthermore, as mentioned earlier, research on IT evaluation in e-business initiatives indicate that many Australian organizations had no measures of success and/or carried out ineffective post-implementation reviews for their investments (Lin et al., 2007; Lin et al., 2008; Marshall & McKay, 2002). Thus, failure to plan for and, derive the benefits from an e-commerce investment can have detrimental consequences on organizational performance.

IT Benefits Realization (BRM)

IT is just one enabler of process change (Grover et al., 1998) and it only enables or creates a capability to derive benefits. The essence of benefits realization is "not to make good forecasts but to make them come true ... and IS/IT on its own does not deliver benefits" (Ward et al., 1996). Benefits realization and identification are also seen as a function of strategic information systems planning (Changchit et al., 1998). Benefits may be considered the effect of the changes or the difference between the current and proposed way that work is done (Ward & Griffiths, 1996). Things only get better when people start doing things differently.

While IT investment evaluation is important, it is insufficient in terms of ensuring that the benefits identified and expected by organizations are realized and delivered (Ward and Griffiths, 1996). As a result, an IT benefits realization methodology is, in most cases, used along with an IT investment evaluation methodology. According to Lin and Pervan (2003), it is simply too risky for organizations to adopt an IT benefits realization methodology without doing a proper evaluation (via an IT investment evaluation methodology). It is more cost effective for organizations to do proper evaluation of their e-business investments and projects before realizing their expected benefits.

Assessing the effective delivery of benefits from these services is very difficult (Ward et al., 1996). Irani (2002) argue that IT cost identification, measurement and control remains a significant challenge for businesses. To overcome this problem, IT projects should be evaluated in the context of accumulated costs and benefits from related initiatives, not solely judged on single initiatives (Galliers et al., 1996). To determine if the desired IT benefits have been achieved in practice, it is necessary to undertake an ex-post evaluation of benefits (Ward et al., 1996). Moreover, if no measurable effects can be identified other than the implementation of the technology itself, then it can be assumed that no benefits have been realized (Ward et al., 1996).

A survey by Seddon et al. (2002) indicates identifying and measuring benefits as the most difficult issue in evaluating IT. In addition, a survey by CIE (Norris, 1996) found that vague statement of benefits, leading to an uncertain allocation of responsibility for managing their delivery, as the number one cause for IT project failure. While the search for benefit identification can contribute to the perceived success of an IT investment such as an e-business initiative, organizations have often found it difficult to evaluate them and as a result tend to use arbitrary values for assessing benefits. Few organizations have a benefits realization approach and, while much attention is paid to justifying investments, little effort is extended to ensuring that the benefits are realized (Ward et al., 1996). Some of the reasons for failure to monitor whether the projected benefits of IT (e.g. e-business) were being realized by organizations include:

- It is difficult to assess benefits after a project has been implemented as benefits are often experienced later (Murphy & Simon, 2002);
- Many organizations have poor IT adoption practices (Fink, 1998);
- It is not necessary as the project was implemented according to plan (Norris, 1996);
- Many organizations tend to give very little attention to the intangible benefits when investment decisions are made (Irani, 2002); and
- It is too costly to undertake the proper post-implementation reviews on benefits (Norris, 1996).

Some of the methodologies for realizing IT investment benefits published in the literature are:

- Active Benefit Realization (ABR) (Remenyi et al., 1997);
- Cranfield Process Model of Benefits Management (Ward et al., 1996);
- DMR's Benefit Realization Model (Truax, 1997);
- Model of Benefits Identification (Changchit et al., 1998); and
- The IT Benefits Measurement Process (Andresen et al., 2000).

IT Maturity

Various stages of growth models have been proposed to describe the evolution of organizational information systems (e.g. Hirschheim et al. (1988), Nolan (1979)). Despite some criticism of these models, they provide an insightful organizing framework for thinking about IT change in organizations (Galliers & Sutherland, 1991). These models were developed to determine the effectiveness of an organization in aligning IT with its business goals and processes as well as in achieving its desired IT outcomes. In addition, these maturity models extend organizations' abilities to evaluate and track impacts of IT investments such as e-business initiatives within their organizations (Schuh & Leviton, 2006).

Nolan's (1979) stages of growth model, one of the earliest IT maturity models, is the most well-known model but has attracted criticisms over the years. The revised stages of growth model by Galliers and Sutherland (1991) is meant to overcome some of the limitations by introducing a means of bringing together a range of key elements associated with the operation and management of an organization generally. The revised model of Galliers and Sutherland (1991) can be represented as six stages, each with its particular set of conditions associated with the seven "S" elements which have provided a rich set of conditions upon which an organization can analyze and measure its IT maturity. The seven elements are strategy, structure, systems, staff, style, skills, and superordinate goals. The six stages of the revised model are: ad hocracy, starting the foundations, centralized dictatorship, democratic dialectic and cooperation, entrepreneurial opportunity, and integrated harmonious relationships. The revised model assesses the organization's capability and readiness to obtain business benefits from IT directly to its level of IT maturity and the results provide crucial information about the organization's ability to effectively utilize IT in order to achieve competitive advantage and business goals.

RESEARCH OBJECTIVE AND METHODOLOGIES

Despite the growing popularity of e-business and its numerous benefits in healthcare, there is a paucity of published literature on the evaluation and benefits realization in e-business investments in the healthcare organizations (Chiasson et al., 2007; Van Akkeren & Rowlands, 2007). An exploratory quantitative case study approach was adopted to identify the relevance, importance and relationships of the concepts (e.g. IEM, BRM, IT maturity) mentioned in the literature (Silverman, 2001). The case study method enables the researcher to examine the IEM and BRM processes and the level of IT maturity of the healthcare organizations studied, particularly in the context of the procurement, supply chain, and distribution of goods and services among healthcare organizations. The method also allows the researcher to better understand the responses given in the interviews through observation. This serves as a method of triangulation of research data and ensures that the questions and answers are properly understood by repeating or rephrasing the questions and through paraphrasing the responses back to the interviewee (Yin, 2002).

Hence, the exploratory quantitative case study approach was utilized to: (1) examine current e-business investment evaluation and benefits realization practices and processes among different types of healthcare organizations; and (2) identify issues and factors that are critical in evaluating and managing e-business investments in healthcare organizations. Interviews were conducted were conducted with participants (i.e. IT managers, IT procurement/supply chain managers, and/or CIOs) from 26 Australian healthcare organizations. All the healthcare organizations that took part in the research have been in business for more than ten years and have adopted at least some sort of IEM and BRM processes. Most of the healthcare organizations participated in this research are hospitals and pharmaceutical companies. These cases were deliberately chosen in order to focus efforts on theoretically useful cases (following the theoretical, non-random sampling strategy by Eisenhardt (1989)). Extensive notes were taken during the interviews. The questions asked related to their e-business investments with respect to the procurement, supply chain, and distribution of goods and services, relevant e-business policies, objectives, and strategies, their level of IT maturity, e-business evaluation and benefits realization practices and processes, e-business benefits, and other e-business adoption and implementation issues.

Questions relating to a particular research theme, for example, level of organizational IT maturity, were examined as a cluster. Divergent views within the same company were assessed in terms of the relative strength of the perspective according to the numbers of responses falling in particular categories. This was done as a form of in-case analysis and to develop general explanations and interpretations (Eisenhardt, 1989). These steps enhance the construct validity, reliability and overall quality of the research (Yin, 2002). Then, qualitative content analysis by Miles and Huberman (1994) was used to analyze the data from the case study. The analysis of the results of the case study was conducted in a cyclical manner and the results were checked by other experts in the field. The external experts were asked to trace the logical flow of the research study, research questions, case findings and analysis and identification of constructs and thereby identifying any gaps in the chain of evidence (Yin, 2002).

RESEARCH FINDINGS

Effective Use of IEM and BRM

Two most cited reasons for undertaking IEM and/or BRM by the participating healthcare organizations were: (1) to improve quality of their e-business systems or initiatives; and (2) to ensure the expected benefits/objectives from their e-business systems or initiatives are met. Respondents were also asked about the effectiveness of the adoption of IT investment evaluation (IEM) and benefits realization (BRM) methodologies or processes for their various e-business investments within their organizations.

Almost half of the healthcare organizations (11) indicated their IEM has been used effectively in ensuring successful e-business investments or projects and about a quarter of the participants (7) reported that their IEM was not used effectively. Only about one-third of the respondents (9) revealed that their BRM was effective in achieving their e-business initiatives and a smaller number of healthcare organizations (7) did not adopt it effectively. With the exception of one healthcare organization, there appeared to be a positive relationship between the use of IEM and BRM in terms of their effectiveness (Figure 1). In other words, those organizations which were able to use IEM effectively in ensuring successful e-business initiatives tend to be able to use BRM effectively. Results of the case study also supported the findings by Lin and Pervan (2003) and Ward et al. (1996) that it is easier for organizations to adopt IEM first before the adoption of BRM.

Figure 1. Effectiveness use of IEM and BRM

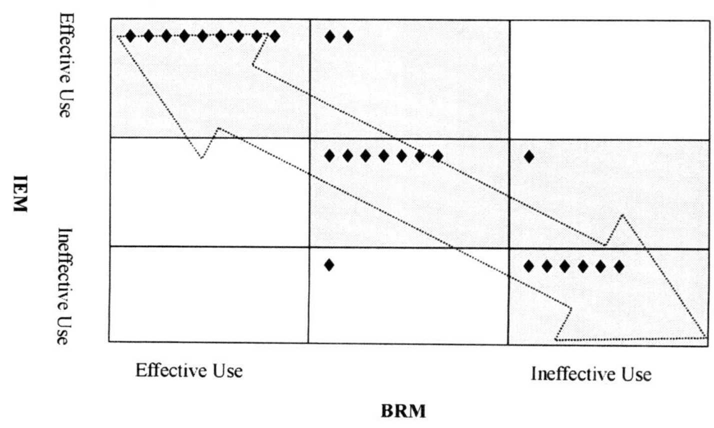

Business Policy, Objective or Strategy and the Level of IT Maturity

Most respondents indicated that it was difficult to just to implement a business methodology or process without first building some sort of technical and infrastructure support/capability. Most case study respondents stated that their business policy, objective or strategy had an impact on their level of IT maturity. Most respondents indicated that the level of IT maturity could only be improved if it is explicitly stated in their business policies, objectives or strategies.

Indeed, results of the case study revealed that there appeared to be a direct connection between a written business policy, objective or strategy in relation to an organization's IT maturity (Figure 2). There were only two exceptions. One healthcare organization was not interested in investing further resources into their ineffective IEM and BRM as they had directed resources at other areas whereas The other healthcare organization already had effective IEM and BRM and they found it too expensive to keep directing IT resources to maintain high level of IT maturity. For example, one IT manager stated: *"......funding and other resources provided through our business strategy and this can affect the level of IT maturity achieved within our organization......"*

The Level of IT Maturity and Effective Use of IEM and BRM

IT maturity was based on Galliers and Sutherland's Revised Stages of Growth Model (Galliers & Sutherland, 1991) which was represented as six stages, each with its particular set of conditions. The original seven elements for IT investments had been examined and tested in the e-business context by Lin et al. (2007) and were therefore reduced to four elements (strategy, style, skills, and overall goals). The case study respondents were also asked to describe about where they thought their organizations stand in terms of its stage of growth for each of the four elements as described in Galliers and Sutherland's Stage of Growth Model (1991). Table 2 showed for each of the four elements what percentage of healthcare organizations are in each stage. It showed that the "average" organizations are in stages 3.5 – 4.5 in most of the four elements. The results also revealed that about 80% of the responding healthcare organizations achieved stages 4 and above for the 'overall goals' element while less than half of the participating healthcare organizations achieved stage 4 and above for the 'strategy' and 'skills' elements.

Results from the case study demonstrate a relationship between more mature organizations

Figure 2. Business policy, objective or strategy vs. the level of IT maturity

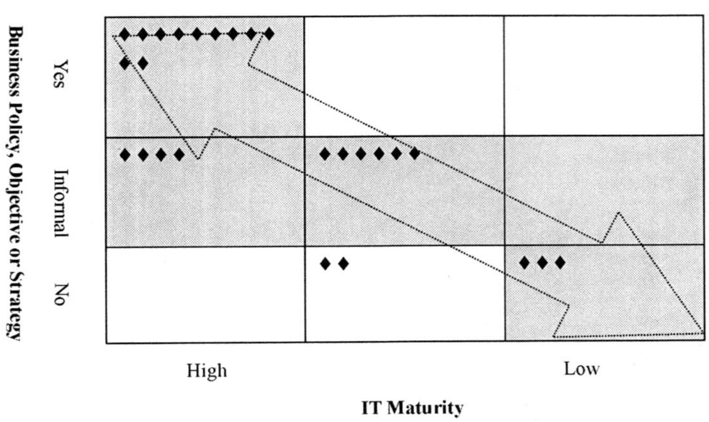

Table 2. Results for stages of growth

	Strategy	*Style*	*Skills*	*Overall Goals*
Stage 1	0%	0%	>1%	0%
Stage 2	14%	>1%	10%	0%
Stage 3	33%	33%	33%	19%
Stage 4	48%	19%	14%	48%
Stage 5	0%	19%	33%	29%
Stage 6	>1%	24%	>1%	>1%
Mean Stage (1-6 stages)	3.5	3.9	3.8	4.2

Figure 3. The level of IT maturity vs. effective use of IEM

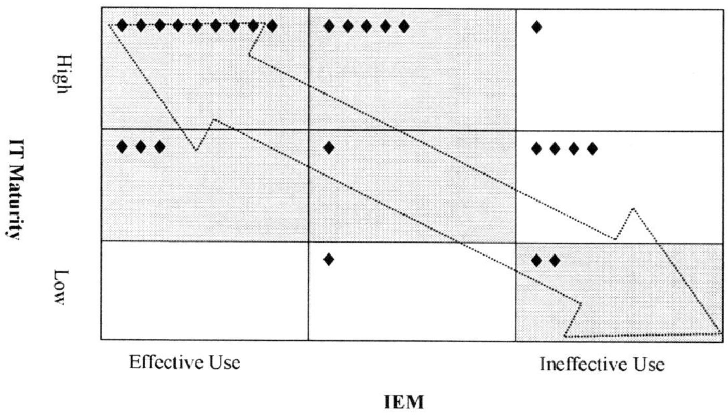

(i.e. higher stages on the four elements) and the effective use of both investment evaluation (IEM) and benefits realization (BRM) methodologies. In other words, the results supported the notion that higher level of IT maturity made easier for healthcare organizations to adopt IEM and BRM more effectively on their e-business investments and projects, and vice versa (Figures 3 and 4). In Figure 3, with some exceptions, the level of IT maturity had a positive relationship with the effective use of IEM. In Figure 4, there are most exceptions than in Figure 4. This is due to the fact that it was more difficult to implement BRM effectively as it requires more time, expertise, and resources than IEM (Lin et al., 2008; Ward & Daniel, 2006).

Effective Use of IEM/BRM and the Level of E-Business Benefits

Results of the case study revealed that effective use of BRM appeared to have more impact on the realization of e-business benefits than the effective use of IEM (Figures 5 and 6). More healthcare organizations had benefited from the effective use of BRM than IEM with respect to the realization of e-business benefits. Many respondents also revealed that the use of IEM alone did not bring about much e-business benefits. According to these respondents, effective use of both IEM and BRM would ensure the realization of e-business benefits. For example, one IT manager stated that benefits only started to flow in after a BRM was used: "*...... an evaluation process was used in conjunction with a benefits realization approach*

Figure 4. The level of IT maturity vs. effective use of BRM

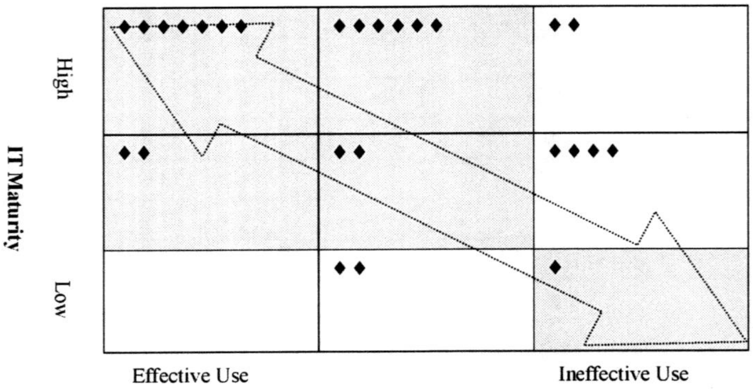

Figure 5. Effective use of IEM vs. the level of e-business benefits

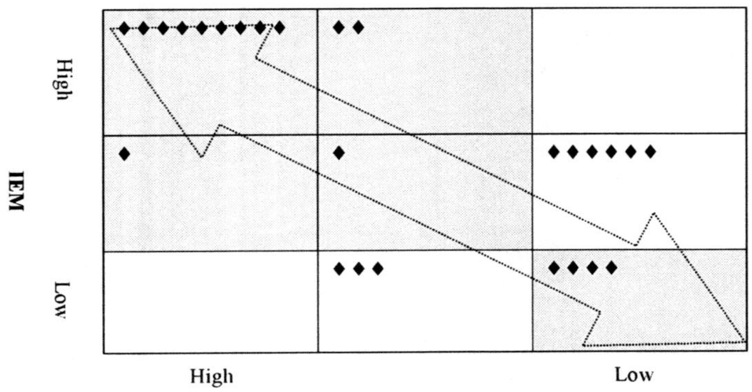

to bring about significant benefits such as cost savings and increased revenue from our e-business investments……the main aim of an evaluation process was to measure the investments…… our benefits realization approach ensured that benefits were delivered."

DISCUSSION AND IMPLICATIONS

Results of the case study revealed that there was a positive correlation between the effective use of IEM and BRM by the participating healthcare organizations (Figure 1). However, only approximately half of the healthcare organizations were able to adopt IEM effectively and just over one-third of healthcare organizations were able to use BRM effectively in ensuring successful e-business investments and projects. The results also indicated that business policy, objective or strategy appeared to affect the level of IT maturity of the participating healthcare organizations (Figure 2).

The level of Healthcare organization's IT maturity, according to most of the case study respondents, was the key determinant in the effective use of IEM and BRM (Figures 3 and 4) as higher levels of IT maturity improved the ability and capability of the participating organizations to adopt the IEM and BRM more effectively. Figures 5 and 6 revealed that effective use of

Figure 6. Effective use of BRM vs. the level of e-business benefits

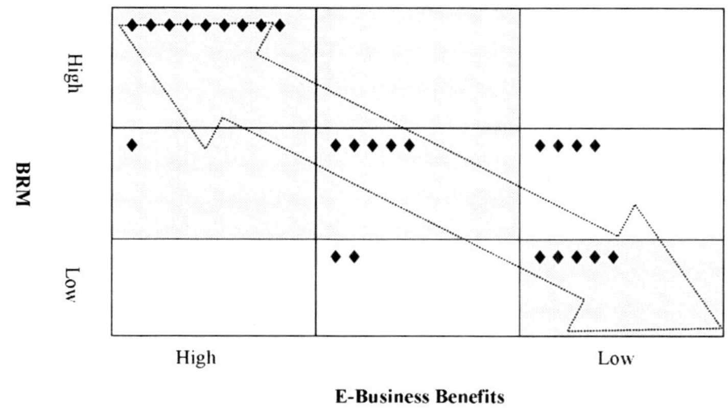

IEM and BRM had a positive relationship with e-business benefits, with BRM appearing to have more impact than IEM.

A key finding of this study is the relationships between the company policy, objective, and strategy, IT maturity, effective use of IEM and BRM, and e-business benefits (Figure 7). Well-thought-out company policies, business objectives and/or strategies can lay a good foundation for building and improving healthcare organizations' level of organizational IT maturity. Capabilities to use IEM and BRM effectively by healthcare organizations stem largely from their levels of organizational IT maturity. As mentioned earlier in the paper, the use of IEM alone does not guarantee the delivery of benefits. Effective use of IEM assists healthcare organizations in the identification of the e-business benefits. On the other hand, effective use of BRM play a more major role in the delivery of e-business benefits. This framework is depicted in Figure 7 and has implications for healthcare organizations. All of the building blocks (company policy, objectives and strategies, IT maturity, IEM, and BRM) are critical for healthcare organizations to achieve appropriate level of e-business benefits. Removal of any foundation blocks will seriously undermine a healthcare organization's ability to realize e-business benefits.

FUTURE RESEARCH DIRECTIONS AND LIMITATIONS

The healthcare industry historically has lagged behind other industries in the adoption of e-business and in the use of evaluation practices or processes but this is expected to change in future years. Healthcare organizations have begun to realize the importance in utilizing e-business tools and in undertaking necessary evaluation processes in order to meet the pressures from regulatory bodies, competition, achieve business performance, and increase the degree of satisfaction of their patients.

Figure 7. A Framework for e-business benefits and satisfaction building blocks

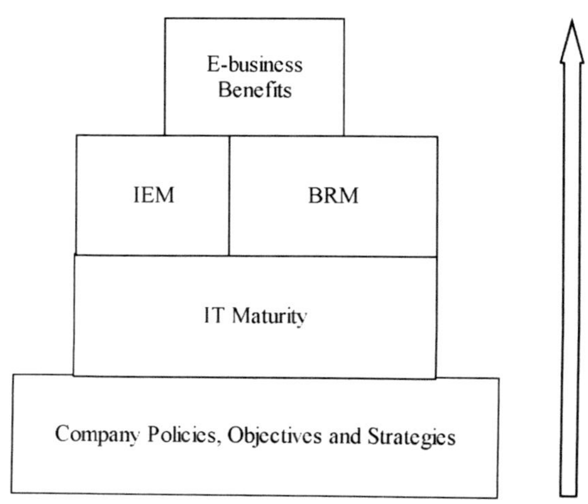

Many different e-business technologies such as electronic records and telemedicine to algorithms to remote monitoring devices can assist in streamlining the systems, saving money, and improving services for the healthcare organizations.

Finally, some limitations in the research need to be acknowledged. First, the inconsistent definitions of what constitutes an IT investment evaluation or benefits realization between studies make research findings difficult to compare and generalize. While there has been a lot of research undertaken in the area of IT evaluation, there has been limited research undertaken in the area of IT benefits realization and, as a result, comparisons are difficult to undertake. Furthermore, our study took place at a particular point in time. Further research could be conducted to capture opinions of benefits realization and investment evaluation at various phases of an e-business projects life cycle and also in terms of their IT maturity. Alternatively, our study could be replicated in a few years time to examine how IT benefit realization and investment evaluation have changed and are being managed in light of other new emerging e-business technologies.

CONCLUSION

Case studies were conducted in 26 healthcare organizations which had undertaken evaluation processes of their e-business systems or projects. Several major issues and factors such as IT maturity and company policy were examined with respect to e-business evaluation and benefits realization practices and processes of healthcare organizations. The findings have the potential to assist healthcare organizations in evaluating and understanding of their e-business systems and projects.

This book chapter has examined the e-business investment evaluation and benefits realization practices of healthcare organizations in Australia. To date, there has not been much research undertaken in this particular area in healthcare organizations, so the findings presented should provide the impetus for organizations to re-consider their approaches to e-business investment evaluation and benefits realization. The inherent difficulties in identifying and assessing e-business investments are often a cause for uncertainty about the expected impact the investments might have on the business. As a result, it is all too easy for healthcare organizations to ignore the evaluation of their e-business investments.

The IT investment in e-business is not just technological gimmickry; it is in fact a whole new way of doing business that replaces traditional models. It not only brings increases operational efficiency and effectiveness, but also a chance to re-engineer the business process. However, its benefits cannot be delivered without taking into account the various issues and factors and the processes of IT investment evaluation and benefits realization. Based on the findings it is recommend that healthcare organizations should focus on improving their level of IT maturity an integral part of their business strategy or policy. In addition, healthcare organizations should conduct an assessment of the e-business available to the organization so that features and costs can be readily identified. Moreover, healthcare organizations should also develop an expected e-business satisfaction and benefits management plan as well as determine if sufficient e-business benefits exist and if organizational culture is supportive of adopting e-business technologies.

REFERENCES

Andresen, J., Baldwin, A., Betts, M., Carter, C., Hamilton, A., & Stokes, E. (2000). A framework for measuring IT innovation benefits. *Electronic Journal of Information Technology in Construction, 5*, 57–72.

Andrew, J. P., Blackburn, A., & Sirkin, H. L. (2000). *The B2B opportunity: Creating advantage through e-marketplaces*. Boston: The Boston Consulting Group.

Bhakoo, V., & Sohal, A. (2008). An assessment into the drivers of e-business adoption within the Australian pharmaceutical supply chain. In *proceedings of the 39th Annual Meeting of the Decision Sciences Institute at Baltimore*, Maryland, 22-25 November.

Brynjolfsson, E., & Hitt, L. M. (2003). Computing productivity: Firm-level evidence. *The Review of Economics and Statistics*, *85*(4), 793–808. doi:10.1162/003465303772815736

Buhalis, D., & Deimezi, O. (2003). Information Technology penetration and e-commerce developments in Greece, with a focus on small to medium-sized enterprises. *Electronic Markets*, *13*(4), 309–324. doi:10.1080/1019678032000135563

Burns, R. B. (1994). *Introduction to research methods*. London: Longman Cheshire Pty Ltd.

Carayannis, E. G., Alexander, J., & Geraghty, J. (2001). Service sector productivity: B2B electronic commerce as a strategic driver. *The Journal of Technology Transfer*, *26*(4), 337–350. doi:10.1023/A:1011130802700

Changchit, C., Joshi, K. D., & Lederer, A. L. (1998). Process and reality in Information Systems benefit analysis. *Information Systems Journal*, *8*, 145–162. doi:10.1046/j.1365-2575.1998.00031.x

Chen, I. J., & Popovich, K. (2003). Understanding Customer Relationship Management (CRM): People, process and technology. *Business Process Management Journal*, *9*(5), 672–688. doi:10.1108/14637150310496758

Chiasson, M., Reddy, M., Kaplan, B., & Davidson, E. (2007). Expanding multi-disciplinary approaches to healthcare Information Technologies: What does Information Systems offer medical informatics? *International Journal of Medical Informatics*, *76*, s89–s97. doi:10.1016/j.ijmedinf.2006.05.010

Clay, K., & Strauss, R. P. (2002). Institutional barriers to electronic commerce: An historical perspective. *Advances in Strategic Management*, *19*, 247–273. doi:10.1016/S0742-3322(02)19008-5

Damanpour, F. (2001). E-business e-commerce evolution: Perspective and strategy. *Managerial Finance*, *27*(7), 16–33. doi:10.1108/03074350110767268

Daniel, E., Wilson, H., & Myers, A. (2002). Adoption of e-commerce by SMEs in the UK. *International Small Business Journal*, *20*(3), 253–269. doi:10.1177/0266242602203002

DeLone, W. H., & McLean, E. R. (2004). Measuring e-commerce success: Applying the DeLone & McLean Information Systems success model. *International Journal of Electronic Commerce*, *9*(1), 31–48.

Devaraj, S., & Kohli, R. (2003). Performance impacts of Information Technology: Is actual usage the missing link? *Management Science*, *49*(3), 273–289. doi:10.1287/mnsc.49.3.273.12736

Eisenhardt, K. M. (1989). Building theories from case study research. *Academy of Management Review*, *14*(4), 532–550. doi:10.2307/258557

Enterprise Ireland. (2004). *IT/e-business status and issues of small and medium sized Irish SMEs*. BSM Ltd Report. Retrieved from http://www.enterprise-ireland.com/ebusiness/eBIT_ICTissues.htm

Fink, D. (1998). Guidelines for successful adoption of Information Technology in small and medium enterprises. *International Journal of Information Management*, *18*(4), 243–253. doi:10.1016/S0268-4012(98)00013-9

Fleiss, J. L. (1981). *Statistical methods for rates and proportions* (2nd ed., pp. 38–46). New York: John Wiley.

Forrester Research. (2010a). *US online retail forecast, 2009 To 2014*. Forrester Research.

Forrester Research. (2010b). *Western European online retail forecast, 2009 To 2014*. Forrester Research.

Freytag, P. V., & Clarke, A. H. (2001). Business-to-business market segmentation. *Industrial Marketing Management, 30*, 473–486. doi:10.1016/S0019-8501(99)00103-0

GS1. (2004). *Hospital pharmaceutical e-commerce looks healthy*. Media Press Release, GS1 Australia.

Galliers, R. D., & Sutherland, A. R. (1991). Information Systems management and strategy formulation: The stages of growth model revisited. *Journal of Information Systems*, *1*(1), 89–114.

Giaglis, G. M., Paul, R. J., & Doukidis, G. I. (1999). Dynamic modelling to assess the business value of electronic commerce. *International Journal of Electronic Commerce*, *3*(3), 35–51.

Goodhue, D. L., Wixom, B. H., & Watson, H. J. (2002). Realizing business benefits through CRM: Hitting the right target in the right way. *Management Information Systems Quarterly Executive*, *1*(2), 79–94.

Grover, V., Teng, J., Segar, A. H., & Fiedler, K. (1998). The influence of Information Technology diffusion and business process change on perceived productivity: The IS executive's perspective. *Information & Management*, *34*(3), 141–159. doi:10.1016/S0378-7206(98)00054-8

Hackbarth, G., & Kettinger, W. J. (2004). Strategic aspirations for Net-enabled business. *European Journal of Information Systems*, *13*(4), 273–285. doi:10.1057/palgrave.ejis.3000511

Heeks, R. (2006). Health Information Systems: Failure, success and improvisation. *International Journal of Medical Informatics*, *75*, 125–137. doi:10.1016/j.ijmedinf.2005.07.024

Hirschheim, R., Earl, M., Feeny, D., & Lockett, M. (1988). An exploration into the management of the information function: Key issues and an evolutionary model. *Proceedings of Information Technology Management for Productivity and Strategic Advantage, IFIP TC-8 Open Conference*, Singapore.

Irani, Z. (2002). Information Systems evaluation: Navigating through the problem domain. *Information & Management*, *40*, 11–24. doi:10.1016/S0378-7206(01)00128-8

Kaefer, F., & Bendoly, E. (2004). Measuring the impact of organizational constraints on the success of business-to-business e-commerce efforts: A transactional focus. *Information & Management*, *41*, 529–541. doi:10.1016/S0378-7206(03)00088-0

Kauffman, R. J., & Walden, E. A. (2001). Economics and electronic commerce: Survey and directions for research. *International Journal of Electronic Commerce*, *5*(4), 5–16.

Klein, H. K., & Myers, M. D. (1999). A set of principles for conducting and evaluating interpretive field studies in Information Systems. *Management Information Systems Quarterly*, *23*(1), 67–94. doi:10.2307/249410

Korchak, R., & Rodman, R. (2001). E-business adoption among US small manufacturers and the role of manufacturing extension. *Economic Development Review (Schiller Park, Ill.)*, *17*(3), 20–25.

Lawson, R., Alcock, C., Cooper, J., & Burgess, L. (2003). Factors affecting adoption of electronic technologies by SMEs: An Australian study. *Journal of Small Business and Enterprise Development*, *10*(3), 265–276. doi:10.1108/14626000310489727

Lee, J., & Runge, J. (2001). Adoption of Information Technology in small business: Testing drivers of adoption for entrepreneurs. *Journal of Computer Information Systems*, *42*(1), 44–57.

Levy, M., & Powell, P. (2003). Exploring SME Internet adoption: Towards a contingent model. *Electronic Markets*, *13*(2), 173–181. doi:10.1080/1019678032000067163

Lin, C., & Huang, Y. (2009). A framework for business-to-business e-commerce evaluation challenges and critical success factors. In Xu, J., & Quaddus, M. (Eds.), *E-business in the 21st Century: Realities, challenges and outlook* (pp. 233–254). World Scientific Publishing Co.

Lin, C., Huang, Y., & Burn, J. (2007). Realising B2B e-commerce benefits: The link with IT maturity, evaluation practices, and B2B e-commerce adoption readiness. *European Journal of Information Systems*, *16*(6), 806–819. doi:10.1057/palgrave.ejis.3000724

Lin, C., & Pervan, G. (2003). The practice of IS/IT benefits management in large Australian organizations. *Information & Management*, *41*(1), 13–24. doi:10.1016/S0378-7206(03)00002-8

Lin, C., Pervan, G., Lin, H.-C., & Tsao, H. (2008). An investigation into business-to-business electronic commerce organizations. *Journal of Research and Practices in Information Technology*, *40*(1), 3–18.

Lin, C., Pervan, G., & McDermid, D. (2005). IS/IT investments evaluation and benefits realization issues in Australia. *Journal of Research and Practices in Information Technology*, *37*(3), 235–251.

Locke, S., & Cave, J. (2002). Information communication technology in New Zealand SMEs. *Journal of American Academy of Business*, *2*(1), 235–240.

Lorenzi, N. M., & Riley, R. T. (2003). Organizational issues = change. *International Journal of Medical Informatics*, *69*(2/3), 197–203. doi:10.1016/S1386-5056(02)00105-3

Malhotra, A., Gosain, S., & Sawy, O. A. (2005). Absorptive capacity configurations in supply chains: Gearing for partner-enabled market knowledge creation. *Management Information Systems Quarterly*, *29*(1), 145–187.

Marshall, P., & McKay, J. (2002). Evaluating the benefits of electronic commerce in small and medium enterprises. *The Australian Journal of Information Systems*, *9*(2).

McKay, J., & Marshall, P. (2004). *Strategic management of e-business*. Queensland, Australia: John Wiley & Sons.

Melville, N., Kraemer, K., & Gurbaxani, V. (2004). Review: Information Technology and organizational performance: An integrative model of IT business value. *Management Information Systems Quarterly*, *28*(2), 283–322.

Miles, M. B., & Huberman, A. M. (1994). *Qualitative data analysis: An expanded sourcebook*. Sage Publications.

Mirani, R., & Lederer, A. L. (1993). Making promises: The key benefits of proposed IS systems. *Journal of Systems Management*, *44*(10), 10–15.

More, E., & McGrath, M. (2002). An Australian case in e-health communication and change. *Journal of Management Development*, *21*(7/8), 612–632.

Murphy, K. E., & Simon, S. J. (2002). Intangible benefits valuation in ERP Projects. *Information Systems Journal*, *12*, 301–320. doi:10.1046/j.1365-2575.2002.00131.x

Nolan, R. L. (1979). Managing the crises in data processing. *Harvard Business Review*, *57*(2), 115–126.

Norris, G. D. (1996). Post-investments appraisal. In Willcocks, L. (Ed.), *Investing in Information Systems: Evaluation and management* (pp. 193–223). UK: Chapman & Hall.

PriceWaterhouseCooper. (2003). *Benefits realization online survey.* PriceWaterhouseCooper. Retrieved from http://www.pwcglobal.com

Rao, S. S., Metts, G., & Monge, C. A. M. (2003). Electronic commerce development in small and medium sized enterprises: A stage model and its implications. *Process Management Journal, 9*(1), 11–32. doi:10.1108/14637150310461378

Reardon, J. L., & Davidson, E. (2007). An organizational learning perspective on the assimilation of electronic medical records among small physician practices. *European Journal of Information Systems, 16*, 681–694. doi:10.1057/palgrave.ejis.3000714

Remenyi, D., Sherwood-Smith, M., & White, T. (1997). *Achieving maximum value from Information Systems: A process approach*. Chichester, UK: John Wiley & Sons.

Rouse, A., & Dick, M. (1994). The use of computerized tools in qualitative Information Systems studies. *Proceedings of the 5th Australasian Conference on Information Systems (ACIS 1994),* Monash University, Melbourne, Victoria, (pp. 209-220).

Sallas, B., Lane, S., Mathews, R., Watkins, T., & Wiley-Patton, S. (2007). An iterative assessment approach to improve technology adoption and implementation decisions by healthcare managers. *Information Systems Management, 24*, 43–57. doi:10.1080/10580530601036802

Schuh, R. G., & Leviton, L. C. (2006). A framework to assess the development and capacity of non-profit agencies. *Evaluation and Program Planning, 29*(2), 171–179. doi:10.1016/j.evalprogplan.2005.12.001

Seddon, P., Graeser, V., & Willcocks, L. (2002). Measuring organizational IS effectiveness: An overview and update of senior management perspectives. *The Data Base for Advances in Information Systems, 33*(2), 11–28.

Silverman, D. (2001). *Interpreting qualitative data* (2nd ed.). London: Sage.

Sohal, A. S., & Ng, L. (1998). The role and impact of Information Technology in Australian business. *Journal of Information Technology, 13*, 201–217. doi:10.1080/026839698344846

Soliman, K. S., & Janz, B. D. (2004). An exploratory study to identify the critical success factors affecting the decision to establish Internet-based interorganizational Information Systems. *Information & Management, 41*(6), 697–706. doi:10.1016/j.im.2003.06.001

Srinivasan, R., Rangaswamy, A. & Lilien, G.L. (2005). Turning adversity into advantage: Does proactive marketing during a recession pay off? *International Journal of Research in Marketing.*

Stamoulis, D., Kanellis, P., & Martakos, D. (2002). An approach and model for assessing the business value of e-banking distribution channels: Evaluation as communication. *International Journal of Information Management, 22*, 247–261. doi:10.1016/S0268-4012(02)00011-7

Standing, C., & Lin, C. (2007). Organizational evaluation of the benefits, constraints, and satisfaction with business-to-business electronic commerce. *International Journal of Electronic Commerce, 11*(3), 107–153. doi:10.2753/JEC1086-4415110304

Stockdale, R., & Standing, C. (2006). An interpretive approach to evaluating information systems: A content, context, process framework. *European Journal of Operational Research, 173*(3), 701–1188. doi:10.1016/j.ejor.2005.07.006

Strassmann, P. A. (2003). Does IT matter? An HBR debate: Web exclusive. *Harvard Business Review*, 7–9.

Stratopoulos, T., & Dehning, B. (2000). Does successful investments in Information Technology solve the productivity paradox? *Information & Management, 38*, 103–117. doi:10.1016/S0378-7206(00)00058-6

Straub, D. W., Hoffman, D. L., Weber, B. W., & Steinfield, C. (2002). Measuring e-commerce in Net-enabled organizations: An introduction to the Special Issue. *Information Systems Research*, *13*(2), 115–124. doi:10.1287/isre.13.2.115.86

Subramani, M. (2004). How do suppliers benefit from Information Technology use in supply chain relationships. *Management Information Systems Quarterly*, *28*(1), 45–73.

Subramaniam, C., & Shaw, M., J. (2002). A study of the value and impact of B2B e-commerce: The case of Web based procurement. *International Journal of Electronic Commerce*, *6*(4), 19–40.

Sugumaran, V., & Arogyaswamy, B. (2004). Measuring IT performance: Contingency variables and value modes. *Journal of Computer Information Systems*, *44*(2), 79–86.

Thatcher, M. D., & Pingry, D. E. (2004). Understanding the business value of Information Technology investments: Theoretical evidence from alternative market and cost structure. *Journal of Management Information Systems*, *21*(2), 61–85.

Timmers, P. (1999). *Electronic commerce*. Chichester, UK: John Wiley.

Torkzadeh, G., & Dhillon, G. (2002). Measuring factors that influence the success of Internet commerce. *Information Systems Research*, *13*(2), 187–204. doi:10.1287/isre.13.2.187.87

Truax, J. (1997). *Investing with benefits in mind: Curing investments myopia*. DMR White Paper. (pp. 1-6).

Tsao, H., Lin, K., & Lin, C. (2004). An investigation of critical success factors in the adoption of B2B e-commerce by Taiwanese companies. *The Journal of American Academy of Business*, *5*(1/2), 198–202.

Van Akkeren & Rowlands, B. (2007). An epidemic of pain in an Australian radiology practice. *European Journal of Information Systems*, *16*(6), 695–711. doi:10.1057/palgrave.ejis.3000715

Vandenbosch, B., & Lyytinen, K. (2004). Much ado about IT: A response to the corrosion of IT advantage by Nicholas G. Carr. *The Journal of Business Strategy*, *25*(6), 10–12. doi:10.1108/02756660410569157

Ward, J., & Daniel, E. (2006). *Benefits management: Delivering value from IS & IT investments*. Chichester, UK: John Wiley and Sons Ltd.

Ward, J., & Griffiths, P. (1996). *Strategic planning for Information Systems*. Chichester, UK: John Wiley & Sons Ltd.

Ward, J., Taylor, P., & Bond, P. (1996). Evaluation and realization of IS/IT benefits: An empirical study of current practice. *European Journal of Information Systems*, *4*, 214–225. doi:10.1057/ejis.1996.3

Wilder, C., Caldwell, B., & Dalton, G. (1997). More than electronic commerce. *Information Week*, *661*, 30–34.

Yin, R. K. (2002). *Case study research, design and methods* (3rd ed.). Newbury Park, CA: Sage Publications.

ADDITIONAL READING

Carr, N. G. (2003). IT Doesn't Matter. *Harvard Business Review*, *8*(1), 4–50.

Cilek, P., Fanko, W., Koch, S., Mild, A., & And Taudes, A. (2004). A Hedonic Wage Model-based Methodology for Evaluating the Benefits of IT Investments in Public-Sector Organisations. *The Journal of Enterprise Information Management*, *17*(4), 269–275. doi:10.1108/17410390410548670

Counihan, A., Finnegan, P., & Sammon, D. (2002). Towards a Framework for Evaluating Investments in Data Warehousing. *Information Systems Journal*, *12*, 321–338. doi:10.1046/j.1365-2575.2002.00134.x

Dhillon, G. (2005). Gaining Benefits From IS/IT Implementation: Interpretations from Case Studies. *International Journal of Information Management*, *25*(6), 502–515. doi:10.1016/j.ijinfomgt.2005.08.004

Dos Santos, B. L. (1994). Assessing the Value of Strategic Information Technology Investments. In Willcocks, L. (Ed.), *Information Management: The Evaluation of Information Systems Investments*. London: Chapman & Hall.

Farbey, B., Land, F., & Targett, D. (1992). Evaluating Investments in IT. *Journal of Information Technology*, *7*, 109–122. doi:10.1057/jit.1992.16

Farrell, D. (2003). The Real New Economy. *Harvard Business Review*, *81*(10), 104.

Hochstrasser, B. (1994). Justifying IT Investments. In Willcocks, L. (Ed.), *Information Management: The Evaluation of Information Systems Investments*. London: Chapman & Hall.

Hu, Q., & Quan, J. J. (2005). Evaluating the Impact of IT Investments on Productivity: A Causal Analysis at Industry Level. *International Journal of Information Management*, *5*(1), 39–53.

Lin, C., & Huang, Y. (2007). A Model of IT Evaluation Management: Organizational Characteristics, IT Evaluation Methodologies, and B2BEC Benefits, LNCS 4655-0149. *Lecture Notes in Computer Science*, (September): 149–158. doi:10.1007/978-3-540-74563-1_15

Lin, C., & Huang, Y. (2007). An Integrated Framework for Managing eCRM Evaluation Process. *International Journal of Electronic Business*, *5*(4), 340–359. doi:10.1504/IJEB.2007.014782

Lin, C., Huang, Y., & Tseng, S. (2007). A Study of Planning and Implementation Stages in Electronic Commerce Adoption and Evaluation: The Case of Australian SMEs. *Contemporary Management Research*, *3*(1), 83–100.

Lin, C., Pervan, G., & McDermid, D. (2007). Issues and Recommendations in Evaluating and Managing the Benefits of Public Sector IS/IT Outsourcing. *Information Technology & People*, *20*(2), 161–183. doi:10.1108/09593840710758068

Lin, K., Lin, C., & Tsao, H. (2005). IS/IT Investment Evaluation and Benefit Realization Practices in Taiwanese SMEs. *Journal of Information Science and Technology*, *2*(4), 44–71.

Liu, Y., & Lin, C. (2008). How Are Public Sector Organizations Assessing their IT Investments and Benefits - An Understanding of Issues For Benchmarking. *International Journal of Advanced Information Technologies*, *2*(2), 86–100.

Serafeimidis, V., & Smithson, S. (2003). Information Systems Evaluation as an Organisational Institution – Experience from a Case Study. *Information Systems Journal*, *13*(2), 251–274. doi:10.1046/j.1365-2575.2003.00142.x

Standing, C., Guilfoyle, A., Lin, C., & Love, P. E. D. (2006). The Attribution of Success and Failure in IT Projects. *Industrial Management & Data Systems*, *106*(8), 1148–1165. doi:10.1108/02635570610710809

Strassman, P.A. (1990). *The Business Value of Computers*. The Information Economics Press: New Canaan.

Tallon, P. P., Kraemer, K. L., & Gurbaxani, V. (2000). Executives' Perceptions of the Business Value of Information Technology: A Process-Oriented Approach. *Journal of Management Information Systems*, *16*(4), 145–173.

van Grembergen, W., & van Bruggen, R. (1998) Measuring And Improving Corporate Information Technology Through The Balanced Scorecard, *Electronic Journal of Information Systems Evaluation*, 1(1), Source: [On-Line] http://is.twi.tudelft.nl/ejise/indpap.html.

KEY TERMS AND DEFINITIONS

B2B E-Commerce: This is also known as Business-to-business electronic commerce. It refers to commerce transactions conducted through the Internet between companies or organizations.

BRM: This is also known as IT Benefits Realization Methodologies. These are approaches that are used to ensure that benefits expected in the IT investments by organizations are realized or delivered.

Electronic Business: It refers to the utilization of information and communication technologies (e.g. the Web, the Internet, intranets, extranets) in support of all business activities. It is a business model that is conducted over the Internet in which clients are able to participate in all phases of a purchase decision. Electronic business can be between two businesses transmitting funds, goods, services or between a business and a customer.

IEM: This is also known as IT Investment Evaluation Methodologies. These are approaches that are used to evaluate organizations' IT investments.

IT Maturity: It refers to an organization's capability to utilize its existing IT infrastructure to obtain business value.

IT Productivity Paradox: Despite large investments in IT over many years, there has been conflicting reports as to whether or not the IT benefits have actually occurred.

Chapter 12
The Agri-Food Industry and the E-Landscape

Kim Bryceson
University of Queensland, Australia

ABSTRACT

This chapter provides a brief introduction to the agri-food industry sector and the role that electronic technologies are increasingly playing in ensuring a highly competitive business environment from production through to the final consumer. Key issues such as the dynamics of sustainable production, the drivers of supply and demand and globalised markets, as well as food safety and food security, are discussed within a "Food for Thought" section which provides the context for a more detailed look at a series of technologies used in various agri-food industry chains. Included for special discussion are those 'e' technologies associated with precision farming, electronic animal identification, robotic weed and pest control, cold chain RFID logistics tracking and e-commerce. The chapter concludes with a look at emerging issues for businesses within the agri-food industry going forward within the global electronic landscape, their significance, and some potential solutions.

INTRODUCTION

In the minds of the public at large, food production has long been associated with on-farm agricultural endeavors - that is, the physical production of crops and livestock. The image that is conjured up tends to be of people on tractors ploughing and sowing the crop seed, or a herd of cattle being mustered in dusty conditions or of backbreaking harvesting of fruit and vegetables by hand. In other words an industry based on hard manual labour. While such production is of course part of the process of putting food products onto retail shelves, it is actually only the start of a very complex set of business processes and systems which have developed over time and involve some of the most innovative technologies developed.

The objectives of this chapter are:

1. To briefly introduce the agri-food industry, its complexity and its importance to society

DOI: 10.4018/978-1-60960-501-8.ch012

through its integral role in sustaining the world's population.

2. To introduce the eLandscape and to provide an up-to-date review of how electronic technologies are being used across the whole agri-food industry from Paddock to Plate in order to address the growing requirement of a global food market.

BACKGROUND

The Agri-Food Industry

The world's population must have food – without nourishment humans cannot survive. Both creating and supplying food is therefore a critical component in any and all approaches to creating a sustainable planet and eradicating hunger (Eriksen, 2007, 2008). The industry primarily involved is a large, multifaceted industry sector known as the agri-food industry that exists worldwide, and involves a range of businesses that create industry specific (e.g. grains, sugar cane, dairy, meat, fruit and vegetables to name a few) agri-food chains that often exist across international boundaries. Businesses in such chains (Ricketts & Rawlins, 2001) include:

- *Input Suppliers* (e.g. Agricultural chemical and fertiliser companies such as Incitec/Pivot, Bayer and Cargill etc);
- *Service Suppliers* (e.g. Banks, R&D organisations, Governments, Consultants etc);
- *Producers* (e.g. Growers of grain, meat, dairy, cotton, fruit, nuts, etc);
- *Wholesalers/Distributors* (Marino Brothers, Sysco, Vestey Group, Shenzhen Agricultural Products Co Ltd etc)
- *Traders* (e.g. Mitsubishi, Australian Wheat Board, Bohemia Nut Company, ConAgra, etc);
- *Processors* (e.g. Peanut Company of Australia, Flour Mills, Parmalat, etc);
- *Manufacturers* (e.g. Companies such as Kraft, Mars, Unilever etc);
- *Retailers* (e.g. Supermarkets such as Tesco, Carrefour, Woolworths as well as smaller retail outlets);
- *Logistics* (e.g. Transport and storage companies such as FedEx, TNT, Euro Pool System, iGPS etc).

Industry deregulation, enterprise restructuring, technology development and uptake, innovation, changing export dynamics and an increasing focus on chain management are all providing both opportunities and stresses for businesses involved in the agri-food industry - as they do for all industry sectors. However there are two major issues for the agri-food industry as a whole that differentiates it from other industry sectors. First and foremost the industry is essentially made up of many different food product chains as outlined above (Thompson et al, 2007; McCulloch et al, 2008) - that is, many specific raw products (e.g. differing varieties of wheat) grown on farms, are transformed through their various chains to manufactured food products (such as bread, biscuits & cakes or pasta dependent on the wheat variety in this example) bought by consumers from the shelves of a retail outlet (IBISWorld Report, 2009). The overall process involves many players and many processes and is regarded as a complex system (Bryceson & Smith 2008). With increasingly competitive and quality conscious global marketplaces for food products, governments and agri-food chain members are beginning to recognise that a whole-of-system approach is the key to successfully addressing the increasingly critical requirements of the agri-food industry as a whole - those of food integrity and traceability - which themselves have become the key to answering consumer demand for safe, clean food with an emphasis on quality (USDA, 2002).

The effectiveness of any agri-food chain will depend on how well its activities integrate and coordinate to create efficiency and value at each link of the chain as well as the value created for final consumers (Porter, 1985). Good chain management integrates user expectations, commercial requirements, and the flow of purchased materials and services: it rewards shareholders by enhancing profitability and providing better returns (Krajewski & Ritzman, 1993; Beamon & Bermudo, 2000; Hausman, 2002). Thus the success of an industry chain depends on good integration, coordination, communication and cooperation (Lee & Amaral, 2002) – all of which can be provided by the appropriate development of electronic technologies and their role as a business infrastructure and facilitator.

The second and most longstanding issue for the agri-food industry is that, it deals in low margin commodities where competitive market forces

Table 1. Technologies of the eLandscape which businesses need to deal with in the 21 century. Examples of specific agri-food uses are in italics. Adapted from Bryceson, K.P. (2006) – The "E" Issues for Agribusiness: The What, Why, How. CABI International

Technology	Function
Telephone, Fax, Mobile (SMS messaging), iPhone (video messaging), Skype, Social networking sites such as Facebook, Twitter, YouTube.	Communication. *Additionally, in recent times mobile phones have been used in the agri-food industry in places where there is no internet access and in developing countries in particular, to obtain market prices and sell product.*
Personal Digital Assistants (PDAs e.g. Palm Pilots, Blackberrys), PCs, Digital Cameras, Computerised office systems & software, Data Warehouses	Data storage and manipulation, word processing, spreadsheets, accounting systems, time management. *Digital photos are often used as proof of product quality at the start and end of a fresh product's journey particularly in an export situation.*
Business Information Systems eg Transaction Processing Systems, Management and Executive Information Systems, Decision Support Systems	Improved managerial information sources leading to better decision making.
Internet, Intranet, WANs, LANs, Email	Communication, Electronic business information flow. *The digital divide definitely exits both in terms of access to the internet in rural areas and a knowledge of how to use it which has efficiency implications for the whole agri-food chain.*
Electronic Transaction Processing – eg Electronic Funds Transfer, EFTPos, BPay, Internet Banking, Web Pages	Electronic money movement, online transaction and marketing.
Electronic Data Interchange (EDI),	Movement of information via electronic networks – primarily used in ordering & purchasing. *Of major importance in food tracking and traceability.*
Material Resource Planning (MRP) & Electronic Resource Planning (ERP)	Complete enterprise wide data & information manipulation resource. *Important for food processing and manufacturing companies to manage inventories.*
Customer Relationship Management (CRM)	Not a technology per se but more a business strategy associated with the collection and use of customer information.
Electronic Demand Forecasting and Supply Chain Management	More accurate forecasting of demand and supply based on real time electronic information flows. *Particularly important to input companies which have to deal with fluctuating climate and weather vagaries complicating orders and order timing*
Vendor Managed Inventory (VMI) and Electronic Range Reviewing	A means of optimizing supply chain performance where the buyer (e.g. Manufacturer or Retailer) provides information on requirements to a supplier of a product who is then responsible for maintaining an agreed inventory either with a consolidator or in house – *e.g. specialist food products and Tescos (via Tescolink – Watson, 2005) in the UK.*
Electronic Marketplaces, eHubs, EWebs, Network orchestrators,	Bring together suppliers and buyers into a virtual trading space. *Public E-marketplaces have not been successful in the agri-industry sector. Private e-marketplaces are more so (e.g. First4Farming)*

have typically resulted in the cost of production being very close to the value created, thus leaving relatively thin profit margins (Boehlje, 1999). In addition, unlike in other industry sectors, the creation of the raw product through agronomic production is directly affected by climate and the resulting weather which can be highly variable and often results in a variable supply of variable quality raw product (Mirander et al, 2001).

Commodity availability as a result of, for example drought, in turn has direct impact on commodity prices world-wide (Boehlje, 1999; Rickets & Rawlins, 2001). Ensuring constant volume, high quality product at the right time and price is a driving business force in domestic and international agri-food systems and has involved the adoption of innovative technologies and management practices both within a business and between businesses in order to remain competitative (O'Keeffe, 1998; Dunne, 2001; Bryceson & Kandampully, 2004; van Hofwegen et al, 2005; Bryceson & Slaughter, 2010).

The eLandscape and the Agri-Food Industry

It is now clear that the internet, and associated electronically enabled business practices form, along with other sophisticated electronic communication devices and associated business systems such as those outlined in Table 1, an electronic landscape or 'eLandscape' (Figure 1) within which

Figure 1. The eLandscape. Source: Bryceson, K.P. (2006) – The "E" Issues for Agribusiness: The What, Why, How. CABI International

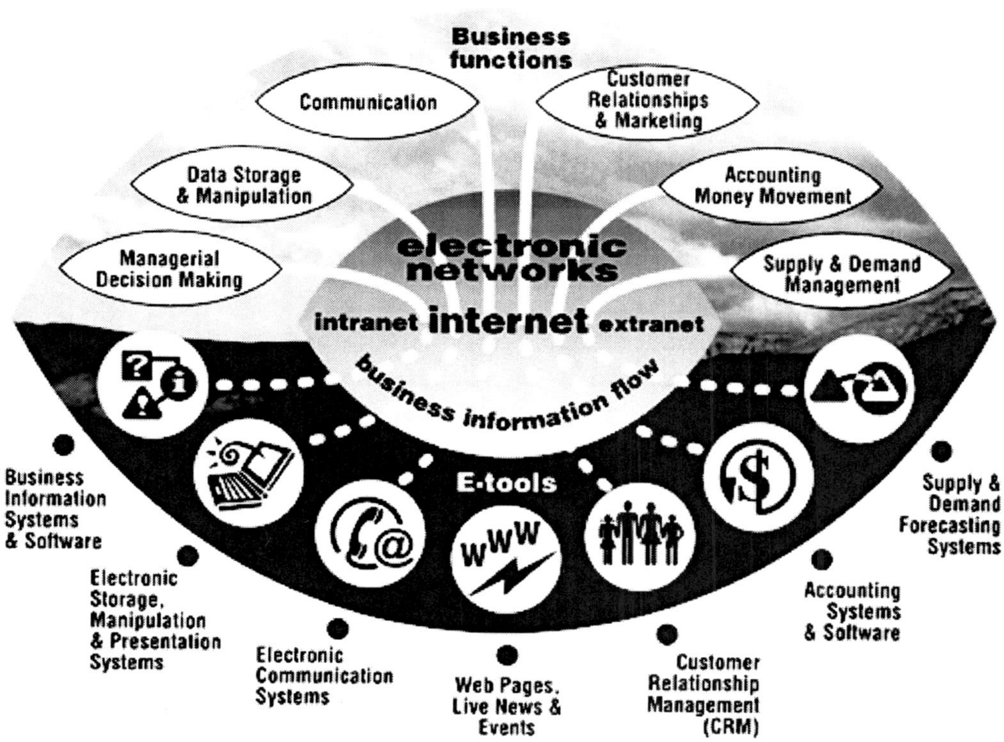

business is carried out (Porter, 2001; Bryceson & Pritchard, 2003; Schiefer, 2003; Bryceson, 2006; Bhaskaran, 2010).

The emergence and adoption of internet technology as an electronic medium of communication and commerce across the board has brought about changes not only in how firms operate and conduct business but how they compete and offer greater value to all stakeholders by proffering an entirely new set of value enhancement options In addition to the impact of new communication technologies on individual businesses in any industry sector, the potential impact on industry chains and the whole industry sector in which they trade, is a major consideration for the likely future success or otherwise of those industry sectors

The reality is that in today's global marketplace, the impact of technology is pervasive in almost all fields of business (Holsapple & Singh, 2000). In the manufacturing industries, technology has gone beyond its initial role as a substitute for manpower to become a value-adding component. In service industries it serves as a unique tool to enhance all aspects of the service sector's endeavors, whether these be operations, marketing, or human resources (Kaplinsky, 1999; Anderson & Lee, 2000, 2001; Power, 2005). In the agri-food industry, in addition to the electronic technologies that provide supporting business infrastructure as already outlined, the specialized nature of the products involved has inevitably meant that specialized technologies have been developed for various parts of the agri-food chain, and these are discussed later in the chapter.

FOOD FOR THOUGHT

Increasing commodity and food prices globally have sharpened the awareness of both policy-makers and of the general public to the complexity and fragility of the global food system - for example, the key issues associated with current agri-food systems revolve around:

- ***The dynamics of sustainable production*** – agriculture is characterised by spatially variable natural resource use and the associated issues of climate variability, a reducing land and water availability for production across the globe, and increasing greenhouse gas pollution. The implication is that unless food production becomes smarter, it will not be sustainable in the long term.
- ***An integrated global food system*** – the overarching trend in agri-food systems as they develop from the localised traditional and poorly organised industry chains to the more advanced industrialised supply chains that exist globally, is the rising need for coordination and the transaction costs that arise as a result (McCulloch et al, 2008). As indicated elsewhere in this chapter, there are marked advantages of organising food production in chains of companies for increased efficiency, flexibility and quality. In particular there has been the increasing requirement for a coordinated flow of information associated both with a food product's traceability from paddock to plate Hammoudi, et al, 2009), the collection of information on its quality and safety for human consumption (Hobbs, 2006), and the flow of information associated with consumer demands needed for both production and inventory management decisions (Schiefer, 2006).
- ***Globalised markets*** - the world is now open for business in both agri-food inputs and outputs, and those players in the agri-food industries with the appropriate assets can, as a result, take advantage of both the new markets that have appeared for traditional products, and of technological and institutional opportunities that exist for

new products. Conversely, those players without such assets (which include both technological and human capital assets) cannot take advantage of the situation - and the gap between the 'haves' and 'have-nots' gets wider. For example, the ability to export fresh tropical fruit as a high value commodity requires post harvest cold chain logistics and monitoring capability (Smith, 2005), and if refrigeration consists of ice blocks in a tub, product quality and shelf life will simply not meet the grade necessary to maintain an export market.

- ***The Key Players*** – In order to survive in increasingly globalized markets, there has been a large increase in vertical integration of businesses in agri-food supply chains since the 1990s - with certain key corporations becoming increasingly dominant. Such enterprises are well serviced with the technologies and business information systems identified in Table 1, and include the household names of Monsanto (agricultural inputs and genetically modified seed), Cargill (food, agricultural and risk management products and services), Nestles (world's largest food and beverage company –originally a food processor), Kraft (world's second largest food and beverage company), Unilever (food, homecare and personal care products) and Carrefour (global distributor and second largest retailer in the world). However it should also be noted that there are a very large number of small agri-food enterprises, particularly in developing countries, which may not have the capability of taking advantage of sophisticated business models or accessing technological innovations in order to add value. In such situations, business based on face to face contact or via the use of minimal technology such as a fax machine or more recently, the mobile phone (Mittal et al, 2010), is still very much the norm.
- ***Food security*** – integrally involved with all of the above, food security is about availability of clean safe food and a person or community's access to it. To be food secure means food is available, food is affordable and food is utilised. Unfortunately, an increasing global population has impacted on both supply and demand for food products and in 2009, FAO estimated that 1.2 billion people were not getting enough regular safe food to eat (FAO, 2009). The issues for the agri-food industry in addressing food security issues are many and complex (Eriksen, 2007, 2008, Fresco, 2009) and include addressing global environmental change (for example through better resource efficiencies) and changing food habits (i.e. by addressing consumer demand). Additionally streamlining food production and supply and developing appropriate systems to deliver and maintain food products is also critical (see next Section and Emerging Issues).

How then can the eLandscape and associated technologies provide some solutions to the issues posed above?

Looking at these drivers with an electronic technologies lens, some clear examples can be brought to mind:

1) The Dynamics of Sustainable Production

Three technologies are briefly reviewed as they impact on the sustainability of production in four major industry chains. The technologies are Precision Farming, Electronic Animal Identification and Robotic Weed and Pest Control. The industry chains benefitting from the technologies are the horticulture (including wine), grains, dairy and cattle industries.

Precision Farming (PF) - PF is a tactical approach to handling sub-paddock (field) variability. It is about mapping and analyzing the inherent variability in field conditions and linking the derived spatial relationships to management action by the tailoring of soil and crop management to match conditions at every location in a field from year to year. The size of each 'location' depends on the particular application and more commonly, on both time and dollar resources available (Bryceson & Marvenek, 1998). PF incorporates the use of technologies not normally associated with agricultural management; these include:

- *Global Positioning System (GPS)* is a satellite navigation system which when accessed via a ground receiver enables every point on earth to have a unique spatial 'address' depicted by a set of co-ordinates e.g. (latitude, longitude) or (easting, northing).
- *Geographic Information Systems (GIS)* which are computer systems designed to allow spatial data storage, manipulation and integration, particularly those data that are geographically referenced.
- *Remote Sensing (RS)* which is the acquisition of information about an object (usually the earth) from some distance away. The classic data sets acquired in this manner are satellite data, aerial photos and airborne video data.
- *Yield Monitoring & Mapping* which involves accurate measurements on a grid basis of production within a paddock. Monitors are on-board the headers and they enable high and low-yield locations to be pin-pointed within paddocks.
- *Variable Rate Technologies (VRT)* allow variable rate application of seed, fertiliser, herbicide and pesticides, based around the GPS recorded locations within paddock.

The core issue in PF is knowing exactly where each location is in the paddock (obtained using the GPS), and having detailed knowledge about the physical and chemical characteristics of that location and how these have been managed. All data for every variable collected at each location (including yield) is thus also geographically located within the surrounding physical landscape of farm and region.

The use of PF technologies to date has mainly been associated with property management planning where the tactical management value of detailed information at the sub-paddock level has been to adjust fertiliser, herbicide and irrigation application rates to address location specific soil properties and conditions (Robert, 2002): additionally, PF technologies have been used to adjust seeding rates for similar reasons, for yield data to be similarly monitored, and more recently, by using wireless sensors and sensor networks to develop environmental management systems, particularly of hazardous situations (Wang et al, 2006) for use in risk management. The strategic management value of collecting this type of detailed information is that it contributes to building a good database of information about the farm which when used in conjunction with other farm regional and catchment management and/or analysis softwares, and can provide the clear, concise and recorded picture of what has happened to a product on farm at source - information that is needed to document such product from a traceability perspective (Dobermann et al, 2004).

Electronic Animal Identification - Livestock identification schemes are the first step in a traceability system for meat and meat products. Many countries are progressing down similar paths in relation to implementing such schemes, including the USA with the United States National Animal Identification Plan (USAIP) (www.usaip.info) and the Animal Disease Traceability Framework (APHIS, 2010); Canada with their National Livestock Identification for Dairy (NLID) (www.nlid.org/English) and the UK with the Cattle Tracing System (CTS) or 'cattle passport' traceability scheme which was implemented in 1998 in compliance with the European Union requirements (Article 6.2 of Regulation [EC] No. 2629/97) (3) (www.defra.gov.uk/animalh/tracing/cattle/cts.htm). In Australia, the National Livestock Identification Scheme (NLIS) is run through Meat and Livestock Australia (MLA) (www.mla.com) working with the various state departments of agriculture/primary industries (e.g. www.dpi.qld.gov.au/nlis/).

Essentially electronic identification of animals involves ear tagging (see Figure 2) each beast with an electronic ear tag at birth so that information such as genetics, weight, husbandry, movements onto and off properties etc. can be recorded against a number linked to each animal until they are slaughtered.

Information from each ear tag is read by a Radio Frequency Identification Device (RFID) reader and downloaded to a computer. In Australia's case this information is then uploaded to the national NLIS database run by the MLA. Saleyards are required to notify the NLIS database of all cattle being sold. Abattoirs are required to notify the database of all cattle slaughtered. In 2010 the whole Australian herd is fully electronically identified for full traceability from birth to death, and the scheme is being rolled out for sheep, goats and, as a result of the equine 'flu epidemic in 2007 which closed down the racing industry in Australia for months, horses.

Figure 2. Australia's National Livestock Identification Scheme (NLIS) RFID tags including ear tags and rumen bolus. Source: Bryceson, K.P. (2006) – The "E" Issues for Agribusiness: The What, Why, How. CABI International

In both the cattle and dairy industry, electronic identification tags as used in the NLIS system have proven additionally useful for herd management purposes – particularly when linked to appropriate herd management software such as Cattleworks (http://www.cattleworks.com/), DairyManagementSystem 21 (http://www.parshwa.com/dairy.html) and CattleLink (http://www.herdlink.com.au/). None-the-less, it should be remembered that electronic identification systems are not a remedy for managing disease outbreaks, they are simply a tool to provide more timely trace-back information. The main issues these systems are attempting to address are biosecurity and surveillance, quarantine and customs and reporting processes, as well as research and development into disease and pest outbreaks.

Robotic weed and pest control - Work in robotics to benefit agricultural efficiency has been ongoing for many years – for example the development of the driverless tractor, which is controlled in 'tracks' in the paddock to the millimetre by GPS locational guidance systems (known as Control Traffic) started in the early 1990s (Lewis, 2003; Tullberg et al, 2007).

However the approach now-a-days is to develop smarter machines that are intelligent enough to work in an unmodified or semi natural environment carrying out a useful task without input from a controller (Blackmore et al, 2005; Pota et al, 2007). A very recent example has been the development of spatially based systems that integrate information from sensors mounted on an Unmanned Aerial Vehicle (UAV) (a small remote controlled aircraft), with Global Positioning System (GPS) locational information to produce large-scale terrain reconstructions and vegetation classifications across vast areas of otherwise unfrequented land (Bryson et al, in press 2010).

The resulting system can be used for a range of environmental monitoring missions such as invasive plant detection and biomass mapping. Early results in terms of weeds show that woody weeds can be distinguished from native vegetation, but discrimination between different species of weed is difficult, due to their similar appearance.

An Integrated Global Food System

Electronic enablement in business has evolved around the premise that information technologies (IT) can perform six data-based functions (capture, transmission, storage, retrieval, manipulation and display) faster and more efficiently than any other type of technology (Alter, 1999). In addition electronic connectivity as it is associated with business-based information flows, revolves around computer networks, particularly the internet – and related technologies. There are many business benefits to being electronically enabled and connected throughout an industry supply/value chain - better efficiencies and better information flows creating better decision making environments and profit generating scenarios being a few (Hoque, 2000; Al-Hakim, 2005). However, in the global food system as it exists today, a particular need has emerged in recent years for well integrated and coordinated reporting and recording systems supported by electronic technologies – the "Paddock-to-Plate" or "Farm-to-Fork" traceability requirement for raw materials and processed food products (Northen, 2001; Hobbs, 2006; Bryceson, 2006).

'Traceability' is defined by the European Union (Article 3 (15) of EU regulation 178/2002), as:

"The ability to trace and follow a food, feed, food producing animal or substance intended to be, or expected to be incorporated into a food or feed through all stages of production, processing and distribution".

To enable full traceability as required by legislation, comprehensive traceability systems or record-keeping systems are required and while these have been largely paper-based in the past, the sheer amount of product now being moved around the world on a regular basis means these systems – even in developing countries, are becoming increasingly electronically based (Setboonsarng et al, 2009).

Food traceability systems are no different from any other product traceability system in terms of the specialised technologies employed (e.g. Barcodes and Barcode scanners as well as Radio Frequency Identification technologies (Bryceson, 2006)) as they are all about maintaining good records. However there are some essential differences in the data required compared to non food product traceability and this is due mainly to biosecurity and food safety information needs which impact on the design of the system – primarily on data capture. For example regulatory issues around food vary between countries, as do import/export requirements, some countries require animal identification, others do not, but in general food processing, storage, transportation and hygiene issues are all issues for which records now need to be kept.

In particular, traceability for food product safety and quality is a necessary part of doing business in the agri-food industry today (Hobbs, 2003, 2006). The increasing number of new food pathogens such as *Escherichia coli* (found in contaminated ground beef, and unpasteurized milk), and *Salmonella enteritidis* (found in eggs), that are now recognized as important causes of food-borne illness; an increasing public health concern about chemical contamination of food, such as the effects of lead in food on the nervous system; plus the size of the food industry and the diversity of products and processes now available, are all contributing to a major push to develop better food traceability systems.

Additionally Golan et al. (2003, 2004) make the point that food companies had been involved in developing food traceability systems for some time before there was a requirement under law. They suggest that food suppliers have had other motives for establishing traceability systems as part of their ongoing business management strategies - for example for managing supply and inventories, for food safety and quality control and for differentiating micro-markets in relation to subtle quality (credence) attributes of food products (Caswell, 1998; Northen, 2001; Kola & Latvala, 2003; Bryceson, 2006).

Globalised Markets and Key Players

With globalised markets come additional opportunities but also additional competition for companies in the agri-food industry. Not only 'Who' do we deal with but 'Where' are they and 'How' do we get our product to that market, are major issues. Two examples of technologies being employed in the agri-food sector as a result of globalization are described – the first is eCommerce and the second is cold chain logistics and RFID tagging of packing containers.

Ecommerce - The 'unique selling point' of the internet is that it provides a timely and efficient method of communication and information flow 24 hours x 7 days a week (24 x 7). Technology advances over the last ten years have enabled businesses to transcend the challenges previously posed by distance and time, and to convert them into potential opportunities. The creative utilization of distance and time, referred to as the 'death of distance' through the impact of technology, constitutes a major economic force that shapes many businesses and countries around the globe by proffering an entirely new set of value enhancement options (Holsapple & Singh, 2000; Ma'ai & Leeming, 2008; Bhaskaran, 2010). Extending the reach of the company and gaining both customers and partnering businesses' acceptance and loyalty as well as focusing on how their needs can be met, is one of the primary preoccupations of managers today.

Ecommerce activities (that is the buying and selling of products over the internet with initiatives being focused on the business to consumer (B2C) model), are now well known and accepted. Just about every company, large or small has a website and most will attempt to sell goods online at some point. In the food industry the key players are large companies all well serviced with electronic technologies and business information systems that enable them to run ecommerce applications of one sort or another where people can buy food and other products online (e.g. Woolworths, 'The Fresh Food People' and the second largest retailer in Australia has a 'Home Shop' facility covering all its products. http://www.homeshop.com.au/). However the biggest barrier to success in this endeavour is that with fresh food, the senses of touch, smell and taste are an important part of the consumer's buying decision and these are not possible to use over the internet. Having said that – there have been some good examples of success – mainly because the companies involved have been selling well known products with a value add proposition such as being organic and/or not available other than through online means, and have full delivery service to the door. For example Silverwood Organics in Queensland, Australia (www.silverwoodorganics.com.au), a small company which sells organic lamb straight from the farm, vacuum packed and with full cold chain logistics to houses in major centres in the State.

Cold Chain Logistics in the food industry can be defined as the maintenance of produce temperature throughout the demand supply chain (from harvest to the consumer) (Commere, 2003). In a globalised market, export cold chain logistics become important and in order to maintain product quality and integrity, involves the seamless movement and management of goods requiring temperature and/or environment controls to maintain product quality, integrity and often, safety.

Effective cold chain logistics is critical to export success (Smith, 2005), as poor cold chain management will negatively impact the product quality of perishable products causing softening, bruising, unwanted ripening, bacterial growth, texture degradation etc. and may lead to the importer rejecting the consignment.

Technology improvements in data capture and processing, product tracking and tracing, temperature control and monitoring and in freight transport transit times, coupled with better communication, the media and the web, have revolutionised cold chain logistics and thus export supply chains which have become more transparent to distribution conditions. Indeed, many large scale customers and/or importers require recorded traces from temperature and moisture probes within the containers and freight trucks to be sure that produce has been kept in appropriate conditions throughout its journey. Such evidence can also be used as part of a company's marketing campaigns – e.g. Ireland's SuperQuinn Retailer (http://www.superquinn.ie/category/FoodTraceability/159.2) and in Carrefour's Filière Qualité labelled range, which is Carrefour's private food traceability system covering hundreds of items in fresh food (www.carrefour.com)

Additionally and associated with logistics in general has been the development of GPS linked RFID tags that are incorporated into plastic food containers and pallets – both as a way of keeping track of food products but also the containers and pallets themselves which are expensive to replace should they not be returned after product delivery (http://www.igps.net/ and http://www.europoolsystem.com/)

EMERGING ISSUES / FUTURE RESEARCH DIRECTIONS

A comprehensive list of researchables is not possible here as this would require a research project in its own right in order to delineate specific projects - however three emerging issues that are clearly associated with agri-food systems in the eLandscape and some ideas for future research projects within these areas can be identified:

1. **Agri-food Issues** – the constant demand for quality, safe product all through the year, everywhere in the world and the stresses that such demand places on sustainable production is a constant issue. Technological innovation, development and implementation in facilitating production that addresses the triple bottom line of people, place and planet is paramount - as is the continuing development of transparent traceability systems from source to customer. Further to this latter point, future research into the role of traceability systems in helping to attain the goal of a food secure world is a key emerging need. Food *in*security is characterised by poor nutrient intake and associated health issues– thus nutrient traceability throughout the food system in order to create interventions that address a *lack* of nutrients in diets, is something to strive for.
2. **EReadiness Issues** – for countries and businesses in those countries to do business supported by electronic infrastructure in order to enable them to create value opportunities in a globalised market is an emerging issue. Some or all of the following have impact on the eReadiness of a country or business and are thus areas of future research in those countries that are not eReady:
    - Technology and ICT infrastructure and technology competencies
    - The political environment
    - The macroeconomic environment
    - Market opportunities
    - Policies towards free enterprise and competition
    - Foreign trade and exchange controls
    - Financing
    - The labour market and infrastructure

 - Access to education and general education levels.

3. **Knowledge Issues** – developing knowledge of both technology and business in players from one end of the agri-food chain to the other is another critical emerging issue. Without a knowledge of the whole industry supply/value chain and the technologies that are now facilitating product and information flows, the imbalance of power and value proposition discontinuities within agri-food chains will continue, and long term sustainability will be compromised. Being aware that the agri-food industry is as technologically complex and business driven as any of the other industry sectors that drive modern human society, and being able to do something with that understanding in order to address key societal issues, is essential.

CONCLUSION

In the Food for Thought section, the current key issues associated with the agri-food industry were highlighted. All five are impacted and are likely to continue to be impacted by the complex, dynamic and competitive global business environment that businesses in the agri-food chain now exist in, and the need for the adoption of modern business practices that have to be backed by innovative business processes and technologies. This chapter has looked at some of the many innovative electronic technologies that are being used in the agri-food industry and has identified that the technologies *per se* are not a barrier to the sector, however the eReadiness of some food producing countries to do business electronically and the technical and business knowledge necessary to use technology to its fullest potential, maybe.

REFERENCES

Al-Hakim, L. (2005). Modelling interdependencies for electronic supply chain management. In *proceedings from 2005 Information Resources Management Association (IRMA) International Conference: Managing Modern Organisations with Information Technology*, 15-18 May 2005, San Diego.

Alter, S. (1999). *Information Systems–a management perspective*. Addison Wesley Longman.

Anderson, D., & Lee, H. (2000). The Internet-enabled supply chain: From the first click to the last mile. *ASCET, 2*. Retrieved on April 20, 2010, from http://web.efzg.hr/dok/INF/bpm/BPM-priprema-03-SCM.pdf

Anderson, D., & Lee, H. (2001). New supply chain business models–the opportunities and challenges. *ASCET, 3*. Retrieved on April 20, 2010, from http://jobfunctions.bnet.com/abstract.aspx?docid=58793

APHIS. (2010). Questions and answers: New animal disease traceability framework. Retrieved on February 7, 2010, from http://www.aphis.usda.gov/publications/animal_health/content/printable_version/faq_traceability.pdf.

Beamon, B. M., & Bermudo, J. M. (2000). A hybrid push/pull control algorithm for multistage, multiline production systems. *Production Planning and Control*, *1*(4), 349–356. doi:10.1080/095372800232072

Bhaskaran, S. (2010). *Using the Internet to enhance small to medium sized food businesses. Rural Industries R&D Report No: 10/004*. Canberra, Australia: RIRDC.

Blackmore, B. S., Stout, W., Wang, M., & Runov, B. (2005). Robotic agriculture–the future of agricultural mechanisation? In J. Stafford (Ed.), *Proceedings of the 5th European Conference on Precision Agriculture*. (pp. 621-628). The Netherlands: Wageningen Academic Publishers.

Boehlje, M. (1999). Structural changes in the agricultural industries: How do we measure, analyse and understand them? *American Journal of Agricultural Economics*, *81*(5), 10–28. doi:10.2307/1244080

Bryceson, K. P. (2006). *E-issues in agribusiness–the what, why, how*. Wallingford, UK: CABI International. doi:10.1079/9781845930714.0000

Bryceson, K. P., & Kandampully, J. (2004). The balancing act: E issues in the Australian agri-industry sector. In *Proceedings of the McMaster World Congress on the Management of Electronic Business*, Hamilton, Ontario, Canada, January 14-16.

Bryceson, K. P., & Marvenek, S. (1998). The role of satellite data and precision farming technologies for on-farm monitoring in Australia. In *Proceedings of the First International Conference on Geospatial Information in Agriculture & Forestry*, 1-3 June, Florida.

Bryceson, K. P., & Pritchard, A. (2003). *The impact of e-business technologies on the grains value chain*. Final Report to the Grains R&D Corporation, GRDC, Canberra, Australia.

Bryceson, K. P., & Slaughter, G. (2010). Alignment of performance metrics in a multi-enterprise agribusiness: A case of integrated autonomy? *International Journal of Productivity and Performance Management*, *59*(4), 325–250. doi:10.1108/17410401011038892

Bryceson, K. P., & Smith, C. (2008). Abstraction and modeling of agri-food chains as complex decision making systems. In Fritz, M., Rickert, U., & Schiefer, G. (Eds.), *EAAA Series on Systems Dynamics and Innovation in Food Networks* (*Vol. 2*, pp. 147–159).

Bryson, M. T., Reid, A., Ramos, F. T., & Sukkarieh, S. (in press). Airborne vision-based mapping and classification of large farmland environments. *Journal of Field Robotics. Special Issue on Visual Mapping and Navigation Outdoors*.

Caswell, J. A. (1998). How attributes of safety and process attributes affect markets for food. *Agricultural and Resource Economics Review*, *27*(2), 151–158.

Commere, B. (2003). *Controlling the cold chain to ensure food hygiene and quality.* Bulletin of the International Institute of Refrigeration (IIR) No: 2003(2).

Dobermann, A., Blackmore, S., Cook, S. E., & Adamchuk, V. I. (2004). Precision farming: Challenges and future directions. In *New directions for a diverse planet, Proceedings of the 4th International Crop Science Congress.* (pp 1-19). 26 Sep – 1 Oct 2004, Brisbane, Australia.

Dunne, A. J. (2001). Supply chain management: Fad, panacea or opportunity? *Agribusiness Perspectives, 48*. Melbourne University.

Ericksen, P. J. (2007). Conceptualizing food systems for global environmental change research. Global environmental change. Retrieved February 9, 2007, from http://www.eci.ox.ac.uk/publications/downloads/ericksen07-foodsystems.pdf

Ericksen, P. J. (2008). What is the vulnerability of a food system to global environmental change? *Ecology and Society, 13*(2), 14. Retrieved from http://www.ecologyandsociety.org/vol13/iss2/art14/

FAO. (2009). *The state of food insecurity in the world.* Retrieved on March 12, 2010, from http://www.fao.org/publications/en/

Fresco, L. (2009). Challenges for food system adaptation today and tomorrow. In *Environmental Science & Policy, Special Issue: Food Security and Environmental Change, Food Security and Environmental Change: Linking Science, Development and Policy for Adaptation, 12*(4), 378-385.

Golan, E., Krissoff, B., Kuchler, F., Calvin, L., Nelson, K. & Price, G. (2004). *Traceability in the US Food Supply: Economic theory and industry studies.* AER-830, USDA/ERS.

Golan, E., Krissoff, B., Kuchler, F., Nelson, K., Price, G., & Calvin, L. (2003). Traceability in the US food supply: Dead end or euperhighway? *Choices Magazine, 18*(2), 36–40.

Hammoudi, A., Hoffmann, R., & Surry, Y. (2009). Food safety standards and agri-food supply chains: An introductory overview. *European Review of Agriculture Economics, 36*(4), 469–478. doi:10.1093/erae/jbp044

Hausman, W. H. (2002). Supply chain performance metrics. In Billington, C., Harrison, T., Lee, H., & Neale, J. (Eds.), *The practice of supply chain management.* Boston: Kluwer.

Hobbs, J. E. (2003). *Traceability & country of origin labelling.* Paper presented at the Policy Dispute Information Consortium, 9th Agricultural and Food Policy Information Workshop, Montreal, April 25 2003. Retrieved on March 25, 2010, from http://www.farmfoundation.org/news/articlefiles/859-hobbs.pdf

Hobbs, J. E. (2006). Liability and traceability in agri-food supply chains. In Ondersteijn, C. J. M., Wijnands, J. H. M., Huirne, R. B. M., & Van Kooten, O. (Eds.), *Quantifying the agri-food supply chain* (pp. 87–100). The Netherlands: Springer. doi:10.1007/1-4020-4693-6_7

Holsapple, C., & Singh, M. (2000). Toward a unified view of electronic commerce, electronic business, and collaborative commerce: A knowledge management approach. *Knowledge and Process Management, 7*(3), 151–164. doi:10.1002/1099-1441(200007/09)7:3<151::AID-KPM83>3.0.CO;2-U

Hoque, F. (2000). *E-enterprise-business models, architecture, and components.* Cambridge, UK: Cambridge University Press. doi:10.1017/CBO9780511584923

Kapplinsky, R. (1999). Globalisation and unequalisation: What can be learned from value chain analysis. *The Journal of Development Studies, 37*(2), 117–146. doi:10.1080/713600071

Kola, J. & Latvala, T. (2003). Impact of information on the demand for credence characteristics. *International Food and Agribusiness Management Review, 5*(2).

Krajewski, L., & Ritzman, L. (1993). *Operations management: Strategy and analysis.* Boston: Addison Wesley.

Lee, H.L. & Amaral J. (2002). *Continuous and sustainable improvement through supply chain performance management.* Stanford global supply chain management forum paper SGSCMF – W1.

Lewis, P. (2003). *Control traffic farming becomes mainstream.* Retrieved on January 12, 2010, from http://www.abc.net.au/landline/stories/s908970.htm

Ma'ai, D., & Leeming, D. (2008). People first network and the potential of ICT in rural development. *Pacific Economic Bulletin, 23*(3), 141–154.

McCulloch, E. B., Pingali, P. L., & Stamoulis, K. G. (2008). Small farms and the transformation of food systems: An overview. In McCulloch, E. B., Pingali, P. L., & Stamoulis, K. G. (Eds.), *The transformation of agri-food systems. Globalisation, supply chains and smallholder farms* (pp. 3–47).

Miranda, M., & Vedenov, D. V. (2001). Innovations in agriculture and natural disaster insurance. *American Journal of Agricultural Economics*, *83*(3), 650–655. doi:10.1111/0002-9092.00185

Mittal, S., Gandhi, S., & Tripathi, G. (2010). *Socio-economic impact of mobile phones on Indian agriculture*. Working Paper 246, Indian Council for Research on International Economic Relations.

Northen, J. R. (2001). Using farm assurance schemes to signal food safety to multiple food retailers in the UK. *International Food and Agribusiness Management Review*, *4*, 37–50. doi:10.1016/S1096-7508(01)00069-6

O'Keefe, M. (1998). Establishing supply chain partnerships: Lessons from Australian agribusiness. *Supply Chain Management*, *3*(1), 5–9. doi:10.1108/13598549810200799

Porter, M. (1985). *Competitive advantage: Creating and sustaining superior performance*. New York: The Free Press.

Porter, M. E. (2001). Strategy and the Internet. *Harvard Business Review*, *79*(3), 63–78.

Pota, H., Eaton, R., Katupitiya, J., & Pathirana, S. D. (2007). Agricultural robotics: A streamlined approach to realization of autonomous farming. In. *Proceedings of the International Conference on Industrial and Information Systems, ICIIS, 2007*, 85–90. doi:10.1109/ICIINFS.2007.4579153

Power, D. (2005). Supply chain management integration and implementation: A literature review. *Supply Chain Management-An International Journal*, *10*, 252–263. doi:10.1108/13598540510612721

Ricketts, C., & Rawlins, O. (2001). *Introduction to agribusiness*. Albany, NY: Delmar.

Robert, P. C. (2002). Precision agriculture: A challenge for crop nutrition management. *Plant and Soil*, *247*(1), 143–149. doi:10.1023/A:1021171514148

Schiefer, G. (2003). New technologies and their impact on agriculture, environment and the food industry. In *EFITA 2003 Conference Proceedings*. (pp. 3-11). Debrecen, Hungary, Schiefer, G. (2006). Information management in agri-food chains. In C.J.M. Ondersteijn, J.H.M. Wijnands, R.B.M. Huirne & O. Van Kooten (Eds.), *Quantifying the agri-food supply chain*. (pp. 137-146). The Netherlands: Springer.

Setboonsarng, S., Sakai, J., & Vancura, L. (2009). *Food safety and ICT traceability systems: Lessons from Japan for developing countries*. ADBI Working Paper 139. Tokyo: Asian Development Bank Institute. Retrieved March 14, 2010, from http://www.adbi.org/working-paper/2009/05/28/3012.ict.food.traceability.system/

Smith, J. N. (2005). Specialized logistics for a longer perishable supply chain. *World Trade Magazine, November 2005*. Retrieved on March 12, 2010, from http://www.worldtrademag.com/Articles/Feature_Article/52de170abaaf7010VgnVCM100000f932a8c0

Thompson, J., Millstone, E., Scoones, I., Ely, A., Marshall, F., Shah, E., et al. (2007). *Agri-food system dynamics: Pathways to sustainability in an era of uncertainty*. STEPS Working Paper 4. Brighton, UK: STEPS Centre.

Tullberg, J. N., Yule, D. F., & McGarry, D. (2007). Controlled traffic farming-from research to adoption in Australia. *Soil & Tillage Research*, *97*, 272–281. doi:10.1016/j.still.2007.09.007

USDA. (2002). Traceability for food marketing & food safety: What's the next step? *Economic Research Service/USDA Agricultural Outlook/January-February 2002*.

van Hofwegen, G., Becx, G., & van den Broek, J. (2005). *Drivers for competitiveness in agri-food chains: A comparative analysis of 10 EU food product chains*. Report No: WP4 EUMercoPol. Retrieved March 14 2010 from: http://www.eumercopol.org/Files/Report%20EU%20Agri-food%20Chains.pdf

Wang, N., Zhang, N., & Wang, M. (2006). Wireless sensors in agriculture and food industry-recent development and future perspective. *Computers and Electronics in Agriculture*, *50*, 1–14. doi:10.1016/j.compag.2005.09.003

Watson, E. (2005). *Tesco admits failures*. Retrieved May 28, 2010, from http://www.foodmanufacture.co.uk/Business-News/Tesco-admits-failures

World Report, I. B. I. S. (2009). *Bread and cake retailing in Australia*. ANZSIC Code: G5124, (p. 31).

Chapter 13
E-Business/ICT and Carbon Emissions

Lan Yi
China University of Geosciences (Wuhan), China

ABSTRACT

The high-tech industry, its associated Information and Communication Technologies (ICT), and subsequent e-business and Internet applications, have had profound effects on our economic lives, social development and environment. This revolutionary change is not only appreciated by computer gurus, but also by private and public organisations. This chapter attempts to focus on the relationship between the e-business/ICT and carbon emission–one of the causes of global warming. Aspects to be considered include energy consumption and CO_2 emission associated with the Internet throughout its lifecycle-from manufacturing of Internet equipment/hardware to its usage and final disposal. On the other hand how Internet technology facilitates environment management and benefits industries via, for example Carbon Trading, will be discussed. The aim of this chapter is to provide a clarification and comprehensive picture of the carbon impact of ICT/e-business to private and public organisations as well as individuals, especially some "behind-the-scene" type of facts. Therefore environmental factors can be taken into account for more informed decision-making during business conduction.

INTRODUCTION

"In 1965, Gordon Moore, one of the visionary founders of Intel, predicted that the high-tech industry could double computing speed and power every two years. This vision, referred to as 'Moore's Law', has been realised" over and over again (SVTC, 2006a). This revolutionary change is not only appreciated by computer gurus, but also the whole society. Through one touch of a mobile phone, one can surf the Internet, send SMS and emails, or shoot 5-mega-pixels resolution photos and share them with the whole world, at a matter of seconds literally. What's more, these

DOI: 10.4018/978-1-60960-501-8.ch013

devices are improving constantly, becoming faster, smaller, cheaper, increasingly powerful and more advanced. However, what shouldn't be ignored also is - they have shorter and shorter lifecycles. Taking pleasure in the revolution exultantly, few of us have noticed what sacrifices have been made to reach this achievement and what side effects are coming along.

Information and Communication Technologies, supported by the high-tech industry, and subsequent e-business and Internet applications, have made profound effects on our lives. The inevitable growing usage has marked it as digital economy era. On the other hand, with the recent Copenhagen Climate Change summit happening during a global recession, how to achieve a low carbon economy has also been on the top agenda of most governments. The further growth of these two trends seem unavoidable, however, are these two going to contribute or contradictory to each other? Study on the interaction between the environment and ICT is relatively new. And very few people really understand the full picture. The obvious links can be exemplified via, just to name a few, teleconferencing reducing carbon emission, online shopping replacing shopping malls, mp3 files substituting CDs, e-commerce deliveries causing additional transport pollution, extra packaging and increasing WEEE (Waste Electronic and Electrical Equipment) issues. Researchers have started to review these impacts from different angles using various methodologies. As can be expected from an emerging field of research, awareness and understanding of this topic is rather limited. Previous studies have presented mixed results.

This chapter attempts to focus on the interactive relationship between the Internet and the environment, energy consumption and carbon emission particularly. Further information regarding the broader linkages between the ICT/e-business and sustainable development (economic growth, environment protection and social development) can be found in the materials provided in the additional reading section at the end of this chapter.

The specific interaction between the e-business/ICT and carbon emission has been mostly looked at from the angle of "the Internet is associated with how much electricity". Rather than assessing the issue from this macroeconomic point of view, which is the most common approach, this chapter attempts to analyse it from a more detailed perspective, at the operational level. The aim is to provide a clearer and comprehensive picture of the carbon impact of e-business/ICT to private and public organisations as well as individuals, especially some "behind-the-scene" type of facts. Therefore they can take the environmental facts into account to make more informed decisions while they conduct business.

Aspects to be considered include (i) energy consumption and carbon emission associated with the Internet throughout its lifecycle, from manufacturing of Internet equipments/hardware to its usage and final disposal, and (ii) how Internet technology facilitates environmental management and benefits the energy industry via, for example Carbon Trading.

BACKGROUND

Mark P. Mills, one of the earliest to study the environmental impact of the Internet, claimed the energy consumed by Internet totalled about 8% of all US electricity use in 1998 and predicted it would grow to half of all electricity use in the next decade (Mills, 1999). These figures were claimed to be "significantly overestimated" by the Energy End-Use Forecasting and Market Assessment Group at Lawrence Berkeley National Laboratory (LBNL) (EUF, 2004). The debate reached an unprecedented level when the US House Subcommittee on National Economic Growth, Natural Resources, and Regulatory Affairs held hearings on "Kyoto and the Internet: the Energy Implications of the Digital Economy" on February 2nd

2000. At that hearing, Jay Hakes, Joseph Romm, and Mark Mills, the most authoritative and active figures in this research field testified. However, no unanimous conclusion was drawn as everybody looked at the problem from different angles and considered different system boundaries. Following this hearing, there were a few rounds of debates and rebuttals between Mills, LBNL etc, which received lots of attention in the media.

In 2005 LBNL published a final report for the PIER (Public Interest Energy Research) programme for the California Energy Commission (Masanet, Price, Can, Brown, & Worrell, 2005). Conclusion was made that despite the ubiquity and importance of PCs in industrialised economies, few data have been published - particularly data that characterise the life-cycle energy consumption and Greenhouse Gas (GHG) emissions of PC. This lack of data is due in large part to the extremely complex nature of the PC production system, which typically involves hundreds of different manufacturing processes, raw materials, and process chemicals, and is often dispersed among a vast global network of suppliers.

Though the results are rather inconsistent, these disputes aroused more interests than ever and triggered researchers and organisations to attempt to find the answer. Even though no conclusions have been made, the scientific value of this topic to researchers as well as the economic interest to industries and governments have been well exhibited. Yi and Thomas have provided a relatively comprehensive review of research on the environmental impact of e-business and ICT till 2007 (Yi & Thomas, 2007), including a review of related papers and thesis published, projects, comparisons of methodologies, conferences and symposiums. In this article, some of the most important development were mentioned and assessed, for example:

- Jokinen et al. examined the relationship between the information society and sustainable development on theoretical and conceptual levels (Jokinen, Malaska, & Kaivo-oja, 1998).
- Miller and Wilsdon published a paper entitled "Digital Futures — An Agenda for a Sustainable Digital economy" in 2001. This paper was one of the major products of the Digital Futures project, one of the most significant projects in this research field. In this paper, findings were summarised in the form of ten headline principles - what they call the ten 'dot-commandments' (Miller & Wilsdon, 2001).
- Matthews et al. compared the environmental and economic performances of traditional retailing and e-commerce logistic networks in the United States and Japan, using book retailing as a case study (Matthews, Williams, Tagami, & Hendrickson, 2002).
- In 2002, Romm published a journal paper entitled "The Internet and the new energy economy" (Romm, 2002), based on work first published in 1999 in a working paper (Romm, Rosenfeld, & Herrmann, 1999). These two papers have been the most frequently quoted and referred papers in this field.

With a glance of these titles, one can see previous efforts have been made from quite different perspectives. It is not feasible to conduct a direct comparison of the results as past work have chosen very different scopes in assessing the problem. Researches range from relatively small-scale case studies to very large pan-European projects. The range of the relationship also spans from strictly e-commerce to all encompassing digital technologies and electronic business platforms, and from a sub-sector of the environment, e.g. energy prospect, to general economic, social and environmental sustainable development. Nevertheless, past efforts have made significant contribution to the possible quantification of the issue in the future and set the background for this chapter.

The complexity of this topic originated from the complication of the Internet structure itself as well as the difficulty to define a clear-cut boundary for the input-output (economic as well as environmental) of the life cycle analysis. One simple network comprises large number of telecommunication equipments and terminals, from hub, router, gateway, server to PCs and peripherals, while every stage of the supply chain could involve energy consumption and CO_2 emission, from production to waste disposal. Furthermore, expertise needed to understand this issue incorporates a range of fields including computing, economics and environment science. Therefore the multi-disciplinary nature of the problem has also made the situation more complex.

Along the life cycle of e-business/ICT related physical existence, the next section of this chapter will provide (i) a review of how relevant e-business/ICT physical elements contribute to energy use and carbon emission in different stages and (ii) why e-business/ICT has contributed to data sharing and transactions on energy derivatives particularly the establishment of carbon trading platform.

ICT RESULTED CARBON EMISSION

This section attempts to break down the Internet associated energy consumption into the following levels along the life cycle of e-business/ICT related physical existence:

- The first section is related to the production which includes energy used in the production of Internet equipment/hardware, e.g. manufacturing of semi-conductors, PCs.
- The second section discusses energy related to and carbon emission resulted from shipment and distribution of components and equipment, e.g. overseas outsourcing.
- The third section presents energy consumption by the usage of hardware.
- The fourth section explains energy and carbon emission related to Internet and e-business applications, e.g. energy saved/used during usage of tele-conferencing and home-working.
- The last section reviews energy consumption and carbon emission related to the final disposal of the hardware.

Production

Along the life cycle of e-business/ICT related physical existence, this section discusses energy used in the production of Internet equipment/hardware, e.g. manufacturing of semi-conductors, PCs.

Researchers have considered different units in studies of computer related resource and energy consumption during the manufacturing stage, with reference to different mixes of processes in different regions. In general, it is agreed that computer related electronics manufacturing is a complicated process involving hundreds of natural resources and chemicals, requiring enormous amount of energy. Some compare the making of a single computer chip to boiling 10 gallons of water to make a cup of tea, as the Silicon Valley Toxics Coalition (SVTC) described. For example, the annual materials usage of a typical semiconductor facility (a semiconductor is an electronic component that is essential in many electrical devices from computers to mobile phones) includes: 832 million cubic feet of bulk gases, 5.72 million cubic feet of hazardous gases, 591 million gallons of deionised water, 5.2 million pounds of chemicals, and 8.8 million kilowatt hours of electrical power (SVTC, 2006b). This energy intensive process also results in large amount of carbon emission.

In 2002 Williams et al studied energy associated with manufacturing a 2-gram microchip in Japan (Williams, Ayres, & Heller, 2002). They stated the fossil fuel and chemical inputs to produce such a chip is estimated at 1600g and secondary materials used in production total 630 times the mass of the final product. This is an order of magnitude much larger than the factor of 1-2 for many other manufactured goods, such as automobiles and refrigerators.

The European Commission study (Atlantic-Consulting, 1998) broke down the assessment to components of a PC – control unit, keyboard, monitor etc and presented the results within the geographic boundary of the European Union as in Table 1 (unit: g CO_2-eqv). These results are however inconsistent. Furthermore data cannot easily be compared to others. Also, what shouldn't be ignored is that a PC or a PC component is just one part of an overall network. And the Internet is a vast collection of different networks that use certain common protocols and provide certain common services (Tanenbaum, 2003). So first of all, what physical elements (components/hardware/devices) are associated with the Internet that are contributing to carbon emission during their lifecycle?

The interconnected networks transmit data packets through guided transmission media (fibre-optic cables, twisted-pair copper wire or coaxial cable), or unguided media (e.g. wireless LAN or a digital satellite channel) (Kurose & Ross, 2003). The Internet's end systems normally include desktop computers (PCs, Macs, and UNIX-based workstations) and mobile computers (portable computers, PDAs and mobile phones with wireless Internet connections), as well as indirect servers (web servers, e-mail servers etc). An increasing number of alternative devices such as thin clients and household appliances, web TVs, set top boxes and digital cameras are also being attached to the Internet as end systems. Furthermore, there are devices which connect network segments between different layers, including repeaters, hubs, bridges, switches, routers and gateways. These switching devices are in common use but they operate in different layers in the OSI (ISO Open Systems Interconnection) reference model (Tanenbaum, 2003).

Also, some computer peripherals are likely to be attached for certain tasks. For example, in relation to e-business for the travelling sector, say a customer buying an e-ticket online, normally would be required to print the reference number/e-ticket or to save the webpage in a mobile. Therefore a printer or a mobile phone is considered to be part of the network.

Although different types of computer networks (e.g. LAN, WAN, MAN…) are available, the hardware and equipment involved are mostly a combination of elements mentioned above, which have carbon emission associated with each individual lifecycle. Nevertheless, the scope of elements involved far exceeds one type of equipment, e.g. PCs.

According to the US census, along with around 26.9 million electronic computers produced in 2006 in the US (US-Census-Bureau, 2007a), over $18,562 million (value of shipments) worth of wireless voice and data networking equipments were also produced, including for example, over $1,081 million worth of enterprise routers (US-Census-Bureau, 2007b). Thus if all Internet related hardware is considered, the contribution to global warming by manufacturing and using these devices will be significantly increased from earlier studies.

Table 1. Emission of CO_2 equivalent from PC components

Components	Material production	Manufacture	Use	Distribution	Disposal	Total
Control unit	2.98E+04	5.32E+04	3.09E+02	1.68E+05	1.34E+03	2.49E+05
keyboard	3.21E+03	2.96E+03	4.35E+01	0.00E+00	6.07E+02	6.66E+03
monitor	1.64E+04	**7.69E+04**	**4.46E+02**	**2.80E+05**	**2.52E+03**	**3.75E+05**

Calculation of these figures would be a confusing and daunting task. It is possible to collect data from a macroeconomics point of view, though it will require data from different sectors including manufacturing of electronic computers, telecommunications equipment, semi-conductors, wiring devices and supplies etc. Nevertheless, when we sit in front of a PC either for personal or business purposes, we should bear in mind the long route taken by this unit and associated environmental impacts.

Shipment and Distribution of Components

Due to cost issues, technology availability and environmental regulations from different countries, e-business/ICT involved hardware are rarely locally manufactured. The material consumption and manufacturing process associated with the Internet hardware are rather energy intensive. On top of that, outsourcing and transportation of the hardware beyond geographical boundaries has made it more difficult to track the life cycle emissions. One PC could have components manufactured in hundreds of different locations and assembled elsewhere.

The reason for the migration and outsourcing of the hi-tech industry has gone beyond being a cost problem. After the discovery of massive groundwater contamination caused by toxic solvents from a computer chip factory in Silicon Valley of Santa Clara County, US in 1980s, hundreds of workers were found to develop cancer, reproductive problems, and other illnesses due to exposure to hazardous chemicals (SVTC, 2006c). The US Environmental Protection Agency had to deem 29 areas toxic Superfund sites, requiring priority environmental cleanup. As of 1996, 20 out of the 29 superfund sites were caused directly by the high-tech electronic industry and 5 were caused by related industries. Therefore, the electronic companies began outsourcing and moving their facilities overseas where labour is cheap and environmental protection laws are weak. "Today, behind any brand name electronic product a consumer buys off the shelf, there may have actually been 30 different companies involved with the making of that product" (SVTC, 2006c).

However, before considering the carbon emission involved in the transportation of computer related products, just how many computer related products are outsourced? Most well known electronic manufacturers such as Intel, AMD and Cisco are actually supported by many components makers worldwide. The quantities of outsourcing products can probably be reflected by the revenues of the overseas computer hardware markets. One of the most important outsourcing countries is undoubtedly China due to its cheap labour cost. China Development gateway, a Chinese NGO originally launched within the World Bank, stated that exports of computer products grew from $200 million in 1990 to $3.78 billion in 1995 (China-gateway, 2004). In 1990 the Chinese computer industry had 191 hardware manufacturers and a total workforce of 100,000 people. By the end of 1995, a total of over 300,000 people worked for 1,000 hardware manufacturers. China's computer industry grew in output value from $921 million in 1990 to $6.37 billion in 1995. When moving such a large quantity of products across the world, no matter what transportation means is used, related carbon emissions are undoubtedly going to be significant. In addition, the market has grown considerably since 1995.

The calculation of freight related emission would require standardisation of indicators, measurements and units. However, just to give an idea of the scale of the problem: flying a fully loaded freight A727 (capacity 19 tonnes) just over 300 kilometres (less than the distance from the Midlands to Brussels) or running an articulated lorry over 9,000 kilometres could generate 10 tonnes of CO_2 equivalent emissions (DEFRA, 2000). So for export from China alone, how much carbon emissions would be involved in moving $3.78 billion worth of computers across the globe every year?

It's nearly impossible to calculate the amount of carbon emission coupled with shipment and outsourcing of Internet components, nonetheless, the above will give some idea of the magnitude of the problem.

Furthermore, Internet supported electronic commerce has facilitated online transactions around the world. The fact that a consumer can order almost anything from anywhere in the world at any time, which will most probably be packed and delivered individually the next day, changes the traditional shipment pattern and correspondingly deteriorates the environment and accelerates carbon emission. This will be discussed further in the Applications section.

Usage of Hardware

Once the hardware are shipped and distributed through warehouses and retailers, the transportation process temporarily stops. The hardware finally reach consumers and are ready to be used. This section provides a simple equation on how to calculate energy consumption during usage of the hardware.

According to the UK's Department for Environment Food and Rural Affairs (DEFRA), powering 86 PCs for a year uses 23,000 kWh of electricity and could generate 10 tonnes of CO_2 equivalent emissions (DEFRA, 2000). To calculate the electricity use of a particular computer or equipment is theoretically straight forward – just multiply its power with the usage time and sum the results from all the units involved. However, the process becomes more complicated as different makes, brands, and modes of operations (e.g. stand by, power save mode…) are involved. LBNL set an equation to calculate the average annual energy used by each piece of equipment as (Kawamoto, et al., 2001):

$$UEC_i = (SPM_i \times (PA_i \times HA_i + PL_i \times HL_i + PO_i \times HO_i) / 7) \times 365/10^3 + ((1\text{-}SPM_i) \times (PA_i \times (HA_i + HL_i) + PO_i \times HO_i) / 7) \times 365/10^3 + EPC_i$$

where

UECi = Unit Energy Consumption for equipment type i (kWh/year)
i = index for equipment type
PAi = Average active mod e power for equipment type i (W)
PLi = Average low- power mode power for equipment type i (W)
POi = Average off mode power for equipment type i (W)
HAi = Hours of operation in active m ode for equipment type i (hours/week)
HLi = Hours of operation in low-power mode for equipment type i (hours/week)
HOi = Hours of operation in off mode for equipment type i (hours/week)
SPMi = Power-management-enabled rate for equipment type i (%)
EPCi = Extra energy for printing or copying for equipment type i (kWh/year)6
365 = days per year
7 = days per week

The results from LBNL are based on a few assumptions including: the average lifetime of all types of network equipment is 4 years; Cisco products are selected as the representative model.

With regards to whether energy use and other environmental impacts in the manufacturing phase of PC equipment outweigh the usage phase, there have been controversial statements made in the past. Williams et al have claimed that for energy, the production phase becomes more relevant than the use phase, a reversal of the situation for products such as automobiles and many household appliances (Williams, Ayres, & Heller, 2002). Williams (2004) stated life cycle energy use of a computer is dominated by production (81%) as opposed to operation (19%). However the European Commission study claims the characterised data for global warming shows that the largest environmental contributions of PC come from the use stage where they are caused by the electricity consumption (4.47E+5, unit: gCO_2-eqv), which is over three times of the second largest contributor - the manufacture stage (1.33E+5, unit: gCO_2-eqv) (Atlantic-Consulting, 1998).

Despite the disagreement among different studies, available research has mainly focused on the energy consumption of PC units, which are only one type of equipment running within the overall network. Other Internet equipment also consumes significant amounts of energy, e.g. the constant running of Internet routers, switches, servers etc. Unlike PCs, these equipments don't really have low-power, standby or power-saving mode. Due to their functions in the Internet structure, these equipments mostly have to be fully powered on 24/7, which means that the associated energy consumption is not negligible.

At the organisational level, it is relatively easy to calculate the electricity consumption and carbon emission associated with PC units. The figure could potential help an organisation's bottom line as well as its environmental performance.

Internet and E-Business Applications

The above section presented energy consumption related to the usage of e-business/ICT physical components. Calculation is rather straightforward. However unlike refrigerators or televisions, energy consumption of e-business/ICT equipments is not just self-resulted. Internet and e-business applications, for example, tele-conferencing, online shopping have significantly changed our life-style – in many ways have contributed to considerable energy saving and reduced carbon emissions compared with traditional life-style. Differentiated from previous section where the focus was on the primary impacts from e-business/ICT involved physical components, this section mainly focuses on the secondary effects created by the ongoing use and application of e-business/ICT.

Some claim that increasingly complex products, such as microchips, require additional secondary materials and energy to realise their lower entropy form, which offsets their dematerialisation contribution – the secondary materialisation proposition (Williams, Ayres, & Heller, 2002). However many still believe that Internet supported applications have been making a positive contribution to resources and energy saving in many areas and consequently to carbon reduction.

For example, the substitution of air travel with tele-conferencing has significantly reduced transport related carbon emissions. DEFRA stated that flying 14 international long haul trips or 110 short haul trips, or travelling 165,000 km on a train could generate 10 tonnes of CO_2 equivalent emissions (DEFRA, 2000). According to the Financial Times, the average transatlantic business flight uses 80,000 to 100,000 pounds of fossil fuel, which could easily be avoided by the use of teleconferencing (Cohen, 2000). Compared with these figures, the electricity consumption of tele-conferencing equipment and associated carbon emission are almost negligible.

Many companies have also started implementing Intranet based home-working schemes, which significantly reduce office space and related energy consumption. The construction, heating and lighting of buildings in the UK accounts for half of all energy consumed, half of carbon emissions and 40% of raw materials. But office space is only actually used for 22% of the average week (Goodman, Alakeson, & Jorgensen, 2004). Many organisations have seen huge savings through tele-working, e.g. British Telecom's flexi-working programme has enabled a 50% reduction in floor space since the mid 1990s; Sun Microsystems' iWork has reduced office space needs by 25% in four years (Goodman, Alakeson, & Jorgensen, 2004); At&T estimates a US $25 million per year saving in real estate through its virtual office programme (Allenby, 2001).

E-commerce related Internet shopping has also replaced physical shopping malls to a certain degree. In contrast to traditional business, some pure online businesses offer even more potential to eliminate retail space completely. C2C (consumer to consumer) e-commerce, for instance EBay, serving over 100 million users with more than 1.4 million items up for auction each day, is maintaining a virtual website rather than a huge retail store. Romm et al predicted the Internet economy could also render unnecessary as much as 3 billion square feet of buildings – some 5% of the US commercial floor space – which would save a considerable amount of construction-related energy (Romm, Rosenfeld, & Herrmann, 1999).

Other Internet facilitated applications, for instance, e-procurement has saved BT 5.8 tons of paper from 100+ page catalogues that were previously circulated and 1 ton of paper from 24,500 order pads that were previously printed (BT, 2003). Furthermore, the internet has made virtualisation of products and digitisation of information possible. Taking the music industry as an example, physical products such as CDs could easily be replaced with musical files downloadable from the Internet. Production as well as related warehouse, transport, etc. could be eliminated. The digitisation of standard business transactions is also changing office paper use (Yi & Thomas, 2006). Transformation of the earlier physical form of these products to virtual format has obviously made a saving in energy across their life-cycles.

Though on the negative side, the "Harry Potter" event – one of the milestone in the e-business history – was strongly criticised because of its environmental impact. On a single Saturday in July 2000, 100 airplanes and 9,000 trucks delivered more than 250,000 copies of 'Harry Potter and the Goblet of Fire' to Amazon.com customers all over the USA. This event had set a record for the online provision of goods in volume, but it probably also set a record for the quantities of empty shipping boxes and packaging that would wind up in landfills, let alone the energy used, transport pollution and carbon emission caused in the deal (Matthews, Hendrickson, & Lave, 2001).

The net impact on energy during the Internet application phase is yet to be further explored. However, in general the positive aspects are believed to outweigh the rebound effects. Realising and analysing each possible detailed aspect, where and how much energy is associated with, is a start.

Disposals of Hardware

During the life cycle of e-business/ICT hardware, the last stage would be the final disposal. Unlike other consumer electronic products such as refrigerators, consumers couldn't be busier throwing away outdated PCs and replacing them with newer versions. The increasingly shorter life cycle of these PCs and other related products (mobile phones, PDAs etc) has induced a unique market with unprecedented amount of demand, compared with products involved in previous major revolutions such as radios and TVs. The enormous appetite for PCs inevitably leads to a high PC disposal rates: for example, an estimated 10,000 PCs become obsolete in California alone each day (CAW, 2004; Masanet, Price, Can, Brown, & Worrell, 2005). On the other hand, despite the massive resources and energy used in manufacturing a relatively small-sized item, the hi-tech industry has designed the computers in such a way that it is more difficult for customers to upgrade than to replace the whole unit.

The EC Directive on Waste Electrical and Electronic Equipment (WEEE) (European-Commission, 2000), was agreed upon in January 2003 and entered into force in January 2007. The Directive aims to minimise the impact of electrical and electronic goods on the environment by increasing re-use and recycling, hence reducing the amount of WEEE going to landfill. IT and telecommunications equipment is among the 10 categories listed and accounts for a considerable portion of the total. WEEE now constitutes one of the fastest growing waste streams in the EU, with current estimates indicating a rate of growth three times that of average municipal waste.

In addition, the process of incineration or recycling of WEEE is not carbon emission free, especially taking into account this amount of WEEE produced. In a report produced by Earthworks and Oxfam America, Farrell et al. stated that PCs alone contain a medley of metals, including gold, silver, aluminium, lead, copper, iron, zinc, and tin. Many of these materials could be salvaged at the end of the computer's life and recycled. But currently, most discarded computers are dumped in landfills or incinerated. Incineration of e-waste releases heavy metals and dioxin into the atmosphere. Some of the larger and older smelters have done extensive ecological damage. Worldwide, smelting adds about 142 million tons of sulphur dioxides to the atmosphere every year, which is 13 percent of total global emissions. And since smelters burn huge amounts of fuel, they also release substantial quantities of GHG, such as CO_2 and perfluorocarbons (Farrell, Sampat, Sarin, & Slack, 2004).

Furthermore, much of the e-waste produced in developed countries is actually shipped to developing countries to recycle. It is estimated that 50% to 80% of the "made-in-USA" e-waste is not recycled domestically but placed on container ships bound for destinations like China due to cheaper labour and lack of environmental standards in these countries (Puckett, et al., 2002). The process of transfer and transportation will cause another round of carbon emission.

CARBON TRADING PLATFORM

The previous section has presented the carbon emissions that Internet and related hardware manufacturing and usage have caused. On the other side, the Internet has enabled effective data sharing and ICT have been making positive contributions on energy issues in various industries. For example:

- The design of intelligent energy efficient products and systems has been facilitated, e.g. intelligent transport systems, environmental management systems, and energy management in housing.
- It can be conducted more effectively to monitor energy flow and other environmental indicators in the supply chain process: from production, distribution to disposal.
- Tools, software/application and platforms for energy trading, especially across countries, have been provided.

In particular, digital technologies supported carbon trading platforms have also created new markets online and is becoming an emerging and fast growing global phenomenon.

Background

Driven by the Kyoto Protocol signed in 1997, governments have committed to reduce carbon dioxide emissions by 20% by 2020. Carbon dioxide is one of the major GHG - the commonly known causes of global warming. Carbon emission has stretched its impact on our economies beyond being a simple environmental issue. On 30 October 2006, the UK government published a 579-page report on the economics of climate change by Sir Nicholas Stern. According to Stern (Stern, 2006), the overall costs and risks of climate change will be equivalent to losing 5% to 20% of GDP. In contrast, the costs of action – reducing GHG to avoid the worst impacts of climate change – can be limited to around 1% of global GDP each year. With no action, each tonne of carbon dioxide we emit will cause at least $85 (£45) of damage.

The British Government has introduced a tax on industrial energy as well as a cap and trade scheme. Companies are granted with an annual allowance of CO_2 emission (the cap). Whoever exceeds the cap will be fined; alternatively the company can buy allowances from other companies who have surplus. In short, industry societies now have to pay for their pollution, e.g. the price of 2005 CO_2 allowance rose to a record of €29.50 a ton on July 7th in the European CO_2 rights market (Levy, 2005). Due to immaturity of the system and policy, the price has fluctuated significantly during the first phase (2005-2007) as well as the recent recession.

The Market and Concept

Carbon is taken by the financial professionals as a form of energy derivatives, which is not a new concept. However, carbon does bear certain unique characters. Since European Commission and national governments still allocate free allowances to industries, only companies that exceed this free allowance needs to buy credits. The accuracy of this initial allocation has significant impact on the subsequent price. The price of carbon also fluctuates under influences of politics, country policies, other energy markets and so on. There are increasing number of regional, national and international systems for trading and transfer of GHG allowances and emission reduction credits. Hasselknippe reviewed a total of 47 systems for GHG trading and transfer included in the International Emissions Trading Association (IETA) Trading Schemes Database (Hasselknippe, 2003).

Various electronic platforms established domestically and internationally include the Climex Alliance (a pan-European market for trading Carbon Emissions Certificates within the EU's Emissions Trading Scheme), the Intercontinental Exchange (ICE), SENDECO2 (the Spanish CO_2 exchange expanding to Italy, Greece and Portugal) etc. The European Union Emission Trading Scheme (EU ETS), the largest multi-national GHG emissions trading scheme in the world and the main pillar of EU climate policy, commenced operation on 1st January 2005. All 27 member states of the European Union participated. In its first year, 362 million tonnes of CO_2 were traded on the market for a sum of €7.2 billion.

The international carbon market in 2005 transacted a total of 799 million tonnes of CO_2 equivalents worth approximately €9.4 billion (Point-Carbon, 2006). The carbon market grew in value to an estimated US$21.5 billion in the first three quarters of year 2006, more than doubling in value over the previous year (Capoor & Ambrosi, 2006). For such a large market to survive and the second phase (2008 ~ 2012) of the ETS scheme to proceed, technological support by ICT will play an important and growing role. Therefore how to integrate all these different systems presents a challenge. The communication and harmonisation of the systems as well as security control, can only be implemented probably with the support of the Internet and other Information and Communication Technologies.

Reduction of carbon emission has become an unavoidable issue for industries and governments. Carbon trading has been developed beyond the single purpose of protecting the environment with the involvement of international financial industry. With more participants and engagements, the mechanism and systems have become more and more complicated. Some even expect the system will further expand from energy industry to cover local government, retail sector, or even individual citizens. One day a carbon credit card for individuals to pay for electricity, gas and petrol might become reality (Sustainable-Development-Commision, 2006). No matter the future of the trading is a global central exchange or over-the-counter mechanism, it wouldn't succeed without the support of ICT – which helps to overcome the barrier of geographical boundaries and the challenge of large amount of data flowing around the globe.

The Case of EU

PriceWaterHouseCooper has reported the state of art of potential expanded use of IT in EU-ETS (PriceWaterHouseCoopers, 2006). It summarised that in the current EU-ETS, IT is restricted to support monitoring, reporting and registry activities. Verifiers and competent authorities use non-standardised software for internal purposes only. For example in the case of the UK, in the Verification Reference Model of EU ETS, IT is used as (i) a template for reporting, (ii) a template for verifiers and (iii) the home developed registry. No software application is currently used in the monitoring stage and the electronic media for reporting is not currently web enabled. The vision of British regulators is to have a web enabled reporting system with different levels of access for regulators, operators, verifiers and the public and a standardised system to be shared across the EU.

The national registry in the UK is accessed via the Internet and has been developed using DEFRA's technical design standards in Microsoft .NET 1.1 and SQL Server. By using web services the registry is able to communicate with the EU Commission's Community Independent Transaction Log (CITL). Currently the communication is secure and processed as real time transactions. The UN Data Exchange Standards (DES) specifies the use of TCP/IP connections using encrypted messages over the Internet in order to protect communications from modification or interception in transit (PriceWaterHouseCoopers, 2006).

During the trading phase, DEFRA (UK) and the French Registry developers, Caisse des ustus et consignations (CDC), are working together to enable trading platforms to interface with the registry systems. DEFRA hopes that the creation of international data exchange standards will facilitate the development of electronic interfaces by end users allowing trading platforms and electronic accounting systems to communicate automatically with Registries (PriceWaterHouseCoopers, 2006).

In order to accommodate growing usage and future demands, transparency, reliability, speed, capacity, adaptabilities, upgradeability and security of these online platforms are critical for the trading to operate effectively and efficiently. Development of a sound IT infrastructure and harmonised interfacing between different systems will be fundamental to the overall success of the programme.

CONCLUSION

ICT and related Internet usages have made profound effects on our economic lives, social development and environment. Research on the environmental impacts of the ICT is relatively new. This chapter attempted to explore the interaction between Internet technologies, energy consumption and carbon emission. The objective is to give a detailed, however not exhaustive, summary of various stages of how the Internet and e-business contribute to the energy issues. Private and public organisations can therefore more clearly understand the environmental impact of their usage of related technologies in their business.

As expected, research that has made a significant difference is limited and is far from finding any definitive conclusions or solutions. Also the depth of research is not enough. However on the positive side, significant interest has been provoked. Awareness of issues involved has also been raised not only with academics but also with policy makers and industries. Some results have paved the way for future research in this field.

With reference to the manufacturing of Internet hardware and equipment, past research has considered different system boundaries with different mixes of processes in different regions. Nevertheless it is agreed that computer related electronics manufacturing is a complicated but clearly energy intensive process. Outsourcing and transportation beyond geographical boundary has made it difficult to track the life cycle of Internet hardware. The usage stage is claimed to involve large amounts of energy due to the constant running of networking hardware but relatively easy for individuals to calculate. Fortunately subsequent Internet supported applications have been making a positive contribution to resources and energy saving in many areas and consequently to carbon reduction. These include home-working, tele-conferencing and dematerialisation of physical products. Finally, the creation of e-waste and its transfer to developing countries has however caused another round of negative ecological impacts.

Additionally, the chapter has also looked at the importance of ICT in facilitating the emerging carbon trading systems. In order to accommodate growing usage and future demand and for the trading to operate effectively and efficiently, the systems have to be transparent, reliable, adaptable, upgradeable, and secure. These critical requirements will all largely depend on advanced ICT and web-enabled technologies.

Understanding of the relationship between the Internet and carbon emission has demonstrated substantial impact on environmental monitoring and protection. Quantitative standardised methodologies are currently under development and will be beneficial in realising the full picture. Nonetheless, presenting a full picture of the interaction hopefully can help both private and public organisations see the facts behind the scene, and therefore manage their purchase of IT equipments and integrate IT into business more sensibly.

SOLUTIONS AND RECOMMENDATIONS

This chapter essentially tries to provide a picture of and reference on how ICT and e-business interacts with the environment, particularly the carbon and energy side along the life cycle of the ICT physical elements. A literature review and survey of the current status is expected to facilitate future research and further understanding.

It is recommended that private and public organisations devote appropriate amount of time and resources to measure the environmental impact such as GHG emissions associated with their general business operations – including ICT and e-business usage. Many industry sectors including mining, utilities, and infrastructure have seen the business value and opportunities of conducting business in an environmentally sound and socially responsible manner.

Organisations are suggested to take on the positive contribution from ICT sector and e-business platforms and to carry out business more sustainably. This includes not only optimising internal operations (such as management of electricity, water, gas use, transportation and waste), but also making the choice of having more sustainable partners. Examples of achieving a greener business might start with a business manager taking a video-conferencing instead of flying across Atlantic, or simply mean reusing or recycling an old PC. Different organisation/company may have different operational strategies, but raised awareness and taking account the environmental factor into business decisions, will only help the planet as well as individual's bottom line.

FUTURE RESEARCH DIRECTIONS

Current dominant approach of analysing the problem is either a micro-level case study approach or a macro-level statistical approach. Traditional assessment approaches are found to be insufficient to accommodate the digital technology revolution and cannot accommodate the challenge of measuring the impacts of ICT on environmental sustainability. New innovative methods need to be created to fill this gap. It is suggested that a more predictive and empirical model, which can be applied within a sector of society, should be more beneficial in the long term. This approach should help simulate potential impacts resulting from changes of indicators, so that positive effects can be promoted and negative ones alleviated proactively, rather than knowing and accepting outcomes passively (Yi & Thomas, 2007).

The challenge of any research does not stop at recognising the problem, but to know: how to solve the problem, alternative solutions with corresponding risks and benefits, and how to choose the best solution. It is not enough to know ICT/e-business has been changing our daily life, economy, transport, air, water, forests, etc., it is not even enough to understand how it is changing everything, ultimately an approach which can influence the behaviour, say of a private organisation for example, is needed.

REFERENCES

Atlantic-Consulting. (1998). *LCA study of the product group personal computers in the EU Eco-label scheme.* Retrieved from http://ec.europa.eu/environment/ecolabel/pdf/personal_computers/lcastudy_pc_1998.pdf

BT. (2003). *E-business and the environment. Better world–our commitment to society*. Report retrieved from www.groupbt.com/betterworld

Capoor, K., & Ambrosi, P. (2006). *State And trends of the carbon market, 2006*. Washington, DC: IETA and the World Bank.

CAW. (2004). *E-waste background. Californians Against Waste.* Retrieved from www.cawrecycles.org/Ewaste/background%20e-waste.html

Chinagateway. (2004). *China's electronics industry.* PRC Ministry of Posts and Telecommunications. Retrieved from http://en.chinagate.com.cn/english/1474.htm

Cohen, A. (2000). Catching the green express from euston. *Financial Times*.

DEFRA. (2000). *Environmental reporting–guidelines for company reporting on greenhouse gas emissions*. Retrieved from http://www.defra.gov.uk/environment/business/reporting/pdf/envrpgas-annexes.pdf

EUF. (2004). Information *Technology and resource use, energy end-use forecasting at Lawrence Berkeley National Laboratory*. Retrieved from http://enduse.lbl.gov/Projects/InfoTech.html

European-Commission. (2000). *Directive on waste electrical and electronic equipment and directive on the restriction of the use of certain hazardous substances in electrical and electronic equipment.* Explanatory memorandum. Brussels.

Farrell, L., Sampat, P., Sarin, R., & Slack, K. (2004). *Dirty metals-mining, communities and the environment*. Earthworks and Oxfam America.

Goodman, J., Alakeson, V. & Jorgensen, B. (2004). *Encouraging green telework: Forum for the future*.

Hasselknippe, H. (2003). Systems for carbon trading: An overview. *Climate Policy*, *3*(2), S43–S57. doi:10.1016/j.clipol.2003.09.014

Jokinen, P., Malaska, P., & Kaivo-oja, J. (1998). The environment in an information society: A transition stage towards more sustainable development? *Futures*, *30*(6), 485–498. doi:10.1016/S0016-3287(98)00054-8

Kawamoto, K., Koomey, J. G., Nordman, B., Brown, R. E., Piette, M. A., & Ting, M. (2001). *Electricity used by office equipment and network equipment in the U.S.: Detailed report and appendices*. Orlando: Lawrence Berkeley National Laboratory.

Kurose, J. F., & Ross, K. W. (2003). *Computer networking: A top-down approach featuring the Internet* (2nd ed.). Pearson Education, Inc.

Levy, A. (2005). *Making pollution pay*. Bloomberg Markets.

Masanet, E., Price, L., Can, S. R., Brown, R., & Worrell, E. (2005). *Optimization of product life cycles to reduce greenhouse gas emissions in California*. California Energy Commission, Public Interest Energy Research Program.

Matthews, H. S., Hendrickson, C. T., & Lave, L. (2001). How much did Harry Potter cost? E-commerce is growing and environmentalists should be happy. Or should they? Retrieved from http://www.oecdobserver.org/news/fullstory.php/aid/415/How_much_did_Harry_Potter_cost_.html

Matthews, H. S., Williams, E., Tagami, T., & Hendrickson, C. T. (2002). Energy implications of online book retailing in the United States and Japan. *Environmental Impact Assessment Review, 22*(5), 493–507. doi:10.1016/S0195-9255(02)00024-0

Miller, P., & Wilsdon, J. (2001). Digital futures-an agenda for a sustainable digital economy. *Corporate Environmental Strategy, 8*(3), 275–280. doi:10.1016/S1066-7938(01)00116-6

Mills, M. P. (1999). *The Internet begins with coal.* Greening Earth Society.

Point-Carbon. (2006). [*-towards a truly global market.*]. *Carbon*, 2006.

PriceWaterHouseCooper. (2006). *Information Technology enhances consistency, transparancy efficiency in EU-ETS.*

Puckett, J., Byster, L., Westervelt, S., Gutierrez, R., Davis, S., & Hussain, A. (2002). *Exporting harm-the high-tech trashing of Asia. The Basel Action Network (BAN), Silicon Valley Toxics Coalition.* SVTC.

Romm, J. (2002). The Internet and the new energy economy. *Resources, Conservation and Recycling, 36*(3), 197–210. doi:10.1016/S0921-3449(02)00084-8

Romm, J., Rosenfeld, A., & Herrmann, S. (1999). *The Internet economy and global warming: A scenario of the impact of e-commerce on energy and the environment. Center for Energy and Climate Solutions (CECS), Global Environment & Technology Foundation.* GETF.

Stern, N. (2006). *The economics of climate change: The Stern Review.* Retrieved from http://www.hm-treasury.gov.uk/independent_reviews/stern_review_economics_climate_change/stern_review_report.cfm

Sustainable-Development-Commision. (2006). Carbon trading for individuals? Retrieved from http://www.sd-commission.org.uk/pages/carbon-trading.html

SVTC. (2006a). Electronics industry-a dazzling industry with a dark side. Retrieved February 15, 2010, from http://svtc.etoxics.org/site/PageServer?pagename=svtc_electronic_industry_overview

SVTC. (2006b). Electronics manufacturing-a deadly process requiring thousands of toxic chemicals Retrieved February 15, 2010, from http://svtc.etoxics.org/site/PageServer?pagename=svtc_electronics_production

SVTC. (2006c). High-tech history: Chronicling the evolution of the high-tech industry. Retrieved February 15, 2010, from http://svtc.etoxics.org/site/PageServer?pagename=svtc_hitechhistory

Tanenbaum, A. S. (2003). *Computer networks* (4th ed.). New Jersey: Pearson Education, Inc.

US-Census-Bureau. (2007a). *Computers and peripheral equipment: 2006 summary (Current industrial reports).* Retrieved from http://www.census.gov/

US-Census-Bureau. (2007b). *Telecommunications: 2006 (Current industrial reports).* Retrieved from http://www.census.gov/

Williams, E. (2004). Energy intensity of computer manufacturing: Hybrid analysis combining process and economic input-output methods. *Environmental Science & Technology, 28*(22), 6166–6174. doi:10.1021/es035152j

Williams, E. D., Ayres, R. U., & Heller, M. (2002). The 1.7 kilogram microchip and material use in the production of semiconductor devices. *Environmental Science & Technology, 36*(24), 5504–5510. doi:10.1021/es025643o

Yi, L., & Thomas, H. R. (2006). E-business and sustainable development. *International Journal of Environment and Sustainable Development*, *5*(3), 262–274. doi:10.1504/IJESD.2006.010896

Yi, L., & Thomas, H. R. (2007). A review of research on the environmental impact of e-business and ICT. *Journal of Environment International*, *33*(6), 841–849. doi:10.1016/j.envint.2007.03.015

ADDITIONAL READING

Abukhader, S. M., & JoÈnson, G. (2003). The environmental implications of electronic commerce: a critical review and framework for future investigation. *Management of Environmental Quality: An International Journal*, *14*(4), 460–476. doi:10.1108/147778303104886685

Arnfalk, P. (2004). The future impact of ICT on environmental sustainability: Evaluation and Recommendations (Fifth Interim Report) Lund: Institute for Prospective Technological Studies (IPTS).

Atkyns, R., Blazek, M., & Roitz, J., & AT&T. (2002). Measurement of environmental impacts of telework adoption amidst change in complex organizations: AT&T survey methodology and results. *Resources, Conservation and Recycling*, *36*, 267–285. doi:10.1016/S0921-3449(02)00082-4

Berkhout, F., & Hertin, J. (2004). De-materialising and re-materialising: digital technologies and the environment. *Futures*, *36*(8), 903–920. doi:10.1016/j.futures.2004.01.003

Caudill, R., Luo, Y., Wirojanagud, P., & Zhou, M. (2001). A life cycle environmental study of the impact of e-commerce on electronic products. Paper presented at the Proceedings of the Institute of Electrical and Electronics Engineering (IIIEE) International Symposium on Electronics and the Environment, USA.

Durieu, X. (2003). How Europe's retail sector helps promote sustainable production and consumption. *UNEP Industry and Environment*, *26*(1), 7–9.

Editorial. (2003). Shopping for a better world: sustainability and retailing. UNEP Industry and Environment, 26(1), 3-6.

Erdmann, L., & Behrendt, S. (2003). The future impact of ICT on environmental sustainability: Script (Second Interim Report). Berlin: Institute for Prospective Technological Studies (IPTS).

Erdmann, L., & Würtenberger, F. (2003). The future impact of ICT on environmental sustainability: Identification and global description of economic sectors (First Interim Report). Berlin: Institute for Prospective Technological Studies (IPTS).

Fichter, K. (2003). E-commerce - Sorting Out the Enviornmental Consequences. *Journal of Industrial Ecology*, *6*(2), 25–41. doi:10.1162/108819802763471762

Goodman, J., & Alakeson, V. (2003). The future impact of ICT on environmental sustainability: Scenarios (Third Interim Report). London: Institute for Prospective Technological Studies (IPTS).

Hesse, M. (2002). Shipping news: the implications of electronic commerce for logistics and freight transport. *Resources, Conservation and Recycling*, *36*(3), 211–240. doi:10.1016/S0921-3449(02)00083-6

Hilty, L. M., Arnfalk, P., Erdmann, L., Goodman, J., Lehmann, M., & Wager, P. A. (2006). The relevance of information and communication technologies for environmental sustainability - A prospective simulation study. *Environmental Modelling & Software*, *21*, 1618–1629. doi:10.1016/j.envsoft.2006.05.007

Hurst, E. (2001). E-commerce and the Environment. Unpublished MSc Thesis, University of London, London.

John, A. (2003). Information Technology and U.S. Energy Consumption: Energy Hog, Productivity Tool, or Both? *Journal of Industrial Ecology*, *6*(2), 13–24.

Macauley, M., Palmer, K., & Shih, J.-S. (2003). Dealing with electronic waste: modeling the costs and environmental benefits of computer monitor disposal. *Journal of Environmental Management*, *68*(1), 13–22.

Matthews, H. S., Hendrickson, C. T., & Lave, L. (2001). How much did Harry Potter cost? - E-commerce is growing and environmentalists should be happy. Or should they? [Electronic Version]. Retrieved 15/02/2010 from http://www.oecdobserver.org/news/fullstory.php/aid/415/How_much_did_Harry_Potter_cost_.html.

Matthews, H. S., Williams, E., Tagami, T., & Hendrickson, C. T. (2002). Energy implications of online book retailing in the United States and Japan. *Environmental Impact Assessment Review*, *22*(5), 493–507. doi:10.1016/S0195-9255(02)00024-0

Miller, P., & Wilsdon, J. (2001). Digital Futures - An Agenda for a Sustainable Digital Economy. *Corporate Environmental Strategy*, *8*(3), 275–280. doi:10.1016/S1066-7938(01)00116-6

Mills, M. P. (1999). The Internet Begins with Coal: Greening Earth Society.

Park, J., & Roome, N. (Eds.). (2002). The Ecology of the New Economy: Sustainable Transformation of Global Information, Communications and Electronics Industries: Greenleaf Publishing.

Plepys, A. (2002). The grey side of ICT. *Environmental Impact Assessment Review*, *22*, 509–523. doi:10.1016/S0195-9255(02)00025-2

Point-Carbon. (2006). [- towards a truly global market.]. *Carbon*, 2006.

PriceWaterHouseCoopers. (2006). Information Technology enhances consistency, transparancy efficiency in EU-ETS.

Reijnders, L., & Hoogeveen, M. J. (2001). Energy effects associated with e-commerce: A case-study concerning online sales of personal computers in The Netherlands. *Journal of Environmental Management*, *62*, 317–321. doi:10.1006/jema.2001.0440

Rejeski, D. (2003). E-Commerce, the Internet, and the Environment. *Journal of Industrial Ecology*, *6*(2), 1–3. doi:10.1162/108819802763471717

Romm, J. (2002). The Internet and the new energy economy. *Resources, Conservation and Recycling*, *36*(3), 197–210. doi:10.1016/S0921-3449(02)00084-8

Romm, J., Rosenfeld, A., & Herrmann, S. (1999). *The Internet Economy and Global Warming: A Scenario of the Impact of E-commerce on Energy and the Environment Center for Energy and Climate Solutions (CECS), Global Environment & Technology Foundation*. GETF.

Roome, N., & Park, J. (2000). Global Sustainability and Information Economy: Old Challenges, New Perspectives. *Greener Management International*, *32*, 24–32.

Siikavirta, H., Punakivi, M., Kärkkäinen, M., & Linnanen, L. (2003). Effects of E-Commerce on Greenhouse Gas Emissions: A Case Study of Grocery Home Delivery in Finland. *Journal of Industrial Ecology*, *6*(2), 83–97. doi:10.1162/108819802763471807

Sioshansi, F. (2000). E-commerce and the Energy Sector: The Pioneers May Not Get It Right; The Procrastinators Are Likely to Become History. *The Electricity Journal*, *13*(5), 42–49. doi:10.1016/S1040-6190(00)00117-2

Stern, N. (2006). The Economics of Climate Change: The Stern Review. Retrieved. from http://www.hm-treasury.gov.uk/independent_reviews/stern_review_economics_climate_change/stern_review_report.cfm.

Sui, D. Z., & Rejeski, D. W. (2002). Environmental Impacts of the Emerging Digital Economy: The E-for-Environment E-Commerce? *Environmental Management*, *29*(2), 155–163. doi:10.1007/s00267-001-0027-X

SVTC. (2003). Just say no to e-waste: background document on hazards and waste from omputers [Electronic Version]. Retrieved 15/02/2010 from http://www.svtc.org/cleancc/pubs/sayno.htm.

SVTC. (2006a). Electronics Industry - A Dazzling Industry With a Dark Side. Retrieved 15/02/2010, 2010, from http://svtc.etoxics.org/site/PageServer?pagename=svtc_electronic_industry_overview

SVTC. (2006b). Electronics Lifecycle - A Wake of Unintended Collateral Damage from Cradle to Coffin Retrieved 15/02/2010, 2010, from http://svtc.etoxics.org/site/PageServer?pagename=svtc_lifecycle_analysis

SVTC. (2006c). Electronics Manufacturing - A Deadly Process Requiring Thousands of Toxic Chemicals Retrieved 15/02/2010, 2010, from http://svtc.etoxics.org/site/PageServer?pagename=svtc_electronics_production

SVTC. (2006d). High-tech History: Chronicling the Evolution of the High-Tech Industry Retrieved 15/02/2010, 2010, from http://svtc.etoxics.org/site/PageServer?pagename=svtc_hitechhistory

Tongia, R., Subrahmanian, E., & Arunachalam, V. S. (2004). *Information and Communications Technology for Sustainable Development: Defining a Global Research Agenda (No. 81-7764-839-X)*. Bangalore, India: Carnegie Mellon University, Pittersburgh & Indian Institute of Science, Banglore.

Tongia, R., Subrahmanian, E., & Arunachalam, V. S. (2005). *Information and Communications Technology for Sustainable Development: Defining a Global Research Agenda*. Bangalore: Allied Publishers.

Velásquez, M., Ahmad, A.-R., & Michael, B. (2009). State-of-the-Art in E-Commerce Carbon Footprinting. *Journal of Internet Banking and Commerce*, *14*(3), 1–21.

Williams, E. (2004). Energy intensity of computer manufacturing: hybrid analysis combining process and economic input-output methods. *Environmental Science & Technology*, *28*(22), 6166–6174. doi:10.1021/es035152j

Wilsdon, J. (Ed.). (2001). *Digital Futures: Living in a Dot-Com World*. London: Earthscan.

Wilsdon, J., & Miller, P. (2001). Digital Futures: an agenda for a sustainable digital economy: Forum for the Future.

Yi, L., & Thomas, H. R. (2006). E-business and Sustainable Development. *International Journal of Environment and Sustainable Development*, *5*(3), 262–274. doi:10.1504/IJESD.2006.010896

Yi, L., & Thomas, H. R. (2007). A review of research on the environmental impact of e-business and ICT. *Journal of Environment International*, *33*(6), 841–849. doi:10.1016/j.envint.2007.03.015

Yi, L., & Thomas, H. R. (2009). A Decision Support System for the Environmental Impact of ICT and E-business. [IJITDM]. *International Journal of Information Technology & Decision Making*, *8*(2). doi:10.1142/S0219622009003387

KEY TERMS AND DEFINITIONS

Carbon Market: Refers to the Emission Trading market, is one of the three mechanisms designed in the Kyoto Protocol. The other two are namely Clean Development Mechanism and Joint

Implementation. The term is also called the cap and trade market, since normally a government sets the limit / cap on the amount of Greenhouse Gas (GHG) that can be emitted. Essentially it is an approach designed to control GHG emission through economic mechanism. Companies are issued with specific permit to emit certain amount of GHG. For those who exceeds the allowance, they either have to pay a fine, or they can buy the allowances from another company who emits less than its assigned allowances. The transfer of the allowances therefore becomes a trade. Transfers and acquisitions of these allowances are tracked and recorded through the registry systems under the Kyoto Protocol. An international transaction log ensures secure transfer of emission reduction units between countries.

DEFRA: The UK government department responsible for policy and regulations on the environment, food and rural affairs.

E-Business: Abbreviated from "Electronic Business", is the application of Information and Communication Technologies in any essential activities of a business throughout its (external and internal) supply chain. The term can be confused with the other concept "e-commerce", which is a subset of e-business and normally only includes conducting business via online platforms.

Greenhouse Gases: The atmospheric gases that contribute to the greenhouse effect, including water vapor, carbon dioxide, nitrous oxide, methane etc. These gases are believed to be able to trap heat therefore affect the temperature of the earth.

ICT: Short for Information and Communication Technologies, is a term used to encompass computing, telecommunication, digital technologies that intend to facilitate the processing, analysing, transfering, exchanging, distribution of information.

Kyoto Protocol: Is an international agreement linked to the United Nations Framework Convention on Climate Change. The Protocol was initially adopted on 11 December 1997 in Kyoto, Japan and entered into force on 16 February 2005. The major feature of the Kyoto Protocol is that it sets binding targets for 37 industrialised countries and the European community for reducing Greenhouse gas (GHG) emissions.

OSI: The Open System Interconnection Reference Model is an industry wide protocol and international agreement for layered communications and computer network. It divides network architecture into seven layers as: the Application, Presentation, Session, Transport, Network, Data Link, and Physical Layers. It is therefore often referred to as the OSI Seven Layer Model.

Peripheral: Is a device that attached to a host computer. It normally does not constitute as the core component but it extends the functions of the main computer. Examples include printer, scanner, speaker, webcam.

Superfund Sites: According to United States Environment Protection Agency (EPA), a Superfund site is an uncontrolled or abandoned place where hazardous waste is located, possibly affecting local ecosystems or people. It is also the name of the fund established by the Comprehensive Environmental Response, Compensation and Liability Act of 1980, as amended (CERCLA statute, CERCLA overview). This law was enacted in the wake of the discovery of toxic waste dumps such as Love Canal and Times Beach in the 1970s. It allows the EPA to clean up such sites and to compel responsible parties to perform cleanups or reimburse the government for EPA-lead cleanups.

SVTC: Created in 1982, Silicon Valley Toxics Coalition is a non-profit organisation engaged in research, advocacy and grassroots organising to promote human health and environmental justice in response to the rapid growth of the high-tech industry. It is based in San Jose, California, the United States.

Chapter 14
E-Business: Definition and Characteristics

Sharon Nachtigal
Royal Holloway, University of London, UK

ABSTRACT

The goal of this chapter is to describe and characterise the domain that is the main focus of this book, or the e-business mode of doing business. In order to be able to discuss e-business and its impact on the modern business and social environments, a working definition of e-business is needed. The definition of the term used here is based on a review of a number of basic concepts essential to an understanding of the business environment in general, and in particular, the changes it has gone through in recent years.

E-business is different from the traditional way of doing business. Practitioners (i.e. executives, developers, and users) need the knowledge on the new way in order to make rational decisions to address the new business environment challenges for the benefit of their firms. Academics need this knowledge in order to progress with new models and approaches on how the new business challenges can be addressed best.

There are a number of different aspects of an e-business, namely: business, operational, and technological. Recent changes in the business world have arisen from the rapid development of Information Technology and changes in the way that information technology is used. These are widely discussed in this chapter, and relevant key terms are given in "Key terms" section following the bibliography list.

Also, the impact of Information Technology (IT) on business needs to be addressed. It is argued here that Information Technology is a significant (if not major) force in developments in the business world, and it is of critical importance for doing business today. The integration of Information Systems (IS)

DOI: 10.4018/978-1-60960-501-8.ch014

into the business environment is then discussed. A variety of e-business models are next presented its characteristics are discussed. The chapter concludes in summary and future research considerations.

The chapter aims at providing the audience of this book with a clear understanding of the term "e-business," and its definition and features, which are based mainly on the fact that e-business is made possible by IT only. As argued above, this understanding is essential in order to discuss various aspects related to this new mode of doing business.

INTRODUCTION

Today, Information Technology (IT) is of fundamental importance to almost all organisations. With the arrival of e-commerce and e-business in the late 1990s, and their growth in the 21st century, the issue of IT security has become acute, as these new modes of doing business necessitate the use of IT and give very large numbers of people access to an organisation's network. According to the European Commission's i2010 report[1] published in 2008, more than half of all Europeans are now regular Internet users, and 60% of public services in the EU are fully available online. The report also reveals that 57% of doctors now send or receive patients' data electronically (as compared to 17% in 2002) and 46% of them receive results from laboratories electronically (11% in 2002), 77% of EU businesses had a broadband connection in 2007 (62% in 2005) and 77% of them use the Internet for dealing with banks (70% in 2005). The next i2010 report published in 2009[2], introduces the term "digital natives" referring to the population aged between 16 and 34 as "the most regular, intensive users ... [which] are veritable users of an interactive borderless space in which content and services are made available for active users to download, exchange, create and recreate, distribute, share and re-use" (p.5). The "e-Business Watch" report in 2006 reveals that 57% of UK firms place orders for supplies online; 16% use Internet communication systems for e-sourcing and procurement; 35% accept orders from customers online; about 70% of firms say that e-business plays a role in their day-to-day operations, and 23% say it has a significant impact[3]. In the US, e-commerce spending increased by 20.1% from $67.2 billion in January-June 2005 to $80 billion during the same period in 2006, and B2B sales reached about $1.8 trillion in 2004, representing about 20% of the total of all B2B sales of $9.1 trillion (Kauffman & Wood, 2007).

Based on the market change and progress figures, and business and technological trends, the necessity to explain and describe the new phenomenon called e-business is quite obvious. Indeed, the goal of this chapter is to provide a definition of the new mode of doing business and to characterise its operational features.

1. E-BUSINESS ORGANISATIONS

In modern business environment, which is both complicated and dynamic, organizations are looking for new ways to compete effectively in their markets (Phillips & Wright, 2009) and satisfy the ever increasing customer demands. Indeed, the ways in which business is conducted are going through very significant changes.

Some companies are already organized in new ways, whereas others are still thinking about it; however, there is a growing general awareness of a new and different way of running a business, namely the e-business approach. Becoming an e-business typically involves adopting new technologies, redesigning processes and coping with various organizational issues (e.g., employees readiness) (Lay, Ong, Yang, & Wang, 2008). In this section we describe the main characteristics of an e-business and provide a definition of the term.

1.1 Basic Concepts and Terms

In order to provide a basis for a discussion of e-business, and also to provide a working definition of an e-business organization, a series of definitions from the business world are first given. A progression from basic terms towards the final definition that forms the conclusion of this section is necessary in order to provide a methodical explanation of the subject under discussion. Because we address a topic spanning both business and technology, the definition we adopt needs to address both business and technological perspectives.

We start by providing a definition of *business* as a concept; followed by other key business-related basic terms, including business functions, business environment, supply chain and value chain. We start with the most basic concept, i.e. that of a *business*. A variety of definitions have been given for this term (including, for example, those in (Laudon & Laudon, 1998; Mercado, Welford, & Prescot, 2001; Pateli & Giaglis, 2003). The various definitions have been reviewed by Pateli and Giaglis (2003). Previous work on this topic has focused on purpose and scope, as well as relationships with other business concepts (such as strategy and business processes, efficiency and effectiveness, management issues) as can be found in (Gupta, Koulamas, & Kyparisis, 2009; Mouzas, 2006; Phillips & Wright, 2009). Some previous studies make use of the term *business model*, which is variously used to mean a "system combined of component pieces", a specification of primary elements and relationships, or a specific architecture to provide value to customers (a detailed review can be found in (Pateli & Giaglis, 2003)). Shafer, Smith and Linder (2005) argue that *business* is fundamentally concerned with creating value and capturing returns from that value, and a model is simply a representation of reality and hence define a business model as a representation of a firm's underlying core logic and strategic choices for creating and capturing value within a value network.

Laudon and Laudon (1998) provide a series of consistent definitions for terms associated with a business organisation, three of which are presented below.

A business organisation is "a complex, formal organisation whose aim is to produce products or services for a profit". (p.39)

A business environment, according to Laudon and Laudon (1998), is associated with the set of external conditions in which a business operates; "the general environment includes government regulations, economic and political conditions, and technological developments, while the task environment includes customers, suppliers and competitors". (p. 43)

As for business functions – these are the "specialised tasks performed in a business organisation (for example, manufacturing and production, sales and marketing, finance and accounting, and human resources activities)". (p. 40)

Given these definitions, it is argued here that a *business* is a formal organisational unit of any size, having a goal (production of a product/service), characterised by specific business functions, performing specialised tasks in order to achieve its goals, and surrounded by an environment. Any organisation that is consistent with the above definition is deemed to be a business.

Traditional businesses usually operate by physical means; business resources and assets (employees, equipment, raw materials, documents, goods, products, etc.) are located in a specific geographic region, occupying specific buildings, using physical means of production, shipping goods, using specific physical links for transmitting documents, and acting within a known and well-defined boundary.

A business, by definition, performs a sequence of actions and operations in order to be able to provide the results of its production to the final consumer. Such a series of actions is known as a supply chain. In this paper we will use the following definition of this term. A *supply chain* is a combination of the core business activities that allows a company to create and deliver its product or service, including all the phases from concept through development and manufacture or conversion into a market for consumption (Poirier & Bauer, 2001).

The supply chain encompasses activities that need to be performed in order to make possible the execution of business functions. These business functions/activities are carried out by *processes* (Lejk & Deeks, 2002). The supply chain starts from an initial stage of developing the product concept, through product development and manufacturing, to the final stages of the business process, in which a consumer purchases the product and/or service. A typical supply chain involves the raw materials suppliers, equipment and parts suppliers, manufacturers/producers which convert the incoming supplies into finished goods, distributors and consumers, which may be either individuals or other businesses. A business customer could be another manufacturer in an extended supply chain, which involves several steps before reaching the final point of business consumption (Poirier & Bauer, 2001). A business supply chain is made up of a chain of integrated and coordinated industrial *processes* (New, 2004). A business organisation not only runs its own supply chains, but is typically also a part of the supply chains of the organisations with which it interacts.

Business activities can be classified into internal (those performed within the organisation) and external (those involving interactions with environmental forces, as presented by the Porter Five Forces model (Porter, Competitive advantage: creating and sustaining superior performance., 1998)).

A well-recognised tool for analysing internal business processes and functions, with the goal of improving (adding value to) these activities by introducing information systems, is the Porter Value Chain model. A value chain is defined by Porter (2001) as a set of discrete, but interconnected, activities (classified as either Primary or Support activities) through which a product or service is created and delivered to customers; these activities have points of connection with the activities of suppliers, channels, and customers.

An organisation's value chain incorporates inbound logistics, manufacturing, outbound logistics, sales and marketing, and service (as primary activities), and human resource management, accounting, administration and infrastructure, product technology and development, and procurement (as support activities). The value chain model makes it possible to view a business as a series of activities that add value to the firm's products or services. Part of the following discussion and analysis is based on this model. Immediately below we provide an explanation of the meaning of each of the activities involved in the model:

- **Inbound logistics:** raw materials, parts, resource handling (receiving, storing, shipping);
- **Manufacturing:** producing the product/ service of the organisation (transformation of raw materials using specific knowledge and/or equipment, following specific procedures);
- **Outbound logistics:** storing the finished goods and/or distributing them to the customer (the final product could also be a service);
- **Marketing and sales:** activities related to promotion, marketing, distribution channels, customer definition, selling and delivering the product/service of the company;

- **Service:** activities performed after the sale has taken place (such as repairing, maintaining and supporting use of the product);
- **Corporate Infrastructure:** support given to the entire value chain, involving finance, administration, and general management;
- **Human resource management:** hiring, training, promoting and compensating employees;
- **Product and technology development:** research and development activities for the product, and the processes and technology needed for its production;
- **Procurement:** the processes (along the value chain) of acquiring raw materials, equipment and information technology.

Porter's value chain model is a widely known and well used tool for identifying the areas that might provide an organisation with additional business and operational capabilities: a value chain analysis makes it possible to separate a business's basic activities into primary and support activities. Performing such an analysis gives managers the opportunity to identify the importance and contribution to the business goals of each activity, and to focus on the specific areas that add most value to a business (and, therefore, to the business goals). Apart from having its own specific value chain, a company is typically also a component of the value chain for other organisations. So, acting together in the market, each in its own environment, companies affect each other through interactions, cooperation and collaboration. However, at least for a traditional organisation, there are still clear boundaries that separate organisations. In general, the value chain is a part of the business supply chain; indeed in some business modes these two terms refer to the same sequence of activities.

A business needs to monitor its own business actions and also its connections and interactions with the external partners of its supply chain. By doing so, a business can analyse not only its internal processes and functions, but also its business environment. Such a supply chain analysis enables a company to improve its operations and also its competitive position in the relevant market. Thus, the supply chain is a key tool for describing, analysing and improving business activities (i.e., business processes).

The adoption of advanced IT is vitally important for any business in the modern environment. The introduction of IT into business processes is made through and between the various elements of the organisational supply chain. IT provides a business with an opportunity to improve its cost/benefit ratio, although there is currently a debate in the academic community with respect to how and to what extent the application of IT within businesses leads to improved organisational performance (Melville, Kraemer, & Gurbaxani, 2004). (Although business investment in these technologies is very high, there is also a debate about the effectiveness of such huge investments, referred as the "productivity paradox" (Bocij, Chaffey, Greasley, & Hickie, 2003; Carr, 2009). We next discuss the role of IT in the business environment.

1.2 Information Technology in Business Activities

In practice, business activities involve transforming documents between stations involved in a specific series of actions in order to complete a specific mission. Documents contain data necessary to perform each of the stages of a specific process. In other words, we use documents as a convenient way of carrying data. Traditionally, documents were paper-based, but advances in information technology enable the business world to use electronic means to transfer business data. These electronic means are increasingly replacing paper for carrying data.

In practice, any kind of interaction between two organisations requires information to be exchanged between them. Obviously, these exchanges (and any associated transformations of data) are performed by means of IT. Such technology is rapidly changing, especially the technology involved in data transformation.

The business world has adopted IT since the very early stages of computer history. Computing and communications technologies, such as mainframes, minicomputers, PCs, LANs and WANs, provide a company with means to support its internal operations. Until relatively recently, companies exchanged data by physical delivery of reports, files, papers, books, etc. The introduction of computer technology has led to the use of computer networks (initially WANs and more recently the Internet), and associated application layer protocols (e.g. EDI and email), in order to communicate data between businesses. Although the Internet was seen from its very early stages as a potential tool for social interactions (ISOC, n.d.) widespread use of the Internet by the business world only took off in the 1990s.

IT has had a very significant influence on the value chain - the value chain serves as a basic tool that enables to understand the influence of information technology on companies (Porter, 2001). Computing and networking technologies have contributed to improving parts of the corporate supply chain (i.e., the value chain). Technologies such as WANs and EDI initiated improvements in external operations by allowing communications with the business environment (e.g. with suppliers); that is, they improved part of the supply chain.

Today, tools such as ERP (Enterprise Resource Planning, a collective name for software that provides integration of major business functions such as production, distribution, sales, finance, and human resource management (Chaffey, 2002), CRM (Customer Relationship Management, an approach to building and sustaining long-term business with customers, (Chaffey, 2002) and SCM (Supply Chain Management, that enables coordination of all supply activities of an organisation from its suppliers and partners to its customers (Chaffey, 2002) are widely used. In a modern business environment, the entire organisational supply chain is considered as a *value system* (Porter, 2001).

Porter (2001) describes the evolution of the use of IT in business in terms of the following five overlapping stages:

1. the first step: automation of discrete transactions (e.g., order entry and accounting);
2. greater automation and functional enhancement of individual activities (e.g., human resource management, sales force operations, and product design);
3. cross-activity integration accelerated by the Internet (e.g., linking sales activities with order processing), linking together multiple activities through CRM, SCM, and ERP;
4. integration of the value chain with the entire value system (the state of the art);
5. in the future: optimisation of the value system in real time.

We are currently seeing the integration of the value chain with the entire value system, that is the *various company value chains are being integrated across entire industry sectors, with tiers of suppliers, channels, and customers*. This integration involves merging tools (e.g., CRM and SCM), and linking the end-to-end applications of various value chain activities and participants (Porter, 2001).

Hence, in today's technology intensive reality, a major opportunity for adding value to a company's supply chain is to further computerise its business functions and activities, i.e. the *business processes*. This means increasingly introducing IS and IT *into* and *between* business activities, i.e. using computer networks to interconnect business functions both inside the organisation and between the organisation and its environment. The business activities that an organisation performs through its supply/value chain, combine the organisation's *business processes*. Hence, the modern interorganisational supply chain involves computerising the participants' business processes.

This was first achieved using computer networks to enable the use of EDI. EDI typically did not offer public access, but was a relatively expensive and proprietary technology, supported by private networks. These private networks were typically controlled by one large organisation (e.g. a manufacturer or supplier), and supported back end activities, such as exchanging invoices or order documents (Zhu & Kraemer, 2005). This kind of business interaction is now achieved in e-business by using the Internet, and is strongly integrated into the value chain at both the front and back ends (Zhu & Kraemer, 2005).

Hence, there are significant differences between the use of EDI and e-business.

We have seen a transition from proprietary networks to IP-based intranets (as an infrastructure for internal organisational activities) and the Internet (as an infrastructure for interactions with suppliers and/or customers). The Internet also makes it possible for organisations to perform all their business connections electronically, including interactions with suppliers and customers. It also provides a channel for advertising, and enables interactions with financial institutions, governmental bodies, potential suppliers, potential customers, partners, competitors, etc.

Businesses now use the Internet to support operations across the entire supply chain, and for interactions with other company's supply chains. The old supply chains can be viewed as information-based value chains that link suppliers, customers and system integrators. By computerising the entire supply chain, the traditional supply chain becomes a supply network. In the modern business environment, e-business technologies enable linking of systems, procedures, and routines of buying and selling organizations, and hence the effective integration of supply chains is one of the organization challenges (Kauremaa, Nurmilaakso, & and Tanskanen, 2009).

In terms of structure, supply chains can be seen as systems of links and nodes (New, 2004). Moreover, by introducing advanced IT into the supply chain, the supply chain becomes, according to Porter's model, a value chain, i.e. a value network or electronic value network.

As a result of changes in the global business environment and technological advances, partly caused by the Internet, flexible collaboration between businesses occurs on a global scale involving the alignment of their business processes (Adam et al., 2005). The new borderless enterprises collaborate with other organisations to produce goods and services, across national boundaries. The importance of opening organization borders has been widely recognized (Adam et al., 2005).

Using Internet-based technologies typically involves using web sites as a means of enabling electronic business. Business webs, or "b-webs", are partner networks of producers, service providers, suppliers, infrastructure companies, and customers, linked via digital channels (Cronin, 2000). While e-commerce (buying and selling over the Internet) involves using the Internet via web sites, e-business (performing all business processes, including buying and selling over the Internet) involves these web sites evolving into portals (web sites that are also entry points to other web sites).

Hence, e-business involves the computerisation of the entire enterprise to automate its business processes across its entire supply chain. By doing so, the organization creates a *Portal Value Network (PVN)* connecting all the participants in the e-business. PVN is a newly suggested here term, and it is identified with the type of network that includes all the parties linked in a specific e-business activity.

When two businesses establish a partnership, say Leymann et al. (Leymann, Roller, & Schmidt, 2002), i.e. when the corresponding service partners are interconnected, cooperation begins, and messages can be exchanged through the established links. However, linked operations have to be executed in a particular order, which needs to be specified. This is achieved by defining a flow model. According to Leymann et al. (2002) since flows often represent business processes, a flow model is sometimes referred to as a business process model, or simply a process model.

From a business perspective, e-business involves the integration of activities, organisations and systems, while from the technological point of view, e-business and e-commerce make use of a vast range of IT concepts and tools. Open Internet standards are used to integrate and automate the value chain by providing a common language for processes to intercommunicate and exchange data (Gloor, 2000).

It follows that the following definition of the e-business can be formulated:

E-business is a way of performing supply chain activities over Portal Value Networks by means of Internet-based Information Technologies.

We use a slightly more detailed version of the above definition in the remainder of this chapter:

An e-business is one that performs its supply chain activities by means of electronic processes only, using Internet-based Information Technology for integration of, and cooperation and interaction with, its participants' PVNs and business processes.

Note that the term "Virtual Business" (v-business) has also been used. There is no widely accepted definition of this term - some authors use v-business to mean any e-business, whereas others define it as a temporary network shared by different independent companies which are linked by information technology in order to share skills, costs, and access to one another's markets (Byrne, 1993). Byrne (1993) argues that this type of networks have neither central office, no hierarchy and nor organisation chart.

The definition of e-business given above is quite broad, and encompasses all types of electronic business, as well as the existing definitions of v-business. As a result we do not use the latter term here.

It is important to note that in this discussion we distinguish between "e-business" and "e-commerce". The term e-commerce refers only to the activities of buying and selling products electronically. That is, e-commerce is just one part of e-business. The discussion here is concerned with e-business.

Guided by our adopted definition of e-business, we now identify the main characteristics of this type of business, and describe various types of e-business organisations.

2. E-BUSINESS CHARACTERISTICS

The main characteristics of e-business organisations are their business models and the fact that they operate using the Internet and Information Technology (IT). All e-business functionality-related properties are derived from these two main characteristics, discussed in sections 2.1 and 2.2 immediately below.

2.1 E-Business Models

In order to analyse e-business, it is important to identify models of its functionality.

To be able to operate as an e-business, an organisation must change its business model, usually in a very radical way (Gloor, 2000). Based on the working definition of e-business given in section 1, it follows that, both practically and theoretically, e-business is a new business model. Unlike a traditional (physical) business, e-business is not focused on static, internally managed and operated chains, but on dynamic business processes, performed through external webs of relationships, that take advantage of the digital arena's power and flexibility (Cronin, 2000). The traditional supply chain has changed into a value network of portals (*Portal Value Networks,* or *PVNs*). Portals provide both the employees and the customers of an organisation with a new way of working and/or buying.

Two types of portals are possible: horizontal and vertical (Gloor, 2000; Zirpins, Weinreich, Bartelt, & Lamersdorf, 2001). A portal that provides functions meeting common user requirements, such as search functions, email, chat, etc. via the Internet, is a horizontal portal. Google and Amazon provide examples of horizontal portals. The objective of a horizontal portal is to provide solutions to a broad range of Internet users (Gloor, 2000). A portal that includes the same functions as a horizontal portal, but that focuses on a specific community and industry, is a vertical portal (Gloor, 2000). Thus the portals of companies such as Blackwells (books) and Dell (computers) are good examples of vertical portals (also known as vortals). They address a very broad audience, but also offer clear and focused functionality aimed at a specific audience. Microsoft's portal is a further example of a vortal, whose target audience includes users with specific requirements. When a company's vortal is directed at business customers, it involves interactions with value chains (i.e. at PVNs, as defined above).

It follows then that one aspect of the business model must be deciding on the type of portal to be provided, i.e. whether it should be horizontal or vertical. Some authors suggest that businesses should build a vortal, not only because vortals improve distribution networks as compared to traditional distribution, but also because a vortal provides production capabilities, and so will enable the exploitation of the Internet as a business tool (Applegate, 2002). An e-business operating via portals (either horizontal or vertical), has to decide upon its business model, which, as stated above, is a different model to that applying to a traditional business.

Applegate (2002) argues that e-business models can be divided into two categories, depending on the type of company involved:

1. Digital businesses, i.e. organisations that are built on the Internet;
2. Businesses that provide the platform upon which digital businesses are managed and operated.

Gloor (2000) provides a more detailed classification of e-business models, based on four fundamentally different types of organisation, as suggested by Malone (in (Gloor, 2000)):

1. **Creators:** producers of goods (physical or information), such as General Electric, Cisco, Dell, Microsoft, and on-line versions of newspapers;
2. **Distributors:** companies that distribute and/or supply goods, such as electronic shops for books or music (e.g. Amazon);
3. **Brokers:** companies that act as intermediaries, such as on-line auctioneers or travel agencies (eBay, Netaction, Thelastminute, and the90minute are examples);
4. **Extractors:** companies that exist only on the Internet, that operate as portals, and whose business model is based on advertising revenue. Examples include Yahoo, MSN, and Google.

This classification appears to be sufficiently detailed for our purposes. Today's e-business environment involves e-businesses which use more than one of the above models. For example, an e-business organisation might be classified as a *creator*, but also operate as a *broker*, and/or might also operate a web site to provide services (e.g., special offers) available only via the Internet (i.e. in the *extractor* role). Airline companies are typical examples of such organisations. There are also companies that do business exclusively via the Internet (defined by Applegate (2002) as *digital* companies) to provide an e-business platform for other (usually creator) organisations, which fall into Applegate's second category. Hence, we propose here an additional two-category classification:

1. type I - a generic e-business organisation (either creator, distributor, broker, extractor, or any combination of these roles);
2. type II - an infrastructure-supportive e-business organisation.

This classification is also consistent with the definition of e-business given in section 1.

E-business organisations can also be classified according to how they collaborate with other organisations. The main current business model types in such a classification are as follows (Androutsellis-Theotokis, Spinellis, & Karakoidas, 2004):

- the old-economy "one-to-one" EDI model (providing direct point-to-point transactions between participants);
- the "one-to-many" model (providing a supplier with access to a group of buyers);
- the "many-to-one" model (providing a buy-side solution for one buyer facing many suppliers);
- the "many-to-many", or net markets, model (involving the aggregation of large groups of buyers and sellers around a central hub offering a wide range of services)

We draw the following conclusions from the above discussions:

- an e-business can operate via either a horizontal portal or a vortal;
- an e-business organisation can perform all its activities in an e-business mode or operate in an e-business mode only to support its main activities;
- an e-business organisation can collaborate with other parties in various ways.

All the types of e-business organisations described above can only operate by deploying Internet-based technologies. The ability to operate using IT only is one of the most important characteristics of an e-business.

2.2 E-Business Technology Infrastructure

It has become common for businesses large and small to make some of their services available on the Web. Using such services typically requires use of a web browser.

E-business operates using the Internet (Porter, 2001; Zhu & Kraemer, 2005), and involves providing broad connectivity between the organisation's front end (namely, sales and customer service) and back end (namely, coordination and procurement) (Zhu & Kraemer, 2005). The Internet makes it possible to exchange real-time information between an e-business organisation and other participants, such as customers and suppliers (Zhu, 2004). The open standard nature of the Internet means that e-business has a number of unique characteristics. E-business use is decentralised, and is often driven by a variety of technological, structural and environmental factors. Hence, e-business is affected both by the technological competence of the organisations, and by organizational and environmental factors, such as international scope, legal protection issues for online transactions, and web functionality (Zhu & Kraemer, 2005).

A typical web-application architecture involves three tiers: web browsers (to provide a ubiquitous user interface), application servers (to manage business activities), and back-end databases (to store the persistent data) (Wasserman & Su, 2007). Although this architecture does not appear to be very complex, implementing it requires a good understanding of both the applications hosted on the application server, and of the appropriate network configuration for a specific e-business organisation (Osborne, 2006).

Some organisations make the change to an e-business mode by also shifting towards a service-oriented architecture (SOA) (Sluiter, 2006), which provides organizations with certain operational benefits. Leymann et al. (2002) argue that web services based on a service-oriented architecture provide a suitable technical foundation for making business processes accessible within and across enterprises.

The emerging semantic web looks set to provide significant advantages to e-businesses, not least because of its potential to host new e-business applications (Thuraisingham, 2005).

Also, conducting business on a platform conforming to open standards facilitates information sharing along the value chain (Zhu, 2004). Doing business by means of networks creates one of the most characteristic features of modern organisations, i.e. transparency (Tapscott & Ticoll, 2003). Employees from different organisations share information, business plans, strategic issues and challenges. In the modern business environment organsations have mistakenly posted on the Web sensitive information (e.g., many different types of legal, medical and financial data (Johnson, 2008). Tapscott and Ticoll (2003) argue that since in a real-time global supply chain, companies and their business partners need to share their competitive and operational secrets, some opacity remains necessary, and transparency isn't always easy. For example, trade secrets and personal data should be confidential (Tapscott & Ticoll, 2003).

There are also significant differences in management issues in two major areas, namely strategic planning and personal privacy and security. These differences result from the fact that e-business is an Internet/Web system, i.e. clients, suppliers; partners can interact with e-business organisations via Internet connections, usually with Web interfaces. Such systems differ significantly from traditional systems in information content, development practices, the developers themselves, and the users (Walter & Scott, 2006).

These Internet-based technologies provide a wide range of services, each with its own capabilities and security vulnerabilities. Hence, security must be taken into consideration while developing an e-business mode of doing business.

3. SUMMARY

In this chapter a conceptual framework that facilitates the development of a clearer picture of an e-business and its characteristics was given. The term *Portal Value Network* was introduced here. This term provides the basis to progress towards the definition and explanation of e-business organisation. E-business has unique features, derived from its use of Information Technology. The term *e-business organisation* was defined. This definition identifies an e-business as a company running its business processes electronically. E-business processes are enabled by the transmission of information flows. E-business, in order to succeed, must provide value to its users (including employees and customers). To deliver the value proposition to all its customers, a business must ensure that it possesses a range of capabilities that underpin the proposed value (Osterwalder & Pigneur, 2002). E-business capabilities are realised through the organisation's underlying e-processes. The correct execution of these e-processes is crucial for the e-business mode of operation.

4. FUTURE RESEARCH

The e-business research field is relatively young, and is still in its initial stages, argue (Gupta, Koulamas, & Kyparisis, 2009). According to the authors (Gupta, Koulamas, & Kyparisis, 2009), the previous research in production and management in e-business during the period 1992-2008 can be classified into four categories:

1. E-auctions
2. Radio frequency identification (RFID)
3. E-business system design
4. C^4 in e-business (competition, collaboration, conflict and coordination issues in e-business).

According to the authors, future research in e-business production and management must focus on solving practical, real-life problems.

Mouzas (2006) suggests that there is a need to learn more about efficiency and effectiveness in e-business and the ways to maximize its value to the stakeholders.

A review of research on e-marketplace (Standing, Standing, & Love, 2010) reveals that there are two issues that the researchers on information systems need to address related to e-business: the nature of electronic markets and their efficiency, and the organizational implications of adopting and managing electronic marketplaces.

The issue of e-business adoption and its organizational impact is also addressed in a study on tourist sector (Andreu, Aldas, Bigne, & Mattila, 2009), where the authors relate also to the supply chain considerations.

To conclude based on the list above, the trend in future e-business research will be focusing on issues such as organizational impact, efficiency and effectiveness and solving practical problems for the e-business management.

REFERENCES

Adam, O., Hofer, A., Zang, S., Hammer, C., Jerrentrup, M., & Leinenbach, S. (2005). A collaboration framework for cross-enterprise business process management. *Preproceedings of the First International Conference on Interoperability of Enterprise Software and Application.* Geneva, Switzerland.

Andreu, L., Aldas, J., Bigne, J. & Mattila, A. (2009). An analysis of e-business adoption and its impact on relational quality in travel agency–supplier relationships. *Tourism Management.*

Androutsellis-Theotokis, S., Spinellis, D., & Karakoidas, V. (2004). Performing peer-to-peer e-business transactions: A requirements analysis and preliminary design proposal. *Preceedings of the IADIS, International e-Commerce 2004 Conference*, (pp. 399-404).

Applegate, L. (2002). *E-business handbook.* The St. Lucie Press.

Bocij, P., Chaffey, D., Greasley, A., & Hickie, S. (2003). *Business Information Systems* (2nd ed.). Financial Times, Prentice Hall.

Byrne, J. (1993). The futurist who fathered the ideas. *Business Week.*

Chaffey, D. (2002). *E-business and e-commerce management. Financial Times.* Prentice Hall.

Cronin, M. (2000). *Unchained value: The new logic of digital business.* HBSWK Publications.

Dahlbom, B. (2001). Postface: From infrastructure to networking, from control to drift. In Ciborra, C. A. (Ed.), *From control to drift-The Dynamics of Corporate Information Infrastructure.* Oxford: Oxford University Press.

Gloor, P. (2000). *Making the e-Business transformation.* London: Springer-Verlag.

Gupta, S., Koulamas, C., & Kyparisis, G. J. (2009). E-business: A review of research published in production and operations management (1992–2008). *Production and Operations Management, 18*(6), 604–620. doi:10.1111/j.1937-5956.2009.01100.x

Internet Society. (2010). *A brief history of the Internet.*

Johnson, M. (2008). Information risk of inadvertent disclosure: An analysis of file-sharing risk in the financial supply chain. *Journal of Management Information Systems, 25*(2), 97–123. doi:10.2753/MIS0742-1222250205

Kauffman, R., & Wood, C. (2007). Revolutionary research strategies for e-business: A philosophy of science view in the age of the Internet. In Wood, C., & Kauffman, R. A. (Eds.), *Economics, Information Systems, and electronic commerce research: Advanced empirical methodologies*. Armonk, NY: M. E. Sharpe.

Kauremaa, J., Nurmilaakso, J.M. & and Tanskanen, K. (2009). E-business enabled operational linkages: The role of RosettaNet in integrating the telecommunications supply chain. *International Journal of Production Economics* .

Laudon, K., & Laudon, J. (1998). *Information Systems and the Internet* (4th ed.). Dryden Press.

Lay, J., Ong, C., Yang, C., & Wang, C. (2008). Assessing and managing employee readiness for embracing e-business. *Proceedings of the 2008 ACM SIGMIS Computer Personnel Research Annual Conference.* Charlottesville, VA: ACM.

Lejk, M., & Deeks, D. (2002). *Systems analysis techniques* (2nd ed.). Addisson-Wesley, Pearson Education Limited.

Leymann, F., Roller, D., & Schmidt, M.-T. (2002). Web services and business process management. *IBM Systems Journal*, *41*(2), 1015–1019. doi:10.1147/sj.412.0198

Melville, N., Kraemer, K., & Gurbaxani, V. (2004). Information Technology and organisational performance: An integrative model of it business value. *Management Information Systems Quarterly*, *28*(2).

Mercado, S., Welford, R., & Prescot, K. (2001). *European Business* (4th ed.). Prentice Hall, Financial Times.

Mouzas, S. (2006). Efficiency versus effectiveness in business networks. *Journal of Business Research*, *59*, 1124–1132. doi:10.1016/j.jbusres.2006.09.018

New, S. (2004). Supply chains: Construction and legitimation. In Westbrook, S. N. (Ed.), *Understanding supply chains*. Oxford University Press.

Osborne, M. (2006). *How to cheat at managing Information Security*. Syngress Publishing, Inc.

Osterwalder, A., & Pigneur, Y. (2002). An e-business model ontology for modeling e-business. *15th Bled Electronic Commerce Conference e-Reality: Constructing the e-Economy*, (pp. 73-86).

Pateli, A., & Giaglis, G. (2003). A framework for undestanding and analysing e-business models. *16th Bled Electronic Commerce Conference - eTransformation (CD-ROM Proceedings).*

Phillips, P. A., & Wright, C. (2009). E-business's impact on organizational flexibility. *Journal of Business Research*, (62): 1071–1080. doi:10.1016/j.jbusres.2008.09.014

Poirier, C., & Bauer, M. (2001). *E-supply chain*. Berrett-Koehler Publishers, Inc.

Porter, M. (1998). *Competitive advantage: Creating and sustaining superior performance*. Free Press.

Porter, M. (2001). Strategy and the Internet. *Harvard Business Review* .

Porter, M. & Millar, V. (1985). How information gives you a competitive advantage. (63), 149-161.

Shafer, S., Smith, J., & Linder, J. (2005). The power of business models. *Business Horizons*, *48*(3), 199–207. doi:10.1016/j.bushor.2004.10.014

Sluiter, J. (2006). Services Oriented Architecture security and deperimeterisation. *Information Security Bulletin*, *11*(1), 65–72.

Standing, S., Standing, C., & Love, P. (2010). A review of research on e-marketplaces 1997–2008. *Decision Support Systems*, *49*(1), 41–51. doi:10.1016/j.dss.2009.12.008

Tapscott, D., & Ticoll, D. (2003). *The naked corporation: How the age of transparency will revolutionise business*. Dow Jones & Company, Inc.

Thuraisingham, B. (2005). Directions for security and privacy for semantic e-business applications. *Communications of the ACM, 48*(12), 71–73. doi:10.1145/1101779.1101812

Walter, Z., & Scott, G. (2006). Management issues of Internet/Web systems. *Communications of the ACM, 49*(3), 87–91. doi:10.1145/1118178.1118181

Wang, W., Bailey, A., Hidvegi, Z., & Whinston, A. (2001). *A framework for proactive, automated and continuous e-commerce control and assurance*. Goizueta Business School Papers Series.

Wasserman, G., & Su, Z. (2007). Sound and precise analysis of Web applications for injection vulnerabilities. *Proceedings of the PLDI'07*. ACM.

Zhu, K. (2004). Information transparency of business-to-business electronic markets: A game-theoretic analysis. *Management Science, 50*(5), 670–685. doi:10.1287/mnsc.1040.0226

Zhu, K., & Kraemer, K. (2005). Post-adoption variations in usage and value of e-business by organizations: Cross-country evidence from the retail industry. *Information Systems Research, 16*(1), 61–84. doi:10.1287/isre.1050.0045

Zirpins, C., Weinreich, H., Bartelt, A., & Lamersdorf, W. (2001). Advanced concepts for next generation portals. *Proceedings of the 12th International Conference on Database and Expert Systems Applications*, (pp. 501-506). Munich, Germany.

ADDITIONAL READING

Battisti, G., Hollenstein, H., Stoneman, P., & Woerter, M. (2007). Inter- and intra-firm diffusion of ICT in the United Kingdom and Switzerland: an internationally comparative study based on firm-level data. *Economics of Innovation and New Technology, 16*(8), 669.

Bowersox, D. J., Closs, D. J., & Cooper, M. B. (2007). *Supply chain logistics management*. New York: McGraw-Hill/Irwin.

Carr, N. G. (2003, May). IT doesn't matter. *Harvard Business Review, 2003*, 5–12.

Disney, S. M., Naim, M. M., & Potter, A. (2005). Assessing the impact of e-business on supply chain dynamics. *Journal of Production Economics, 89*(2), 109–118. doi:10.1016/S0925-5273(02)00464-4

Ketchen, D., & Hult, G. T. (2007). Bridging organization theory and supply chain management: The case of best value supply chains. *Journal of Operations Management, 25*(2), 573–580. doi:10.1016/j.jom.2006.05.010

Klassen, M., Gupta, P., & Bunker, M. P. 2009. *Comparison shopping on the internet.* Int. J. Bus. Inf. Syst. 4, 5 (May. 2009), 564-580. DOI= http://dx.doi.org/10.1504/IJBIS.2009.025207

Lee, H. L. (2004). The triple-A supply chain. *Harvard Business Review, 82*(10), 102–112.

Meron˜o-Cerdan, A.L., Soto-Acosta, P., 2005. *Examining e-business impact on firm performance through website analysis.* International Journal of Electronic Business 3 (6), 583–598.

Meron˜o-Cerdan, A.L., Soto-Acosta, P., 2007. *External Web content and its influence on organizational performance.* European Journal of Information Systems 16 (1), 66–80.

Miles, R. E., Miles, G., & Snow, C. C. (2005). *Collaborative entrepreneurship: How communities of networked firms use continuous innovation to create economic wealth.* Stanford, CA: Stanford Books.

Miles, R. E., Miles, G., & Snow, C. C. (2006). Collaborative entrepreneurship: A business model for continuous innovation. *Organizational Dynamics, 35*(1), 1–11. doi:10.1016/j.orgdyn.2005.12.004

(2005). of e-brokerage service providers. *Journal of the Academy of Marketing Science*, *33*(3), 360–375.

Overby, E., Bharadwaj, A., & Sambamurthy, V. (2006). Enterprise agility and the enabling role of information technology. *European Journal of Information Systems*, *15*, 120–131. doi:10.1057/palgrave.ejis.3000600

Ray, G., Barney, J. B., & Muhanna, W. A. (2004). Capabilities, business processes and competitive advantage: choosing the dependent variable in empirical tests of the resource-based view. *Strategic Management Journal*, *25*, 23–37. doi:10.1002/smj.366

Saini A, Johnson JL. *Organizational capabilities in e-commerce: an empirical investigation*

Soto-Acosta, P., Meron˜o-Cerdan, A.L. (2006). An analysis and comparison of Web development between local governments and SMEs in Spain. *International Journal of Electronic Business*, *4*(2), 191–203.

Soto-Acosta, P. (2009, May). Meron˜o-Cerdan,A.L.,2009. *Evaluating Internet technologies business effectiveness. Telematics and Informatics*, *26*(Issue 2), 211–221. doi:10.1016/j.tele.2008.01.004

Zhu, K., & Kraemer, K. L. (2004). Information Technology Payoff in E-Business Environments: An International Perspective on Value Creation of E-Business in the Financial Services Industry. *Journal of Management Information Systems*, *22*(1), 17–54.

KEY TERMS AND DEFINTIONS

Business Processes: Combination of business activities that an organisation performs through its supply/value chain.

Business: Formal organization that performs a sequence of actions (i.e. *supply chain*) in order to provide the results of its activities to the final customer.

Customer Relationship Management (CRM): An approach to building and sustaining long-term business with customers.

E-Business: Performs its supply chain activities by means of electronic processes only, using Internet-based Information Technology for integration of, and cooperation and interaction with, its participants' PVNs and business processes.

E-Commerce: The activities of buying and selling products electronically.

Electronic Data Interchange (EDI): The exchange of business information by using digital devices.

Enterprise Resource Planning (ERP): A collective name for software that provides integration of major business functions such as production, distribution, sales, finance, and human resource management.

Local Area Network (LAN): At least two computers/devices sharing a common communications channel in a relatively small geographic area.

Portal Value Network (PVN): Network connecting e-business partners through their portals

Portal: Web site that provides possibilities to enter other websites information by means of directories, search engines, links, etc.

Service-Oriented Architecture (SOA): A set of design principles that enables to provide web-based application services to consumers.

Supply Chain Management (SCM): Enables coordination of all supply activities of an organisation from its suppliers and partners to its customers

Supply Chain: a combination of the core business activities that allows a company to create and deliver its product or service, including all the phases from concept through development and manufacture or conversion into a market for consumption.

Value Chain: Defined by Porter (2001) as a set of discrete, but interconnected, activities through which a product or service is created and delivered to customers.

Wide Area Network (WAN): At least two computers/devices sharing a common communications channel within a wide geographic area.

ENDNOTES

1 The Commission's i2010 report is available at: http://ec.europa.eu/i2010

2 The report is available at: http://ec.europa.eu/information_society/eeurope/i2010/docs/annual_report/2009/sec_2009_1060_vol_1.pdf

3 Source: e-Business Watch (2006), available at: http://www.ebusiness- watch.org/ key reports/documents/ EBR06.pdf

Chapter 15
Web Services and E-Business Technologies:
Security Issues

David Anyiwo
Bowie State University, USA

Sumana Sharma
Bowie State University, USA

ABSTRACT

The large scale deployment of Web services and e-business technologies in general – is affected by numerous implementation (for example, interoperability and integration) and security issues. The focus of this chapter is mainly on exploring the latter from the perspective of initiatives for mitigating security risks in Web services (for example, WS-S specifications) and e-business technologies (for example, security standards, business continuity planning, and cyber legislation). Best practices and recommendations for addressing security risks in the rapidly evolving Web environment are presented. Future research trends associated with the topic of security in Web services and e-business technology applications are also discussed.

INTRODUCTION

We are witnessing the evolution of Web 2.0! This marks the birth of a new Web, drastically different from the traditional Web, which is characterized by static screens and passive browsing. The hallmark of Web 2.0 (O'Reilly, 2007) is 'true' user interaction with Web sites (giving users the ability to modify content) and the use of the Web as a computing platform (that allows users to run software applications through their browser). The latter set of applications is commonly referred to as Web services. Enterprises are increasingly deploying Web services and other e-business technologies in their quest to adapt quickly to the rapidly chang-

DOI: 10.4018/978-1-60960-501-8.ch015

ing business context owing to the ushering in of this new era. While Web services have delivered enormous benefits to the enterprises that employ them, and have indeed revolutionized e-business and other Internet-enabled services, they have also introduced greater complexity in the service environment. They are highly vulnerable to attack, and are still plagued by inadequate levels of enterprise system integration and interoperability (Anyiwo, 2006). Securing disparate systems and applications in a very complex service environment continues to pose significant challenges.

In spite of the recent advances in agent-based Web technologies, Web services are still hindered by security and interoperability (especially, semantic interoperability) problems. Although the Xtensible Markup Language (XML) has enabled some levels of structural and syntactic integration and interoperability in Web services, achieving semantic interoperability in a services-oriented architecture (SOA) is still very difficult.

This chapter provides an overview of the underlying opportunities as well as the security problems and other related challenges facing Web services, and other e-business technologies. In the second section of the chapter, the security risks affecting Web services and e-business technologies are presented. The current approaches in the mitigation of the security and other risks associated with each of these technologies are considered in the next section. The chapter also provides a list of recommendations for managing security risks in the Web 2.0 era as evidenced by the academic and practitioner literature, and explores emerging issues in semantic Web service security as well as the direction for future research.

WEB SERVICES SECURITY ISSUES

As the leading US analyst firm, Gartner, predicted a skyrocketing increase in Web services applications in 2003, it simultaneously identified Web services security as one among the top 11 security issues facing companies in that same year (Parry, 2003). Seven years later in 2010, the concern has only increased, mainly due to the fact that we have continued to see Web services applications being developed and deployed even though we have not exactly been able to address all the security challenges that confront these services. More recent reports also indicate that attackers are focusing their attention on interactive Web 2.0 elements with some 95 percent of user-generated comments on blogs, message boards, and chat rooms being either spam or containing malicious links (Websense, 2009). However, as O'Neill et al. (2003) point out, Web services needs to be secure in order to enjoy widespread deployment.

So what are some of the main issues associated with securing Web services? A survey of the literature reveals that many of the issues identified are in fact common to all distributed computing technologies and includes factors such as authentication, authorization, confidentiality, integrity, non-repudiation, availability and end to end security. These are in fact some of the main security issues facing Web service deployments. In addition, there could also be issues which arise from the nature of Web services themselves, such as diversity of standards specifications, need for new XML formats to structure security data, interoperability of requirements and online security elements, to name a few (Gutiérrez et al. 2004).

In their latest industry report, the Secure Enterprise 2.0 Forum (a group of organizations and individuals composed of executives at Fortune 500 companies who have embraced the trend of Web 2.0 tools and services in enterprise) has identified the top eight Web 2.0 vulnerabilities (Perez 2009). These are described in Table 1.

Hasan and Duran (2005) have grouped security challenges facing Web services deployment into three main categories: (1) interoperability and policies (since a plethora of encryption mechanisms and variety of platforms are involved); (2) message security (since a Simple Object Access Protocol (SOAP) message has to

Table 1. Top eight Web 2.0 vulnerabilities (Source: Secure Enterprise 2.0 Forum)

	Vulnerability	Description
1	Insufficient authentication controls	In many Web 2.0 applications, content is trusted in the hands of many users, which means that there is a greater chance that a less-experienced user will make a change that will negatively affect the overall system.
2	Cross Site Scripting (XSS)	In a stored cross site scripting (XSS) vulnerability, malicious input sent by an attacker is stored in the system then displayed to other users. Especially at risks are blogs, social networks, and wikis because hackers can easily steal the session cookies and use them for unauthorized purposes.
3	Cross Site Request Forgery (CSRF)	In CSRFs, victim visit what appear to be innocent-looking Web sites, but which contain malicious code which generates requests to a different site instead. As compared to legacy applications, Web 2.0 applications are more vulnerable to this type of attack.
4	Phishing	Although this phenomenon is not just restricted to web 2.0 technologies, it is more common in the web 2.0 era because the plethora of dissimilar client software makes it harder for consumers to distinguish between the genuine and the fake web sites thereby enabling more effective phishing attacks.
5	Information Leakage	The growing tendency to access information from anywhere at any time often leads people to inadvertently share information that their employer would have considered sensitive. Even the compromise of small snippets of non-sensitive information can result in a considerable breach of data privacy in the future.
6	Injection Flaws	Web 2.0 technologies tend to be vulnerable to new types of injection attacks including XML injection, JavaScript injection etc, because such applications tend to use and rely on these technologies.
7	Information Integrity	In a business environment, having systems open to many users allows a malicious user(s) to post and publish inaccurate information which destroys the integrity of the data.
8	Insufficient Anti-automation	Programmatic interfaces of Web 2.0 applications let hackers automate attacks more easily. These can be avoided by using Captchas or similar mechanisms.

naturally cross domain boundaries and has to go through intermediaries, hereby rendering traditional security models that work within a controlled environment ineffective); and (3) identity and trust (both the client and the server need to know that they can rely on the other party and both need to be authenticated).

The topic of security issues facing Web services can be also be studied from the perspective of the users of such services. While most discussion of security is based on the perspective of the services themselves, Yee (2006) has explored this issue from the perspective of the end user and proposed ways to ensure privacy of individuals when using Web services.

Integration and interoperability issues relate to Web services as such applications require the integration of services from other companies, vendors and third parties, as well as the interoperability of the differing computer systems involved. Integration refers to the capability of services to be joined together on demand to create composite services. Integration, achieved through loose coupling, is a means of ensuring that factors such as language, platform, etc are decoupled from the service itself. Interoperability, on the other hand, refers to the capability of two or more computer systems to share data and resources, even though they are made by different manufacturers.

Given that Web services are based on open standards, such as SOAP and Universal Description, Discovery and Integration (UDDI), they are regarded as an open standards way of supporting interoperability. However, interoperability issues are the greatest stumbling block that are preventing us from truly realizing the SOA vision (Kanneganti & Chodavarapu, 2008). In simple terms, this issue arises from the fact that the enterprise systems required to interoperate were built by diverse organizations using diverse structures and a diverse set of vocabularies. We are therefore presented with substantial heterogeneity in syntax, structure and semantics when it comes to interoperation between these systems (Nagarajan et al., 2006).

Of these, it is semantic heterogeneity (where the same real world entity is represented using different terms or the same term is used to describe different real world entities) that has come to the forefront as the greatest impediment to successful deployment of Web services. Having one common vocabulary across differing domains is not a feasible solution but the challenge in fact is to share across different vocabularies.

Laskey (2005) proposes some approaches that could be of interest as we deal with this issue. These include: (1) using annotations that connect structured documents with ontologies describing related domains; (2) using upper ontologies to encode higher level concepts and provide consistent relationships for more detailed ontologies to leverage; (3) ontology of mapping concepts to provide basis for generating and capturing specifics of matching circumstances; (4) transitive reasoning to leverage specific relationships in making connections to previously unmapped entities; and (5) polymorphic reasoning to leverage entities related to one ontology when working with another ontology with which there are formal relationships.

Cloud Computing Security Issues

Cloud computing is a related technology trend that enables businesses to access and run traditional business applications, through the cloud (metaphor for Internet), and thereby prevent expensive installation and maintenance costs. The application (for example, Human Resources (HR), Customer Relationship Management (CRM), Accounting or any customer-driven system) can be made available in only a few days time frame. This is clearly a drastic reduction in the amount of time (several weeks or months) that is typically required to install and configure an application at a business' end. The main attraction of cloud computing is its pricing model which works on a pay-as-you-go basis and therefore consumers are only expected to pay based on the amount of service used.

Cloud computing, in its present form, is essentially a world where there are no standards whatsoever (Williams, 2010). Therefore issues such as security and compliance are of great concern. As more and more data about individuals and companies are hosted on it, the cloud becomes particularly attractive to cyber crooks (Ashford, 2010). Other issues include, managing trust, multi-tenancy, encryption and compliance. To address these issues, Novell and the Cloud Security Alliance have recently announced a vendor-neutral "Trusted Cloud Initiative" for developing standards and certification of cloud security, compliance, identity management and other best practices (Babcock, 2010). The National Institute of Science and Technology (NIST) has also stated that it wants to act as a catalyst in the formulation and adoption of cloud standards by government and industry (Mell & Grance, 2009). NIST recognizes that over-specification may inhibit innovation in this dynamic domain and so it is only pushing for minimal standards at this time.

E-Business Technology Security Issues

It is clear that the identification and structuring of the security risks underlying the organization's information assets will remain as the primary focus of IT security practitioners in all types of enterprises. In this section, we explore some of the current and emerging security risks associated with e-business technologies.

Internal Threats: Insider threat continues to remain as one of the top challenges confronting organizations. According to the 2010 CyberSecurity Watch Survey (Carnegie Mellon, 2010) conducted by *CSO* magazine, the leading resource for security professionals, the most costly or damaging attacks to an organization's security are caused by insiders. 67% of the respondents polled admitted that internal security breaches are more costly than external breaches.

External Threats: Hackers too present a significant menace especially today, given their increasing use of more and more sophisticated technologies (Barber & Integralis, 2001) for breaking into enterprise systems. We are also seeing more and more coordinated attacks (as compared to attacks from an individual hacker) wherein a large number of hackers launch an attack on one or more systems simultaneously. A case in point is the Estonian Cyber War that swamped Web sites of Estonian organizations, including the Estonian Parliament, banks, ministries, newspapers and broadcasters, all in the same time frame (Traynor, 2007).

Countermeasures to address internal and external threats: Organizations must implement different defensive and preventative countermeasures to address the different security challenges facing them. For example, solutions for addressing insider threats should take into consideration the observations that typically emerge from the analysis of most insider IT sabotage problems, such as, (1) most insiders who committed the IT sabotage were disgruntled due to unmet expectations, and (2) stressful events, including organizational sanctions, increased the likelihood of insider IT sabotage (see, Moore et al., 2008, for a detailed review). Such general observations should be incorporated in refining an organization's security policy and thereby improve its effectiveness.

While most security efforts aimed at addressing the issue of external threats posed by hackers have been largely defensive, we are also now seeing approaches (for example, honeypots) that allow organizations to take the offensive (Spitzner, 2003). In addition, most organizations are engaging in periodic penetration tests of their systems and investing in programs related to increasing employees' awareness of security issues and challenges.

Mitigating Security Risks in E-Business Technologies

Emerging trends in cyber legislation: As instances of cyber crime are growing, many nations are enacting cyber legislation to combat this problem. In the year 2000, India implemented the Information Technology Act to counter cyber crimes, which was amended in December 2008 (Nappinai, 2010) to take into consideration the advent of newer weapons of cyber warfare. Recently, China introduced three new articles to its criminal code based on which online criminals could now be sentenced to 7 years of imprisonment, an increase of 4 years over the previous recommended punishment of 3 years (Fletcher, 2010). The push for tougher cyber legislation is not unexpected given that there could be extremely high monetary losses if such attacks were to materialize. According to a recent research conducted by the U.S. Cyber Consequences Unit, a non-profit research institute that assesses the economic impact of cyber attacks, a single wave of cyber attacks on critical infrastructures could exceed $700 BILLION, which is the equal of 50 hurricanes hitting America at one time (PRNewswire, 2010).

It is important to note that cyber laws are of consequence not only to companies who conduct their business online but also to individuals who navigate the Internet in any manner, such as through posting of reviews, use of social networking sites, etc. Trout (2007) discusses the legal aspects of cyber laws and the implications of such laws for businesses and consumers operating in cyberspace.

Information Security Governance and Compliance: Information Security compliance standards and models are an important step in mitigating the risks associated with e-business technology. Numerous relevant standards apply, such as the USA's Health Insurance Portability and Accountability Act (HIPAA) security rule, which is related to the safeguarding of electronic-protected health information; ISO/IEC 27002, which provides best practice recommendations on information security management for use by those who are responsible for initiating, implementing, or maintaining Information Security Management Systems (ISMS); Control Objectives for Information Technology (COBIT), which provides managers, auditors, and IT users with a set of generally accepted measures, indicators, processes and best practices to assist them in maximizing the benefits derived from the use of information technology and developing appropriate IT governance and control in a company, etc.

Even though newer standards for information security governance are being formulated, compliance with the proposed standards remains a major problem. Siponen & Willison (2009) researched various information security management standards and concluded that this problem is due to (1) the generic nature of standards and their indifference to the fact that organizations are different and have differing security requirements; and (2) guidelines are often validated by appeal to common practice and authority, which is not a sound basis for important international information security guidelines.

The same opinion is also voiced by Höne & Eloff (2002) that while standards are a good starting point for determining what the information security policy should consist of, they should not be relied upon exclusively for guidance. They contend that it is far more important that the information security policy fits in with the organization's culture and must therefore be developed with this in mind.

Business Continuity Strategies: Businesses rely on business continuity planning (BCP) to ensure that their ability to respond and deliver services is not hindered in the event of a disaster. Business continuity planning is "planning which identifies an organization's exposure to internal and external threats and synthesizes hard and soft assets to provide effective prevention and recovery for the organization, whilst maintaining competitive advantage and value system integrity" (Herbane, Elliot, & Swartz, 1997).

BCP is regarded as a safety net for businesses because it in essence forces them to conduct a rigorous threat and risk assessment as well as develop and test plans that avoid or mitigate the risks identified. As firms' dependence on IT is increasing, there is a greater need than ever before to integrate business continuity planning with IT security planning (Cerullo & Cerullo, 2004). With data now being universally regarded as an organization's most valuable asset, data security should be designed in order to reduce exposure to the likelihood of a disaster and minimize the impact of a disaster if it does occur (Myers, 1996). We are also witnessing the emergence of standards being designed especially for BCP. In December 2006, the British Standards Institution (BSI) released a new independent standard for BCP — BS 25999-1. This was followed by the proposal of the BS 25999-2 "Specification for Business Continuity Management", that specifies requirements for implementing, operating and improving a documented Business Continuity Management System (BCMS).

Recently the United States Congress enacted Public Law 110-53, "Implementing Recommendations of the 9/11Commission Act of 2007." Title IX of this law is being regarded as the next step in a complex pattern of laws and regulations requiring the private sector to develop and maintain business continuity plans. The ultimate goal of this law is to achieve SOX (Sarbanes-Oxley) type compliance for private sector's business continuity plans which will need to be certified by the DHS (Firestorm, 2008).

Mitigating Security Risks in Web Services Applications

Web servicesare based on technologies such as Web Services Description Language (WSDL), SOAP, and UDDI, which by themselves do not provide any security mechanisms to protect the information or assets that are involved in the Web services application. Given this situation, there is clearly a need for development of a risk management framework that enables us to identify and address the risks confronting such applications.

Given the nature of Web services, mechanisms such as the Hypertext Transfer Protocol Secure (HTTPS) or secure HTTP are not as effective in securing Web services applications as they centre on implementing point to point security. According to Michael Barrett, the Chief Information Security Officer at eBay Inc.'s PayPal Division, these basic standards need to be rethought, reevaluated and rewritten (McMillan, 2007). Clearly, in the case of Web services, we need more dynamic, message oriented security. The Web Services Security (WS-Security or WSS) specification is a flexible and feature-rich extension to SOAP that applies security to Web services.

WSS specifies how integrity and confidentiality can be enforced on messages and allows the communication of various security token formats like Security Assertion Markup Language (SAML), Kerberos, and X.509. WS-S specification addresses different issues such as authentication (process of identifying a user/application based on its credentials); authorization (process of validating whether an already authenticated user has access to a particular resource); integrity (means that the message was not tampered in transit); and confidentiality (which guarantees that only authorized individuals have access to a message). Furthermore, WS-S enables security information to be stored directly in the message header and it therefore stays in the message even if the message gets routed to more than one end point (Hasan & Duran, 2005).

RECOMMENDATIONS

Management of security in the new era ushered in by Web 2.0 is a highly demanding and complex task. That a vast volume of highly valuable information is now being migrated to the "cloud" for the purpose of seamless access by Web services and other e-business technology implementations should lead us to rethink our existing security strategies and standards. As we research the extant literature and review case studies associated with Web services and e-business technology implementations, several general directives in the form of best practice measures emerge.

1. In their quest to achieve compliance with security standards, organizations should not neglect the realities of their own context and refine their security strategies to meet the ground reality of their business; in the same vein, agencies responsible for formulating security standards must be sensitive to the difference in context that applies to different organizations and must use this to guide the development of their compliance rules and measures.
2. The importance of employee training must be recognized and proper training should be imparted to thwart the risk of insider threats, as much as possible.
3. Rather than cling to the traditional defensive security measures, innovative approaches (such as honeypots outlined earlier in the chapter) must be tried and tested by organizations as they try to secure their assets.
4. Organizations involved in creating and distributing Web-based applications must follow agile and iterative development methodologies that incorporate user feedback in the newer versions of the applications being designed.
5. Implementing the triad of confidentiality, integrity and authentication is the key to the success of Web 2.0 technology imple-

mentations and the determinant of whether or not the Web services application will be in demand.

6. Given that information will be traveling over the Web, high quality data encryption is essential, an issue that is much simplified today owing to fast, inexpensive, and on-the-fly data encryption techniques.
7. Maintenance is important and regularly testing and evaluating applications is important, especially when they have been changed or updated.
8. Last but not the least, it is important to develop a security culture, whereby implementation of security measures and standards, and adhering to best practices is not perceived as burdensome, but where the importance of such directives is defined and effectively communicated by the management.

FUTURE RESEARCH DIRECTIONS

The advent of Web 2.0, the new era marking Internet-based computing presents us with diverse research opportunities.

First, to truly realize the vision being posited by Web 2.0, research needs to focus on how new Web services can be created from already published services without any human mediation and how service discovery, acquisition, and monitoring can be automated.

Second, research needs to focus on creating a common, shared language and understanding between service integrators, vendors and customers of all sizes, so that the Web services applications can be effectively deployed.

Third, a better understanding of the issues and challenges associated with securing Web services, service-oriented architectures, and cloud computing platforms is needed. We are also presented with research opportunities in the area of development of security standards and specifications. Given that the Web 2.0 is a totally different context than Web 1.0, the older security models are challenged because the older assumptions no longer apply.

Fourth, future research also needs to focus on the related topic of formulating effective organizational level security policy to deal with the new open context, where valuable informational assets are being increasingly migrated to the cloud or the Internet.

Fifth, we need to rigorously understand chief interoperability issues to ensure that valuable services can be quickly deployed.

Last, but not the least, we are presented with research opportunities in the area of formulating cyber legislation where complexities are compounded by the fact that the jurisdiction of territorial laws is vague at best.

What are described above are only a few of the many research opportunities that will enable us to leverage the power of Web 2.0. It is clear that most research projects will require cross disciplinary research involving disciplines such as Information Systems, Computer Science, Business/Management, and Law, to name a few. We are indeed fortunate that we are witnessing this critical revolution and can participate in shaping it.

CONCLUSION

Security and implementation challenges continue to affect the deployment and use of Web services, cloud computing and other e-business technologies in general. The security issues mostly stem from factors such as authentication, authorization, confidentiality, integrity, non-repudiation, availability and end to end security. On the other hand, implementation challenges arise because such applications often require the integration of services from other companies, vendors and third parties, as well as the interoperability of the differing computer systems involved.

With respect to mitigating risks of Web services applications, the Web Services Security (WSS or WS-Security) specification is a flexible and feature-rich extension to SOAP that applies security to Web services. WS-S specification addresses different issues such as authentication, authorization, integrity and confidentiality. Risks to the broader spectrum of e-business technologies can be managed through approaches such as Cyber legislation, Information Security Governance and Compliance and Business Continuity Planning. As organizations deal with the security implications of Web 2.0 offerings they need to realize that a cookie cutter approach is not beneficial when formulating security standards, as in order to be successful the standard must take into consideration the contextual realities that differ from organization to organization. Organizations need to engage their employees through formal training thereby helping them become aware of security loopholes and vulnerabilities of e-business technologies in general; the focus needs to remain on building an overall security culture or mindset. Numerous research opportunities can be explored within the context of Web 2.0. Some of these include, building a common, shared language that can be used by vendors, integrators and customers alike, complete automation of Web services creation and monitoring, development of security standards, formulation of and cyber legislation, to name a few.

REFERENCES

Anyiwo, D. (2006). *Perspectives on Semantic Web Services security.* Paper presented at the 1st Annual Web Services Security Conference, College Park, Maryland.

Ashford, W. (2009). Cyber-criminals to favour financial scams in 2009, says McAfee. *Computer Weekly*. Retrieved February 26, 2010, from http://www.computerweekly.com/Articles/2009/01/20/234325/cyber-criminals-to-favour-financial-scams-in-2009-says.htm

Babcock, C. (2010). Cloud security alliance to tackle cloud standards. *Information Week*. Retrieved March 2, 2010, from http://www.informationweek.com/news/software/hosted/showArticle.jhtml?articleID=223101102&cid=RSSfeed_IWK_All

Barber, R., & Integralis, A. (2001). Hacking techniques: The tools that hackers use, and how they are evolving to become more sophisticated. *Computer Fraud & Security*, (3): 9–12. doi:10.1016/S1361-3723(01)03014-7

Beachboard, J., et al. (2008). Improving Information Security risk analysis practices for small- and medium-sized enterprises: A research agenda. *Issues in Informing Science and Information Technology, 5.*

Cerullo, V., & Cerullo, M. (2004). Business continuity planning: A comprehensive approach. *Information Systems Management, 21*(3), 70–80. doi:10.1201/1078/44432.21.3.20040601/82480.11

Dimitrakos, T., et al. (2002). CORAS-a framework for risk analysis of security critical systems. *ERCIM News, 49.*

Firestorm. (2008). *Public law 110-53, Title IX.* Retrieved March 2nd, 2010, from http://www.publiclaw110.com/?p=3

Fletcher, O. (2010). China working to toughen up hacking laws. *Infoworld.* Retrieved February 2, 2010, from http://www.infoworld.com/d/security-central/china-working-toughen-hacking-laws-681

Gutiérrez, C., et al. (2004). Web Services security: Is the problem solved? *Information Security Journal: A Global Perspective, 13*(3), 22–31.

Herbane, B., Elliot, D., & Swartz, E. (1997). Contingency and continua: Achieving excellence through business continuity planning. *Business Horizons, 40*(6). doi:10.1016/S0007-6813(97)90063-X

Höne, K., & Eloff, J. (2002). Information security policy—what do international information security standards say? *Computers & Security, 21*(5), 402–409. doi:10.1016/S0167-4048(02)00504-7

Kanneganti, R., & Chodavarapu, P. (2008). *SOA security*. Greenwich, CT: Mannings Publications.

Laskey, K. (2005). Semantic interoperability for Web Services: The needs and challenges for the demanding consumer. *W3C Workshop on Frameworks for Semantics in Web Services*.

McMillan, R. (2007). *Researchers: Web 2.0 security seriously flawed*. Retrieved June 10th, 2010, from http://www.pcworld.com/article/131215/researchers_web_20_security_seriously_flawed.html

Mell, P., & Grance, T. (2009). *Effectively and securely using the cloud computing paradigm*. Retrieved March 02, 2010, from http://csrc.nist.gov/groups/SNS/cloud-computing/

Moore, A., Cappelli, D., & Trzeciak, R. (2008). The big picture of insider IT sabotage across U.S. critical infrastructures. In S. Stolfo, et al. (Eds.), *Insider attack and cyber security: Beyond the hacker.* New York: Springer Science + Business Media, LLC. Also published as an SEI technical report, CMU/SEI-2008-TR-009. Retrieved from http://www.cert.org/archive/pdf/08tr009.pdf.

Myers, K. (1996). *Total contingency planning for disasters: Managing risk, minimizing loss, ensuring business continuity*. New York: John Wiley & Sons.

Nagarajan, M., et al. (2006). Semantic interoperability of Web Services–challenges and experiences. *Proceedings of the Fourth IEEE International Conference on Web Services* (ICWS 2006), Chicago, IL.

Nappinai, N. S. (2010). Cyber crime law in India: Has law kept pace with emerging trends? An empirical study. *Journal of International Commercial Law and Technology, 5*(1), 22–28.

O'Neill, M. (2003). *Web Services security*. New York: McGraw Hill Publications.

O'Reilly, T. (2007). What is Web 2.0: Design patterns and business models for the next generation of software. *Communications & Strategies, First Quarter, 17*.

Parry, E. (2003). *Gartner releases top 11 list of security issues. SearchCIO.* Retrieved February 26, 2010, from http://searchcio.techtarget.com/news/article/0,289142,sid182_gci891984,00.html

Perez, S. (2009). *Top 8 Web 2.0 security threats*. Retrieved June 10th, 2010, from http://www.readwriteweb.com/enterprise/2009/02/top-8-web-20-security-threats.php

PRNewswire. (2010). *Security legislation will help protect U.S. economy, Says VirnetX researcher*. Retrieved March 1st, 2010, from http://markets.chron.com/chron/?GUID=11788469&Page=MediaViewer&Ticker=VHC

Siponen, M., & Willison, R. (2009). Information Security management standards: Problems and solutions. *Information & Management, 46*(5), 267–270. doi:10.1016/j.im.2008.12.007

Spitzner, L. (2003). *Honeypots: Tracking hackers*. Reading, MA: Addison-Wesley Professional.

Traynor, I. (2007, May 17). Russia accused of unleashing cyber war to disable Estonia. *The Guardian*.

Trout, B. (2007). *Cyber law: A legal arsenal for online business*. World Audience, Inc.

Websense. (2009). *State of Internet security*. Q1-Q2 2009. Retrieved June 10th, 2010, from http://www.websense.com/assets/reports/report-wsl-state-of-internet-security-q1-q2-2009.pdf

Williams, A. (2010). *Cloud computing security: What the big guns have in store*. Retrieved March 2nd, 2010, from http://www.readwriteweb.com/cloud/2010/03/cloud-computing-security-what.php

Yee, G. (2006). Measuring privacy protection in Web Services. *Proceedings of the ICWS'06, International Conference on Web Services,* Chicago, IL, (pp 647–654).

ADDITIONAL READING

Annapaola, M., Marco, P., & Paolo, T. (2008). Automated composition of web services: the ASTRO approach. *IEEE Data Eng. Bull, 31*(3), 23–26.

Bertoli, P., et al. (2009). Continuous Orchestration of Web Services via Planning. In Gerevini, A. et al. (Eds.): *Proceedings of the 19th International Conference on Automated Planning and Scheduling*, ICAPS 2009, Thessaloniki, Greece, September 19-23, 2009. AAAI 2009, ISBN 978-1-57735-406-2

Buyya, R., et al. (2009). Modeling and simulation of scalable cloud computing environments and the CloudSim toolkit: Challenges and opportunities. *7th High Performance Computing and Simulation Conference* (HPCS 2009). Leipzig, Germany, June 2009

Buyya, R. (2009). Cloud computing and emerging IT platforms: Vision, hype, and reality for delivering computing as the 5th utility. *Future Generation Computer Systems, 25*(6), 599–616. doi:10.1016/j.future.2008.12.001

Chinnici, R., et al. (2004) *Web Services Description Language (WSDL) Version 2.0. Part 1: Core Language*, W3C Working Draft 26.

Curbera, F. (2002). Unraveling the Web services web: an introduction to SOAP, WSDL, and UDDI. *Internet Computing, 6*(2), 86–93. doi:10.1109/4236.991449

Erl, T. (2005). *Service-oriented architecture: Concepts, technology, and design*. Upper Saddle River, NJ: Prentice Hall, ISBN 0-13-185858.

Gottschalk, K., Graham, S., Kreger, H., & Snell, J. (2002). Introduction to Web services architecture. *IBM Systems Journal, 41*(2), 170–177. doi:10.1147/sj.412.0170

Grossman, R. L., & Gu, Y. (2009). On the Varieties of Clouds for Data Intensive Computing. *IEEE Data Eng. Bull., 32*(1), 44–50.

Hao, Y., Ding, Y., & Wen, G. (2001). SOAP Protocol and its Application [J]. *Computer Engineering, 27*(6), 128–130.

Little, M. (2003). Transactions and Web Services. *Communications of the ACM, 46*(10), 49–54. doi:10.1145/944217.944237

Narayanan, S., & McIlraith, S. A. (2003). Analysis and simulation of Web services. *Computer Networks, 42*(5), 675–693. doi:10.1016/S1389-1286(03)00228-7

Newcomer, E., & Lomow, G. (2004). *Understanding SOA with Web Services*. Reading, MA: Addison-Wesley Professional.

O'Reilly, T. (2005), *What is Web 2.0*? Retrieved March 02, 2010 from http://www.oreilly.com/go/web2.

Paolucci, M., Kawamura, T., Payne, T. R., & Sycara, K. P. (2002). *Proceedings of the First International Semantic Web Conference on The Semantic Web,* Volume: 2342, Publisher: Springer-Verlag, 333-347.

Schou, C., & Shoemaker, D. (2007). *Information Assurance for the Enterprise: A Roadmap to Information Security*. New York, NY: McGraw-Hill/Irwin.

Shirin, S., & McIlraith, S. A. (2009). Optimizing web service composition while enforcing regulations. *International Semantic Web Conference,* 601-617.

Sycara, K., Poulucci, M., Ankolekar, A., & Srinivasan, N. (2003). Automated Discovery, Interaction and Composition of Semantic Web Services. *Journal of Web Semantics*, *1*(1), 27–46. doi:10.1016/j.websem.2003.07.002

Voorsluys, W., Broberg, J., Venugopal, S., & Buyya, R. (2009). Cost of Virtual Machine Live Migration in Clouds: A Performance Evaluation. *Proceedings of CloudCom*, *2009*, 254–265.

Yonah, A., & Swetnam, M. S. (1999). *Cyber terrorism and Information Warfare*. Dobbs Ferry, NY: Oceana Publ. Inc.

Zhang, L. J., Zhou, Q., & Chao, T. (2004). A Dynamic Services Discovery Framework for Traversing Web Services Representation Chain. *Proceedings of the ICWS*, *2004*, 632–639.

KEY TERMS AND DEFINITIONS

Cloud Computing: Cloud computing is a model for enabling convenient, on-demand network access to a shared pool of configurable computing resources (e.g., networks, servers, storage, applications, and services) that can be rapidly provisioned and released with minimal management effort or service provider interaction.

Extensible Markup Language (XML): Is used to tag the data. XML is an open industry standard managed by the World Wide Web Consortium. It enables developers to describe data being exchanged between PCs, smart devices, applications, and Web sites. In XML, the data is separate from the format and style definitions, and can therefore be easily organized, programmed, edited, and exchanged between any Web sites, applications, and devices.

Service Oriented Architecture (SOA): Is an application architecture in which all functions, or services, are defined using a description language and have interfaces that can be invoked to perform business processes. Although SOA and Web Services are sometimes used interchangeably, they are in fact very different. While SOA is the design, Web Services are technologies such as SOAP and XML. SOA is independent of technologies and can be implemented using technologies other than Web Services too (for e.g., by using Java).

Simple Object Access Protocol (SOAP): Is a simple XML-based communication protocol that enables applications exchange information over HTTP. It is platform- and language- independent, allowing disparate applications to communicate across various operating systems, with different technologies, and programming languages.

Universal Description, Discovery and Integration (UDDI): In contrast to WSDL is used for 'listing' the web services available. It comprises of a Web-based distributed directory that enables businesses to list themselves on the Internet and discover each other, similar to a traditional phone book's yellow and white pages.

Web 2.0: Web 2.0 is a term given to describe a second generation of the World Wide Web that is focused on the ability for people to collaborate and share information online. It was coined by Darcy DiNucci in 1999, but was popularized when John Battelle and Tim O'Reilly used in the first Web 2.0 conference. In essence Web 2.0 basically refers to the transition from web 1.0 which comprised of static HTML Web pages, which were passively browsed by users, and signifies a more dynamic Web which enables user interaction with Web Content (as opposed to passive browsing) and is based on serving Web applications to users.

Web 3.0: Web 3.0 or the Semantic Web is emerging from the programs initiated by the Defense Advanced Research Projects Agency (DARPA) and other organizations, including the World Wide Web Consortium (W3C) and the European Union at the turn of the 21st century. The Semantic Web enables the semantic integration and linkage of related Web documents to machine-readable ontologies.

Web Services Description Language (WSDL): Is used for 'describing' the services available. WSDL was developed jointly by Microsoft and IBM.

Web Services: The term Web services describes a standardized way of integrating Web-based applications using the XML, SOAP, WSDL and UDDI open standards over an Internet protocol backbone. Web services allow different applications from different sources to communicate with each other without time-consuming custom coding, and because all communication is in XML, Web services are not tied to any one operating system or programming language.

Chapter 16
E-Business and Security

Sharon Nachtigal
Royal Holloway, University of London, UK

ABSTRACT

This chapter is concerned with a major problem for any e-business organization, the security of its Information Systems. A review of information security characteristics and components is presented, followed by a detailed discussion of e-business security issues. Based on a structured approach for describing e-business functionality, e-business characteristics relevant to information security are identified. The major e-business security challenges are considered and e-business security issues are discussed and requirements are identified in different aspects of the realm. The current perimeter security approach appears to be inadequate to the modern business environment. Hence, a different approach is needed. A few alternative approaches are discussed and a review of previous and future research on e-business security is presented. Hence, the chapter aims to contribute both to academics and to e-business executives by providing the information security insight and awareness to the e-business unique security issues and challenges.

INTRODUCTION

E-business has unique features, derived from its use of Information Technology. The definition identifies an e-business as a company running its business processes electronically (Nachtigal, 2007).

DOI: 10.4018/978-1-60960-501-8.ch016

Today's organization faces a business environment which is very different from that of the 1980s; the possibility of interacting with suppliers, customers, competitors, partners, and financial and government institutions via the Internet changes the way operations are performed, the business and financial relationships, and the ways of dealing with money. As a result, there are new security vulnerabilities and threats, and so this new environment requires a different set of security goals. Knowledge about secure e-business mode of doing business appears to be essential in order to implement it effectively.

BACKGROUND: INFORMATION SECURITY ISSUES

In this section the basic information security concepts and terms are presented. The presentation of the information security basics is essential in order to proceed with the unique e-business security challenges, which are discussed in the rest of this chapter.

The business environment is constantly changing along with (and due to the) advances in technology. Indeed, business environment benefits from advances in Information Technology (IT) by expanding and/or modifying its business activities, but also these advances in IT produce a new range of threats. A *threat* can be defined as any circumstance or event that has the potential to harm a system (Slade, 2006).The various threats to information and Information Systems (IS) also constantly change, depending on a specific business environment, IT environment, specific business goals and functionality. Hence, it is not possible to provide a list of all possible threats, but their classification into four categories (Bishop, 2005), such as:

1. disclosure (unauthorised access to information);
2. deception (acceptance of false data);
3. disruption (interruption or prevention of correct operation); and
4. usurpation (unauthorised control of some part of a system).

Although the level of information security varies according to specific threats relevant to a specific organisation, there are common general *security goals*. *Confidentiality, Integrity,* and *Availability* (CIA) are commonly considered to be the fundamental goals of IT security (Slade, 2006); they are sometimes also referred to as objectives, requirements, or properties (see, for example, (Furnell, 2005; Harris; Tettero, 2000)).

Sinclaire (2005) based his findings on a review of a range of MIS research literature published over the period 2002 - 2004 in four highly ranked journals[1] for IS research publications. The author makes a distinction between *information security* and *information privacy*. Based on the definitions of *information privacy* as the ability of the individual to control personal information about one's self (Stone et al cited in Sinclaire, 2005), and the Panko (cited in Sinclaire, 2005) definition of *IT security* as providing confidentiality, integrity, and availability, Sinclaire (2005) presents the following statistics regarding research on security. A total of 24 articles were reviewed, of which 74% addressed information security and 26% addressed information privacy; 29% of the information security research articles address planning, and 71% address protection. Analysis of research within the protection category reveals that 50% pertain to authentication/verification, 25% are about types of threats, 17% address standards, and eight percent pertain to firewalls. Analysis of information privacy research reveals that 67% address user perceptions and the remaining 33% pertain to surveillance issues. In this context *confidentiality* refers to the holding of sensitive data in confidence, limiting access to an appropriate set of individuals (Slade, 2006).

Integrity involves maintaining the accuracy and reliability of information and systems, and preventing unauthorised modification of data, thereby ensuring that data is correct as defined by the process designer (Harris; Tettero, 2000).

Availability refers to the state when the system, resources, and data are in the place needed by the user, at the time the user needs them, and in the form the user needs (Slade, 2006).

Information and/or Information Systems Security is concerned with assuring that the security requirements triad (CIA) is achieved. The CIA triad can be violated by various *attacks*. An *attack* is an act that tries to bypass security (Slade, 2006). A list of *attack categories*, including some of the more widely discussed classes of attack, is provided below:

1. malware (a collective term for many varieties of deliberately malicious software (Furnell, 2005; Slade, 2006), including:
   - viruses (self-replicating and propagating programs, usually operating with some form of input from the user, although generally the user is unaware of the intent of the virus (Slade, 2006)); An example of such an attack in 2008 was inflected to an U.S. military base in Afghanistan "which affected nearly three quarters of the computers on the base" (Mullrine, 2008)
   - worms (replicating programs that can spread between systems autonomously, without the need to infect a carrier in the manner that a virus does (Furnell, 2005)); A very new example of such malware in September 2010 is the new worm that arrives via emails with the subject line "Here You Have" and contains a link to a site that will download a malicious file to the victim's PC. According to McAfee analysis, "the virus attempts also to stop and delete security services"[2].
   - Trojans (programs that pretend to be something else in order to enter a system and encourage people to use them, typically resulting in unexpected and unwanted effects (Furnell, 2005)); An example of a Trojan horse is SpySheriff[3], which masquerades as an anti-spyware program but when executed essentially takes over a user's computer, preventing access to files and the internet.
   - spyware (a program that reports on the contents, status, or operation of a computer to a remote system or user (Slade, 2006)); Spyware can steal confidential business information, leaving companies vulnerable in several ways. For example, an online advertisement is one of the ways to install a spyware program into a user's computer - many spyware programs display advertisement[4];
2. denial of service (DoS or DDoS) (a form of attack in which legitimate access is prevented or impeded as a direct result of activities originating from unauthorised parties (Furnell, 2005));
3. social engineering (the use of fraud, spoofing, or other social or psychological measures to get legitimate users to break security policy (Slade, 2006));
4. insider attacks (attacks involving an employee or other trusted individual, generally one with a higher than normal access (Slade, 2006));
5. impersonation attacks (an attempt to gain access to a system by posing as an authorised user (Slade, 2006));
6. hacking (breaking security systems by either skilled or unskilled persons (Slade, 2006));
7. exploitation of implementation errors (i.e. an attack taking advantage of weaknesses that exist as a result of errors in development, yielding a system which is not consistent with security policy requirements).
8. harassment (sending unwanted threatening or injurious messages to an opponent directly, either in person or through a medium such as email (Denning, 2006));
9. cyberterrorism (the use of information technology by terrorist groups and individuals to further their agenda*);

10. political or industrial net espionage (network-enabled espionage (Boni & Kovacich, 2000)).

There is a wide variety of attack techniques, which may include, for example, one or more of the following:

1. An SQL injection (an attack that manipulates parameters that are used in SQL statements (Osborne, 2006));
2. A buffer overflow (made possible as a result of a common programming error in which excessive input exceeds the memory space allocated to it, potentially causing the program to execute arbitrary code or switch operational control to an arbitrary memory location (Slade, 2006));
3. spoofing/masquerade (an attempt to gain access to a system by posing as an authorised user (Slade, 2006));
4. packet sniffing (traffic monitoring used by an attacker within a network to gather information about the network (Liska, 2003);
5. abuse of cookies (using the information contained in cookies for illegitimate purposes);
6. DNS (Domain Name System) hacking (an attack on the DNS infrastructure that has the potential to affect a large portion of the Internet (Chakrabarti & Manimaran, 2002));
7. phishing (posting of a fraudulent message to a large number of people via spam or other general posting, asking them to submit personal and financial information (Slade, 2006));
8. SMiShing (a compound of "phishing" and "SMS", where mobile phone users receive text messages containing a Web site hyperlink, which, if clicked, downloads a Trojan horse to the mobile phone[†]);
9. vishing (the telephone equivalent of phishing[‡]);
10. spamming (indiscriminately sending unsolicited, unwanted, irrelevant, or inappropriate messages, especially commercial advertising, in mass quantities (Slade, 2006)); and several others.

As a result of a successful attack, damage can be caused to information and information systems. Some of the possible types of damage include software corruption/modification, hardware malfunction, data corruption/modification/exposure/theft, identity theft, intellectual property theft, financial loss, damage to reputation, national-level infrastructure disaster.

Information security *safeguards* include methods, tools, techniques, policies, procedures, and standards to ensure the organisation security goals. These include a wide variety of access control models, cryptographic algorithms, antivirus software, firewalls, IDSs (Intrusion Detection Systems), IPSs (Intrusion Prevention Systems), organisation security policies, security standards (e.g. HIPAA and SOX) and other protective and preventive means.

Given this brief review of information security field, this chapter will progress with the detailed discussion of e-business security issues, which is provided immediately below.

E-BUSINESS SECURITY ISSUES

The following section will present issues relevant to what is widely accepted as one of the main barriers to the successful deployment of e-business, i.e. security (Jarvis, 1999; Jones, Beynon-Davies, & Muir, 2003; Katsikas, Lopez, & Pernul, 2005). Security issues are classified here into two groups; a broader group covers Internet-related security problems in general, and a more specific group includes Web application vulnerabilities.

Internet-Related Security Problems

Interconnectivity, one of the main characteristics of an e-business, widens the scope of potential attacks. Security breaches and unreliable e-services can result in a range of types of business losses, including e-business productivity losses, asset losses, and reputation damage (Wang, Hidvegi, Bailey, & Whinston, 2001).

New threats and problems arise when using Internet technology, and especially when a company adopts the e-business mode. A particular danger in running an e-business arises from its "openness" to the environment, and the various connections and communication channels it shares with the external world. E-business involves performing business interactions (transmitting documents represented by information flows) between organization portals using Internet technology.

Use of a web browser brings with it various threats and vulnerabilities for the end user (business or private customer). The range of threats includes issues such as (Joshi, et al. 2001):

1. violation of user privacy by abusing information held about a user (such as login name or computer name);
2. cookies, stored on the client machine and exchanged between the Web client and the Web server to maintain connection information, can be used for the purpose of gathering potentially sensitive user information;
3. executable downloads, such as Java applets and ActiveX controls, can be a source of vulnerability at the client;
4. push technology, used by many sites to deliver web content to customers, give rise to serious security vulnerabilities (the content provider could, for example, send malicious code or cause Denial of Service (DoS) attacks);
5. attacks on network servers (Web servers, application servers, mail servers, etc.).

Poorly configured networks are subject to confidentiality, integrity and/or availability threats (Osborne, 2006).

Traditional authentication methods (such as passwords), when used for access control in an online environment, are subject to a range of vulnerabilities (e.g., arising from keystroke loggers, which record the keys struck by a user).

Because of the global access it provides, the Internet is very vulnerable to malicious actions, while worms, viruses and other types of malware continue to evolve both with respect to the mechanisms they use and their infection speed (Jungck & Shim, 2004). Existing server and network infrastructures include large numbers of tools and mechanisms designed to mitigate known threats. Finally, "day zero" vulnerabilities (those that reveal themselves only when they are first exploited) need different approaches and solutions (Jungck & Shim, 2004).

Jungck and Shim (2004) use the January 2003 SQL slammer worm as an example of an attack that took advantage of vulnerabilities in firewalls and IDSs (Intrusion Detection Systems). In order to allow service provision, argue the authors, many firewalls left open the port that Slammer attacked, and most IDSs left that port unmonitored.

According to Jungck and Shim (2004), it is very difficult to update every system to address every known threat; it is difficult for administrators to manage such a process, and for organizations to afford the necessary updating activities. The authors also emphasize that increasing Internet bandwidth is another obstacle to information security, because the growth in bandwidth (e.g. in optical networks) has exceeded the ability of processors to filter the traffic. Closing the gap between bandwidth and processor capabilities is only made possible by either slowing traffic to a speed at which security applications can be performed, or by abandoning security monitoring to meet network performance goals.

It is thus evident that information security problems span a wide range of issues, including lost flexibility for high-speed operation, changing protocol threats, and problems related to traditional designs. According to the authors, the problem is not about silicon, systems, and applications, but in the standards that are written without security in mind, the complexities that are being introduced by not adhering to standards, and the network topologies pasted together without a holistic view and systematic approach (Jungck & Shim, 2004). Internet businesses often face major operational uncertainties arising from system complexity, rapid development, interconnectivity, and a lack of familiarity with the new technology based economy (Wang et al., 2000; Wang et al., 2001).

Many e-business organizations suffer from problems with the information systems that facilitate e-business (Anderson, et al. 2005). Such problems include poor security, flawed controls, inadequate management controls, and poorly designed and unreliable back-end systems (Anderson et al., 2005; Wang et al., 2001). Wang et al. (2001), argue that management lacks reasonable assurance whether or not their digital operations are effective and efficient, or whether information generated by e-processes is reliable for decision making, or whether e-operations are compliant with the applicable laws and regulations. The lack of these types of confidence is resulted in highly publicized security breaches and failures in e-service. Effective design of e-business processes is essential to avoid security defects (Anderson et al., 2005). Because of the critical reliance of e-business on correct execution of its e-processes, tools to verify e-process design and implementation need to be developed (Wang et al., 2001).

From a slightly different perspective, digital copyright has become a major concern for businesses that engage in online content distribution using approaches such as pay-per-view, subscription, trading, etc. (Kwok, et al. 2004). Advances in Internet technology have enabled digital service providers to sell their digital content via computer networks. Digital content can be easily copied, altered, and distributed to a large number of recipients, which could cause significant revenue loss to media companies (Liu, et al. 2003). Intellectual property protection is a growing concern for content owners (Kwok, et al. 2004).

Privacy issues and identity of the customers are considered (Katsikas et al., 2005) to be an important aspect of e-business, as well as communications- and transaction-related issues (Aissi, et al. 2002; Katsikas et al., 2005).

Attacks on e-business information system can occur at various levels of the system. Examples of such attacks include:

- denial-of-service attacks;
- damage caused by malware;
- attacks arising from exploits of the Web Services application interface;
- script-based injections causing damage to an application and/or its data;
- network interception arising from protocol weaknesses;
- defeat of encryption arising from faulty cryptographic key management or poor selection of encryption methods;
- database administrators stealing sensitive database content or configuration parameters;
- application programmers inserting undetected malicious code in application software, causing problems such as widespread security failures or subversion of critical business transactions.

Next session will discuss the information security issues that arise while using the Web application and technologies.

Web Application Vulnerabilities and Challenges

Web services have great potential for both application developers and users (Gehling & Stankard, 2005). However, certain features of web services make them particularly vulnerable to a number of possible attacks. Gehling and Stankard (2005) observe that attacks on web applications and web services are the fastest growing new category of attack, and describe the following vulnerabilities particular to web services:

1. the web services format was designed to bypass existing security measures, to be platform independent, and to support any application call structure; hence, additional security measures are needed;
2. although the flexibility of SOAP (Simple Object Access Protocol) and other technologies makes communication among applications easy, it also simplifies interception and manipulation of messages;
3. SOAP messages are transparent to firewalls, which mean that perimeter security controls are likely to be bypassed.

Much of today's online activity (such as online shopping, online banking, and online business) makes use of web servers, which are also frequently attacked. One class of weaknesses that makes such attacks possible are the well known and commonly occurring SQL (Structured Query Language) command injection vulnerabilities (Wasserman & Su, 2007).

Such attacks are made possible by inadequate checking of untrusted user inputs (substrings) during communication between the application layer and the back-end database in the web-application architecture (Wasserman & Su, 2007). These attacks typically involve manipulation of SQL statement parameters in input fields, such as customer number, address, or search phrase (Osborne, 2006). Attacks taking advantage of such vulnerabilities can result in loss of confidentiality, integrity and authenticity for important business data.

Many security flaws in e-business applications arise from design-related flaws (Woon & Kankanhalli, 2007). According to the reports of a survey conducted by AtStake[§], security design flaws were found in 70% of all analysed defects. Although part of the flaws were of low business impact or were not easily exploitable, nearly half (47%) of the remaining serious defects could have been caught and fixed inexpensively during the design stage (Jaquith, 2002). There are several reasons for the presence of such flaws, including:

- many designers and developers are not trained in general security principles (Woon & Kankanhalli, 2007);
- designers and developers do not consider security as an explicit application requirement (Mead & Stehney, 2005; Woon & Kankanhalli, 2007).
- security issues are addressed only after a vulnerability is discovered (Sherwood, Clark, & Lynas, 2005; Woon & Kankanhalli, 2007);
- marketing pressures lead to shortcuts in application design and development, resulting in less robust software that may include security flaws (Woon & Kankanhalli, 2007);
- there are many different e-commerce applications written in a variety of languages running on a range of different systems (De Win, Van den Bergh, Matthijs, De Decker, & Joosen, 2000);
- generic e-commerce applications have specific security requirements, which some existing security architectures (e.g. SESAME[5], and CORBA[6]) are not able to meet (De Win et al., 2000);
- e-business operation is very complex, and so are the e-business applications that implement these operations.

Also, once operational use of an application has commenced, it is very difficult and expensive to significantly improve its security (Mead & Stehney, 2005).

The business and financial relationships of the modern organizations, as well as the ways of dealing with money, have changed due to the possibility of interacting with suppliers, customers, competitors, partners, and financial and government institutions via the Internet. As a result, there are new security vulnerabilities and threats, and so this new environment requires a different set of security goals.

The security models, mechanisms and techniques currently used, support the perimeter paradigm by meeting the "pre-e-business reality" security objectives. The safeguards and methods reviewed above are designed to support security policies that presume, according to the perimeter paradigm, that organisation should be surrounded by a fence separating it from the external world. This leads to the conclusion that the perimeter security approach is not appropriate for the e-business mode of doing business. The use of the perimeter security approach is thus a major problem for the Information Systems Security (ISS) discipline today.

Leyman et al. (2002), describing modern security measures, argue that new approaches are needed to support dynamic business processes and their management, which provides:

1. the ability to prescribe how Web services are used to implement activities within a business process;
2. a way of deciding how business processes are represented as Web services; and enhances
3. the ability to decide which business partners perform what parts of the business process.

In a process-based security paradigm (Nachtigal & Mitchell, 2006; Nachtigal, eBPSM—A New Security Paradigm for e-Business Organisations (e-Business Process Security Model), 2007) was suggested for e-business organizations as a new alternative to the perimeter security approach. According to this newly suggested approach, e-business mode of doing business is built upon performing intra- and inter-organisational electronic processes (e-processes). E-business operation is dependent on proper execution of its comprising e-processes. Therefore, secure execution of these e-processes is crucial for any e-business organisation. Other researchers argue that robust e-process design and implementation is the first line of defence (Wang, et al., 2001).

E-business implies different security challenges as compared to the traditional business environment. Hence, the CIA triple is not sufficient for this new type of business. The security goals set must be extended to include additional goals (see, for example, e-business security requirements septet as suggested in (Nachtigal, E-Business security design using process security requirements septet, 2007).

E-business is a relatively new discipline – and so is the e-business information security. Also, by its nature, it is dynamic and ever changing – and so are the threats and ways to violate the e-business security. Efforts to discover and improve the ways in which its security is designed and implemented, are evident both in research and in practice. We next present a review on e-business security research.

E-BUSINESS SECURITY RESEARCH

This section contains a review of prior research on security for e-business and e-commerce, although the distinction between these two modes of business is often not clearly made.

Ngai and Wat (Ngai & Wat, 2002) conducted a literature review of 275 research articles on e-commerce published in nine leading journals over the period 1993-1999. This review was used to devise a classification scheme for e-commerce research. They classify articles under three main headings, namely by year of publication, by the percentage of the total number of articles in selected journals, and by topic. According to the findings of this review, among the technology-related articles, security was the most popular topic - about 33% of the technology-oriented articles were on e-commerce security (Ngai & Wat, 2002).

In recent years (i.e. since the late 1990s) research in e-commerce security has focussed on two broad areas (Wareham, Zheng, & Straub, 2005), namely research on improving e-business security through policy measures, and research on technical security measures for e-commerce. Wareham et al. (2005) came to this conclusion based on an analysis of 582 articles on e-commerce research published over the period 199 7- 2003 in both academic and professional journals. Following Ngai and Wat (2002), the authors give the distribution of primary topics identified in their sample. The topics were grouped into four major domains, namely information technology and infrastructure, applications and industry themes, business issues, and other social issues. The findings of the study performed by Wareham et al. (2005) have significant differences to those of the Ngai and Wat (2002) review of the period 1993 - 1999. Security was found to be one of the most underserved research areas, represented by only 2.4% of the articles. The authors point out the surprisingly low percentage (16.6%) of research on information technology and infrastructure, and argue that IT researchers have to understand the technical underpinnings of the central issues before proceeding with enlightened social, business, and application research (Wareham, Zheng, & Straub, 2005).

Prior research on improving e-business security is presented here using the classification provided in (Wareham, Zheng, & Straub, 2005), i.e. we divide our discussion into technical and organisation-related issues.

Technical Security Research

Security for e-commerce (which under our definition forms part of e-business) is often regarded as purely a communications security problem (Gollman, 2003), with the use of cryptography as a suggested solution. Gollmann (2003) argues that cryptography protects the information transmitted between two points, i.e. it protects the communications media.

Digital watermarking (Cox, et al., 2008; Kutter & Hartung, 2000) is one technique that can be used to protect on-line content. A watermark can serve various purposes, including content protection, fingerprinting, ownership assertion, authentication and integrity verification, and usage control (Memon & Wong, 1998). Various watermarking techniques have been suggested. Proposed applications include using digital watermarking for intellectual property protection in electronic commerce transactions (Kwok, Yang, & Tam, 2004), for identity authentication in e-business activities (Wang, Yuan, & Archer, A contextual framework for combating identity theft, 2006), and for automatically protecting e-business data (Pons & Aljifri, 2002; Pons & Aljifri, 2005). The specific requirements on the watermarking technique vary with the application (Memon & Wong, 1998).

Although Memon and Wong (1998) argue that watermarking is undoubtedly important for protecting various forms of digital content, there are difficulties that still need to be solved. Watermarking techniques can be classified as either fragile or robust. While fragile watermarks are easily corrupted by image-processing procedures, robust watermarks resist common image-manipulation procedures, and are useful for ownership assertion purposes. However, devising robust digital watermarking schemes is a very difficult problem because of the numerous image manipulation techniques that a robust watermark must be able to survive (Memon & Wong, 1998).

Two classes of security service, namely access control services and communications security services, are usually considered to be the most crucial for an e-business organisation (Joshi et al., 2001). Access control prevents unauthorised use of resources, and communications security ensures the confidentiality and integrity of transmitted data. Joshi et al. (2001) argue that new access control mechanisms for Web-based applications are needed, because the existing access models are insufficient. The authors present a comparative assessment of security models used for Web-based applications and workflow systems, and claim that existing access models do not support dynamic changes in the content and context of information, or allow monitoring of the system state or transactional activities. Such features should be present in access models (Joshi et al., 2001).

Organisation-Related Security Research

Chua, et al. (2005) conducted a survey designed to test whether or not existing e-commerce research focuses primarily on certain specific stakeholders. The survey was based on publications in seven out of the top nine e-commerce journals. The results demonstrate that academic e-commerce researchers concentrate their attentions on two stakeholder groups, specifically customers and the internal organisation (i.e., managers and employees) of a Net-Enhanced Organisation (NEO). Other stakeholders, such as suppliers, indirect stakeholders, investors, and regulators, receive disproportionately less research interest (Chua et al., 2005). The authors argue that at least four stakeholder groups, namely investors, suppliers, regulators, and indirect stakeholders, will increasingly demand the attention of NEOs, and therefore IS and e-commerce research needs to reposition itself (Chua et al., 2005).

Privacy is a security-related issue relevant to e-business. The findings of a study conducted by Sinclaire (Sinclaire, 2005) reveal that there has been limited research in the area of information security and privacy, particularly at the organisational level.

Detailed requirements definition and analysis, as the first step in e-business systems design, has been proposed (e.g., in (Androutsellis-Theotokis, Spinellis, & Karakoidas, 2004)) to support collaboration between the parties in e-business activities. Androutsellis-Theotokis et al. (2004) discuss various requirements with respect to their importance for e-business collaboration, such as workflow and collaboration orchestration, authentication and access control, logging and non-repudiation, data storage security, availability, integrity, and anonymity. Herrmann and Pernul (1999) argue that security and integrity are two security requirements that are relevant for e-business.

Jones et al. (Jones, Wilikens, Morris, & Nasera, 2000) consider e-business requirements in terms of the trust and dependability of business partners.

E-business security management research covers a wide range of issues. As elsewhere, some authors use the terms e-commerce and e-business interchangeably, without making a distinction between these two concepts. Koskosas and Paul (2003) suggest a socio-organisational approach to IS security management in terms of a framework of security goal setting. The framework illustrates three important issues in the process of security goal setting, namely trust, culture, and risk communication.

Wang and Chen (2005) discuss the issue of e-business management in terms of three types of flow. The authors present an approach in which the object of e-business management is the set of e-business activities. According to Wang and Chen (2005), generally speaking, e-business management is a dynamic object formed by the circulation of three flows, namely information flow, fund flow and material flow (or logistics). Hence, e-business management should focus on guiding the coordination of these three flows.

Bodin et al. (Bodin, Gordon, & Loeb, 2005) propose a method for optimally allocating a budget for maintaining and enhancing the security of an organisation's information systems. The method is based on computing ratings according to confidentiality, data integrity, and availability criteria.

Rees et al. (Rees, Bandyopadhyay, & Spafford, 2003) propose a framework for creating security strategy and policy for applications. This framework, known as PFIRES (Policy Framework for Interpreting Risk in E-business Security), was initially developed for e-commerce activities, but has been generalised to handle security policy for all types of organisations engaged in computing and Internet operations. The authors (Rees, Bandyopadhyay, & Spafford, 2003) argue that PFIRES offers a possible starting point for understanding the impact of a security policy on an organisation, and is intended to guide organisations in developing, implementing, and maintaining security policies.

In summary, practice and research show that modern organizations need to adapt to the new business reality, which is dominated by IT development, and technological tools alone are not capable to cope with various security threats inherent in the information technology. Organizations need policies and guidelines as a part of a comprehensive vision and approach to achieve their information security goals.

CONCLUSION

E-business processes are enabled by the transmission of information flows. E-business, in order to succeed, must provide value to its users (including employees and customers). To deliver the value proposition to all its customers, a business must ensure that it possesses a range of capabilities that underpin the proposed value.

The Internet-based technologies used by e-business provide a wide range of services, each with its own capabilities and security vulnerabilities. Even if we assume perfect protection from firewalls, strong defence from unbreakable cryptographic algorithms and installation of all available software patches, if e-business processes are not designed or implemented properly then the e-business will still be exposed to losses (Wang etal.,2001). Also, generic e-business applications are created using a variety of technologies, which may not work in all environments.

Prior research on e-business and e-commerce security has focussed mainly on technical security measures, and on improving security through policies. Unfortunately, relying solely on technology solutions for e-business security is not enough. We conclude that a process- based security approach is a rational way to address e-business information security. In order to protect e-business information, we need a methodological approach to come to decisions regarding the degree and type of security protection that should be provided to information. Future research should be conducted in order to continue on addressing the issues raised in the review above.

REFERENCES

Aissi, S., Malu, P., & Srinivasan, K. (2002). E-business process modeling: The next big step. *Computer*, *35*(5), 55–62. doi:10.1109/MC.2002.999776

Anderson, B., Hansen, J., Lowry, P., & Summers, S. (2005). Model checking for design and assurance of e-business processes. *Decision Support Systems*, *39*, 333–344. doi:10.1016/j.dss.2003.12.001

Androutsellis-Theotokis, S., Spinellis, D., & Karakoidas, V. (2004). Performing peer-to-peer e-business transactions: A requirements analysis and perliminary design proposal. *IADIS, International e-Commerce 2004 Conference, December*, (pp. 399-404).

Bishop, M. (2005). *Introduction to computer security*. Addison-Wesley.

Bodin, L., Gordon, L., & Loeb, M. (2005). Evaluating information security investments using the analytic hierarchy process. *Communications of the ACM*, *48*(2), 78–83. doi:10.1145/1042091.1042094

Boni, W., & Kovacich, G. L. (2000). *Netspionage- the global threat to information*. Butterworth-Heinemann.

Chakrabarti, G., & Manimaran, A. (2002). Internet infrastructure security: A taxonomy. *Network*, *16*(6), 13–21.

Chua, C., Straub, D., Khoo, H., Kadiyala, S., & Kuechler, D. (2005). The evolution of e-commerce research: A stakeholder perspective. *Journal of Electronic Commerce Research*, *6*(4).

De Win, B., Van den Bergh, J., Matthijs, F., De Decker, B. & Joosen, W. (2000). Security architecture for electronic commerce applications. *SEC*, 491-500.

Denning, D. (2006). *Information warfare and security* (12th ed.). Addison-Wesley.

Furnell, S. (2005). *Computer insecurity*. London: Springer-Verlag.

Gehling, B., & Stankard, D. (2005). E-commerce security. *InfoSecCD '05: Proceedings of the 2nd annual conference on Information security curriculum development* (pp. 32-37). New York: ACM Press.

Gollman, D. (2003). *Computer security*. John Wiley & Sons.

Harris, S. (2008). *CISSP all-in-one exam guide* (2nd ed.). McGraw-Hill/Osborne.

Herrmann, G., & Pernul, G. (1999). Viewing business-process security from different perspectives. *International Journal of Electronic Commerce*, *3*(3).

Jaquith, A. (2002). *The security of applications: Not all are created equal.*

Jones, P., Beynon-Davies, P., & Muir, E. (2003). E-business barriers to growth within the SME sector. *Journal of Systems & Information Technology*, *7*(1-2), 1–26. doi:10.1108/13287260380000771

Jones, S., Wilikens, M., Morris, P., & Nasera, M. (2000). Trust requirements in e-business. *Communications of the ACM*, *43*(12), 81–87. doi:10.1145/355112.355128

Joshi, J., Aref, W., Ghafoor, A., & Spafford, E. (2001). Security models for Web-based applications. *Communications of the ACM*, *44*(2), 38–44. doi:10.1145/359205.359224

Jungck, P., & Shim, S. (2004). Issues in high speed Internet security. *Computer*, *37*(7), 36–42. doi:10.1109/MC.2004.58

Katsikas, S., Lopez, J., & Pernul, G. (2005). *Trust, privacy and security in e-business: Requirements and solutions*. (LNCS 3746), (pp. 548-558).

Kauffman, R., & Wood, C. (2007). Revolutionary research strategies for e-business: A philosophy of science view in the age of the Internet. In Kauffman, R., Wood, C., & Kauffman, R. A. (Eds.), *Economics, Information Systems, and electronic commerce research: Advanced empirical methodologies*. Armonk, NY: M. E. Sharpe.

Koskosas, I., & Paul, R. (2003). A socio-organisational approach to Information Systems security risks. [IJRAM]. *International Journal of Risk Assessment and Management*, *4*(2/3). doi:10.1504/IJRAM.2003.003527

Kutter, M., & Hartung, F. (2000). Introduction to watermarking techniques. In Petitcolas, S. K. (Ed.), *Information hiding techniques for steganography and digital watermarking*.

Kwok, S., Yang, C., & Tam, K. (2004). Intellectual property protection for electronic commerce applications. *Journal of Electronic Commerce Research*, *5*(1).

Leymann, F., Roller, D., & Schmidt, M.-T. (2002). Web services and business process management. *IBM Systems Journal*, *41*(2), 1015–1019. doi:10.1147/sj.412.0198

Liska, A. (2003). *The practice of network security-deployment strategies for production environment.* Prentice Hall PTR, Pearson Education Inc.

Liu, Q., Safavi-Naini, R., & Sheppard, N. (2003). Digital rights management for content distribution. *ACSW Frontiers '03: Proceedings of the Australasian information security workshop conference on ACSW frontiers,* (pp. 49-58). Darlinghurst, Australia: Australian Computer Society, Inc.

Mead, N., & Stehney, T. (2005). Security Quality Requirements Engineering (SQUARE) methodology. *SESS '05: Proceedings of the 2005 workshop on Software engineering for secure systems-building trustworthy applications,* (pp. 1-7). New York: ACM Press.

Melville, N., Kraemer, K., & Gurbaxani, V. (2004). Information Technology and organisational performance: An integrative model of it business value. *Management Information Systems Quarterly, 28*(2).

Memon, N., & Wong, P. (1998). Protecting digital media content. *Communications of the ACM, 41*(7), 35–43. doi:10.1145/278476.278485

Nachtigal, S. (2007). eBPSM—a new security paradigm for e-business organisations (e-business process security model). *ICEC '07: Proceedings of the Ninth International Conference on Electronic Commerce,* (pp. 101-106). Minneapolis: ACM.

Nachtigal, S. (2007). E-business security design using process security requirements septet. *ICETE - International Joint Conference on E-Business and Telecommunications, SECRYPT.* (pp. 408-413). Barcelona, Spain: INSTICC Press. Co-sponsored by IEEE. In cooperation with ACM SIGMIS.

Nachtigal, S., & Mitchell, C. (2006). Modelling e-business security using business process. *ICETE- SECRYPT 2006, International Conference on Security and Cryptography,* (pp. 459-464). Setubal, Portugal: INSTICC Press, in cooperation with IEEE Systems, Man and Cybernetics (SMC) Society.

Ngai, E., & Wat, F. (2002). A literature review and classification of electronic commerce research. *Information & Management, 39*, 415–429. doi:10.1016/S0378-7206(01)00107-0

Osborne, M. (2006). *How to cheat at managing information security.* Syngress Publishing, Inc.

Pons, A. & Aljifri, H. (2002). An active watermarking system. *IACIS, Issues in Information Systems.*

Pons, A., & Aljifri, H. (2005). Data protection using watermarking in e-business. *Journal of Database Management, 14*(4).

Rees, J., Bandyopadhyay, S., & Spafford, E. (2003). PFIRES: A policy framework for information security. *Communications of the ACM, 46*(7), 101–106. doi:10.1145/792704.792706

Schneier, B. (2004). *Secrets and lies.* Wiley.

Sherwood, J., Clark, A., & Lynas, D. (2005). *Enterprise security architecture: A business-driven approach.* CMP Books.

Sinclaire, J. (2005). Current research in information security and privacy. *Proceedings of the 2005 Southern Association of Information Systems Conference.*

Slade, R. (2006). *Dictionary of information security.* Syngress.

Tettero, O. (2000). *Intrinsic information security. Embedding security issues in the design process of telematics systems.* Telematics Institute Fundamental Research Series.

Thuraisingham, B. (2005). Directions for security and privacy for semantic e-business applications. *Communications of the ACM, 48*(12), 71–73. doi:10.1145/1101779.1101812

Walter, Z., & Scott, G. (2006). Management issues of Internet/Web systems. *Communications of the ACM, 49*(3), 87–91. doi:10.1145/1118178.1118181

Wang, W., Bailey, A., Hidvegi, Z., & Whinston, A. (2001). *A framework for proactive, automated and continuous e-commerce control and assurance.* Goizueta Business School Papers Series.

Wang, W., Hidvegi, Z., Bailey, A., & Whinston, A. (2000). E-process design and assurance using model checking. *Computer, 33*(10), 48–53. doi:10.1109/2.876292

Wang, W., Hidvegi, Z., Bailey, A. & Whinston, A. (2001). Model checking-a rigorous and efficient tool for e-commerce internal control and assurance. *Knowledge Emory* (DIA-2001-007).

Wang, W., Yuan, Y., & Archer, N. (2006). A contextual framework for combating identity theft. *IEEE Security and Privacy*, *4*(2), 30–38. doi:10.1109/MSP.2006.31

Wang, X., & Chen, D. (2005). Research on e-business management thought. *Proceedings of the ICEC'05.* Xi'an, China: ACM Press.

Wareham, J., Zheng, J., & Straub, D. (2005). Critical themes in electronic commerce research: A meta-analysis. *Journal of Information Technology*, *20*, 1–19. doi:10.1057/palgrave.jit.2000034

Wasserman, G., & Su, Z. (2007). Sound and precise analysis of Web applications for injection vulnerabilities. *Proceedings of the PLDI'07.* ACM.

Woon, I., & Kankanhalli, A. (2007). Investigation of IS professionals intention to practise secure development of applications. *International Journal of Human-Computer Studies*, *65*(1), 29–41. doi:10.1016/j.ijhcs.2006.08.003

ADDITIONAL READING

Ben Aissa, A., Abercrombie, R. K., Sheldon, F. T., & Mili, A. 2009. *Quantifying security threats and their impact.* In Proceedings of the 5th Annual Workshop on Cyber Security and information intelligence Research: Cyber Security and information intelligence Challenges and Strategies (Oak Ridge, Tennessee, April 13 - 15, 2009). F. Sheldon, G. Peterson, A. Krings, R. Abercrombie, and A. Mili, Eds. CSIIRW '09. ACM, New York, NY, 1-6. DOI= http://doi.acm.org/10.1145/1558607.1558637

Fan, S., He, Z., Zhang, Y., Zhang, H., & Su, R. 2008. *An m-Business Model Based on Session Initiation Protocol.* In Proceedings of the 2008 international Symposium on Electronic Commerce and Security (August 03 - 05, 2008). ISECS. IEEE Computer Society, Washington, DC, 250-253. DOI= http://dx.doi.org/10.1109/ISECS.2008.66

Garcia, D. Z., & Felgar de Toledo, M. B. 2008. *Ontology-Based Security Policies for Supporting the Management of Web Service Business Processes.* In Proceedings of the 2008 IEEE international Conference on Semantic Computing (August 04 - 07, 2008). ICSC. IEEE Computer Society, Washington, DC, 331-338. DOI= http://dx.doi.org/10.1109/ICSC.2008.35

Goel, S., & Chen, V. (2008, September). Can business process reengineering lead to security vulnerabilities: Analyzing the reengineered process. *International Journal of Production Economics*, *115*(Issue 1), 104–112. doi:10.1016/j.ijpe.2008.05.002

Kerschbaum, F., & Robinson, P. (2009, April). Security architecture for virtual organizations of business web services. [Elsevier.]. *Journal of Systems Architecture*, *55*(Issue 4), 224–232. doi:10.1016/j.sysarc.2008.10.001

Kim, B., Chen, P., & Mukhopadhyay, T. 2009. *An Economic Analysis of the Software Market with a Risk-Sharing Mechanism.* Int. J. Electron. Commerce 14, 2 (Dec. 2009), 7-40. DOI= http://dx.doi.org/10.2753/JEC1086-4415140201

Kim, C., Tao, W., Shin, N. and Kim, K. *An empirical study of customers' perceptions of security and trust in e-payment systems.* Electronic Commerce Research and Applications

Kovač, D., & Trček, D. 2009. *Qualitative trust modeling in SOA.* J. Syst. Archit. 55, 4 (Apr. 2009), 255-263. DOI=http://dx.doi.org/10.1016/j.sysarc.2009.01.002

Li, L., & Tao, L. 2009. *Security Study of Mobile Business Based on WPKI*. In Proceedings of the 2009 Eighth international Conference on Mobile Business (June 27 - 28, 2009). ICMB. IEEE Computer Society, Washington, DC, 301-304. DOI= http://dx.doi.org/10.1109/ICMB.2009.58

Nurse, J. R., & Sinclair, J. E. 2009. *BOF4WSS: A Business-Oriented Framework for Enhancing Web Services Security for e-Business*. In Proceedings of the 2009 Fourth international Conference on internet and Web Applications and Services (May 24 - 28, 2009). ICIW. IEEE Computer Society, Washington, DC, 286-291. DOI= http://dx.doi.org/10.1109/ICIW.2009.48

Pimenidis, E., & Georgiadis, C. K. 2009. *Web services security evaluation considerations*. Int. J. Electron. Secur. Digit. Forensic 2, 3 (Jul. 2009), 239-252. DOI= http://dx.doi.org/10.1504/IJESDF.2009.027520

Tsai, W., Huang, B., Liu, J., Tsaur, T., & Lin, S. 2010. *The application of Web ATMs in e-payment industry: A case study*. Expert Syst. Appl. 37, 1 (Jan. 2010), 587-597. DOI= http://dx.doi.org/10.1016/j.eswa.2009.05.058

Volume 9, Issue 1, January-February 2010, Pages 84-95. Elsevier. Special Issue: Social Networks and Web 2.0

Wolter, C., Menzel, M., Schaad, A., Miseldine, P., & Meinel, C. (2009, April). Model-driven business process security requirement specification. [Elsevier.]. *Journal of Systems Architecture*, *55*(Issue 4), 211–223. doi:10.1016/j.sysarc.2008.10.002

Xi, L., Xiaoyu, Z., Aimin, Z., & Lijuan, Y. 2010. *Research on the Evaluation in Alliance E-business Security*. In Proceedings of the 2010 international Conference on E-Education, E-Business, E-Management and E-Learning (January 22 - 24, 2010). IC4E. IEEE Computer Society, Washington, DC, 74-78. DOI= http://dx.doi.org/10.1109/IC4E.2010.30

Zhou, H., & Dai, S. 2008. *PKI-Based E-Business Security System*. In Proceedings of the 2008 3rd international Conference on innovative Computing information and Control (June 18 - 20, 2008). ICICIC. IEEE Computer Society, Washington, DC, 203. DOI= http://dx.doi.org/10.1109/ICICIC.2008.421

KEY TERMS AND DEFINITIONS

Availability: is concerned with keeping data in the place and in the form it should be kept and providing it to the users at times they need it.

Confidentiality: is concerned with keeping sensitive data in confidence and not exposed to unauthorized persons.

E-Business Process Based Security: a security paradigm according to which the e-business organization security is achieved by securing its electronic processes.

E-Business Security: is concerned with secure execution of its electronic processes.

E-Business: is a business that performs its supply chain activities by means of electronic processes only, using Internet-based Information Technology for integration of, and cooperation and interaction with, its supply chain participants' webs and business processes.

Information Security: is concerned with protecting assets of information and information systems by assuring that the security goals are achieved.

Information System Security: is a set of actions and measures that enable an information system to meet the defined security requirements, and to operate in a reliable manner.

Information Systems Security Management: is the process of using business resources for planning, organising, leading, and controlling the security of an information system and ensuring its reliable operation.

Integrity: Is concerned with the accuracy and reliability of information and systems, and preventing unauthorised modification of data.

Perimeter Security: Deals with surrounding an organization with security safeguards fence in order to separate it from the outside environment.

Security Requirements/Goals: the commonly accepted CIA (Confidentiality, Integrity, and Availability) triad.

ENDNOTES

* NCSL, National Conference of State Legislatures
Retrieved from http://www.ncsl.org/programs/lis/cip/cyberterrorism.htm.

† Webopedia, retrieved from http://www.webopedia.com/TERM/C/crack.html.

‡ The same as [2]

§ www.atstake.com

1 Communications of the ACM, Information & Management, Information Systems Management, and the Journal of Management Information Systems

2 Can be found at: http://www.avertlabs.com/research/blog/index.php/2010/09/09/widespread-reporting-of-here-you-have-virus/?utm_source=feedburner&utm_medium=feed&utm_campaign=Feed%3A+McafeeAvertLabsBlog+%28McAfee+Avert+Labs+Blog%29

3 See information presented at http://www.software.com/article/antivirus/Introduction-To-Antivirus-Trojan-Horse

4 Wikipedia, the free encyclopedia; retrieved from http://en.wikipedia.org/wiki/Spyware.

5 A Secure European System for Applications in a Multi-vendor Environment

6 Common Object Request Broker Architecture

Chapter 17
E-Business Efficacious Consequences:
The Etiquettes and the Business Decision Making

Wilhelmina Djoleto
The CulRitzWil Co., USA

ABSTRACT

It stands with no contention that a society without virtues and values would be a muddled one, and etiquette is a systemic rectitude that helps shape society. The once eccentric Internet now epitomises regularized modern society and has paved way for new diverse business processes and operations that necessitate critical decision making. These proliferating business processes have been termed e-business or e-commerce, both of which have been used interchangeably in the literature. We consider two groups in the United States of America – business organisations and Higher Education Institutions; specifically the different higher education systems in the United States. Traditionally white institutions are of choice, as members of the Internet society in this chapter. A reconnoitring of the etiquettes of the Internet and e-business vis-à-vis decision making is presented and readers are driven through the elements of etiquettes that govern e-business and how these impact businesses as a whole. It would not be over-amplified to state that this component of e-business is important in ways that translate into institutions' and organisations' efficacies. The Higher Education Institutions and organisations vary in size and ownership, each institution or organisation deals with the essence of e-etiquette and the data show direct relationships between e-etiquette, decision making and the success of organisations and institutions and e-societal members.

DOI: 10.4018/978-1-60960-501-8.ch017

INTRODUCTION

The Internet is an epitome of a society which in this chapter, referred to as *eSociety* (society, Internet society and eSociety will be used interchangeably in this chapter) has paved way for diverse business processes and operations, to forge a global culture (Djoleto 2008a; Djoleto, 2008b; Okunbor & Okunbor, 2005). These thriving Internet business processes have been termed Enterprise Electronic Business (e-business) or Enterprise Electronic Commerce (e-commerce), both of which have been used interchangeably in the literature (Djoleto-Okunbor, 2008c; Schneider, 2003).

In this research, we have chosen two groups in the United States of America – (1) Business Organisations and (2) Higher Educational Institutions (HEIs), specifically, Traditionally White Institutions (TWIs), as members of the Internet society. TWIs are one sect of institutions of higher education in the United States of America, originally established for the purpose of educating whites only, until the mid to late twentieth century when schools and universities began witnessing desegregation of people of every colour in schools as a result of a 1954 United States of America supreme court ruling, that segregation is unconstitutional in the USA. Diversity, in recent times, has been witnessed in educational institutions in the USA; and attests to the overall global trend in diversity (Allen, Epps, & Haniff, 1991).

Enterprise Electronic Business, has become the normalized effervescent source of the way organisations do business in an effort to achieve efficiency and efficacy, is without disputation (Djoleto-Okunbor, 2009; Henderson & Venkatraman, 1999; Aird, 2001; Goral, 2003a; Fraumeni, 2001; Kennedy, 2000; Olsen, 2000; Periasamy, Ho, Poh and Bok, 2002). As a result, organisations have made significant investments in this venture, ranging from Information Technology (IT) – the bedrock, to e-business (Djoleto, 2009). It is to be noted that academic institutions or HEIs, have also invested in various forms of IT (Djoleto-Okunbor, 2008b). While the literature is full of empiricism of how e-business implementations have aided organisations and institutions to attain almost full enhancement of their business and decision-making processes, little emphases have been made specifically on the etiquettes of Enterprise e-business/e-commerce or the Internet society and its environs, which spurs this expository; eEtiquette in the context of this chapter parallels with the definition of etiquette from the Oxford Dictionary (2003) as the formal rules of correct or polite behaviour in society or among members. eEtiquette will be used to denote Internet or e-business etiquette, however, in this chapter, both will be used mutually.

Since its inception in the 1950s when its patronage was limited to the United States Department of Defence, till present, with its augmented patronage, the eccentric Internet has become a society and has thus, been flooded with diverse business models (Djoleto 2008a; Djoleto, 2008b; Djoleto, 2009; Djoleto-Okunbor, 2008a). The speedy increase in Information Technology spectacled in the late twentieth century and the commencing of the twenty-first century heightened the infiltration of applied IT in every facet of business organisations, e-business being an aspect (Burke, Nakayama, Ralston &Tolani, 2000; Cherian, 1999; Djoleto, 2008c; Djoleto, 2009; Djoleto-Okunbor, 2009; Fraumeni, 2001; Gil-Garcia, 2005; Gil-Garcia, 2006; Karake-Shalboub, 2002; Kraemer & King, 2003; Lee, Park & Ahn, 2000; Molla & Licker, 2001; Technology Forecast 2002-2005). A myriad of business transactions are conducted through the Internet and other Telecommunication channels would be an understatement (Armstrong & Sambamurthy, 1999; Djoleto-Okunbor, 2008a). The emergence of Enterprise e-business, though, impacts organisations in both positive and negative ways; eEtiquette must play a role in this ensue and is an important component of Enterprise e-business in ways that allow institutions and organisations to be productive, agile, innovative and reputable, if practised.

Djoleto (2008a), Djoleto (2008b) and Djoleto (2009) stipulated that ever since their commencement, the empiricism documented in the literature shows that e-business have richly and positively enhanced the organisational decision-making process with speedy improved and prescience communications and collaboration, improved data management and dexterous support of organisational or broadly, societal members. And, organisations that have not yet clinched to Enterprise e-business prescience struggle to compete in the competitive global market, as the process of such decision-making, has critical impact on organisational leadership and the effectiveness and the efficiency of organisations' image. (Cherian, 1999; Djoleto, 2008a; Djoleto-Okunbor, 2008a; Kopp, 2000).

Partitioned into sections and subsections, this chapter, specifically, espouses the notion that in a venture to immensely enhance Enterprise e-business, institutions and organisations must orchestrate ways to commence taking critical practice of engulfing eEtiquette in their businesses – as e-business in this modern era underpin the way organisations engage in their day-to-day business operations (Djoleto, 2008a; Djoleto 2008b; Djoleto, 2009; Djoleto-Okunbor, 2008b;); and etiquette helps shape society. It is to be noted that while, traditional societies are making effort to lessen ethnocentrism, cynicism and prejudice (Gupta & Chanttopadhyaya, 1998), the same stands true, that the Internet society must invest immense effort in combating ethnocentrism, cynicism and prejudice for this global society [eSociety] to thrive. The important question is – *what impact and to what extent does the eEtiquette component have on the governing of e-businesses*? We reconnoitre the prospects to how organisations and HEIs have embarked on the rectitude of Enterprise e-business solutions, what role the eEtiquettes governing this eSociety have played on their consequences and how to mitigate unintended negative consequences that could be apocalyptical as a result of eEtiquette oversight. The degree of the savoir-faire nature of organisations and HEIs as members of the Internet society with respect to e-business etiquettes is of immediate interest. Furthermore, to present real life cases, some organisations' and institutions' eEtiquette practices will be cited. Telephone interviews were conducted to collect data on the organisations and institutions presented in this chapter. It is no tautology to state that readers will be driven through this expository and perpetuity demonstrated by the prospects of e-business delineated throughout this chapter.

When tackling issues such as etiquettes of Enterprise e-business or the Internet society, pertinent factors like social, cultural, ethical, trust, security and legal issues come to play (these factors must be of critical importance to institutions and organisations as they reflect on their savvy nature); these factors will be referred to as eSocial, eCultural, eEthical, eTrust, eSecurity and eLegal respectively and these terms will be used interchangeably with the former in the text from here forward. It is imperative, therefore, for these institutions and organisations to understand the etiquettes that govern e-business, without which leads to fermented e-business society. Enterprise e-business and its etiquettes must be judiciously practised to meet the needs of all societal members and the environs thereof; enormous venture must be made to resist any acts that will jeopardise the image of this eSociety, thereby maintaining its integrity and reputation and minimising probable business mayhem.

BACKGROUND

In this section, we *drive* readers through the exploration of areas of significance deemed critical; such areas include business decision-making, e-business, business etiquettes and their efficacious consequences. Additionally, in an endeavour to illustrate our intended ensue; these components will be married in this chapter; that is – good

Enterprise e-business etiquettes impacts organisations' and institutions' successes and failures.

United States Higher Education System

Noted in the introduction, Allen et al. (1991) elucidated that institutions in the United States have more diversified in recent years. This was not the case in times prior to the 1954 ruling by the supreme court of the United States that segregation is unconstitutional in America, an effort to diminish desegregation and to provide quality higher education to all Americans in the US. And prior the 1954 ruling, the US congress had passed two Land-Grant Acts also known as the First Morrill Act (1862) and Second Morrill Act (1890) to require states with all-white and non-white systems of higher education systems to provide land-grant for both systems. As a result HEIs whether all-white or non-white receive federal and state funding. Some of these funding are specifically designated for IT implementation. (For further reading, see Djoleto, 2008a, Djoleto, 2008b; 2008c; Djoleto-Okunbor, 2008a; 2008b; 2008c; Djoleto-Okunbor 2009).

It is to be noted that while diversity has definitely evolved in its higher educational systems, the US also maintains various higher educational systems – Traditionally White Institutions, Public Institutions, Private Institutions, Four-Year Institutions, Two-Year Institutions, Traditionally Hispanic Institutions and Historically Black Colleges and Institutions (American Council on Education, 2001; Djoleto, 2008a; 2008b; 2008c; Djoleto-Okunbor, 2008a; 2008b; 2008c; Djoleto-Okunbor 2009). The latter two are also known as minority institutions and were established educate people other than whites at a time when desegregation did not literally exist in the US. However, as mentioned in the introduction and in the preceding paragraph, all the higher educational systems are now diversified and admit all people (Djoleto, 2008a; 2008b; 2008c; Djoleto-Okunbor, 2008a; 2008b; 2008c; Djoleto-Okunbor 2009).

In previous cross-sectional research, Djoleto (2008a) and Djoleto (2009) reconnoitred the Impact of Ecommerce on Historically Black Colleges and Universities using mixed model method, sampling 716 higher level administrators from 55 of these universities across the U.S. and data was collected via U.S. mail, website and electronic mail. The sample was obtained using cluster stratified random sampling (see Djoleto, 2008a; 2008b; 2008c; Djoleto-Okunbor, 2008a; 2008b; 2008c; Djoleto-Okunbor 2009).

Efficacious Consequence as a Function of Decision Making

Decision-making is a process that occurs on a daily basis. Business decision-making can be strategic (by higher echelon business executives) or tactic (by managerial level employees). Plausible superior-decision making is achieved when all stakeholders in the business environs are critically factored into the decision making equation. Such strategies call for good leadership and while it is supposed that the overarching goal of organisations and institutions is to have a positive presence in society or eSociety for that matter, these strategies call for decision makers or leaders who will lead to that effect. To that effect, business decision making at organisations and HEIs vis-à-vis their Enterprise e-business with a focus on their eEtiquette will be considered.

Regardless of the perspective of the decision maker or makers, the ultimate goal of the decision maker or makers is to resolve to a system at which juncture would resolve to an effective and efficient system (Djoleto 2008a; Djoleto, 2008b; Djoleto, 2009; Turban, Aronson, Liang & Sharda, 2007). On one hand, the resulting system must be proven to be accomplishing the task it was and is intended to do in order to be dubbed effective; and on the other hand, the system would be hailed as efficient if it is doing task it was or is intended to do the right way. Failure to accomplish this ensue, the system becomes an albatross. Figure 1 has been developed to depict the ensued business organisations' and institutions' effective and efficient Enterprise e-business process.

Enterprise E-Business: At a Glance

Contrary to traditional face-to-face business dating back to the advent of communication, e-business which evolved in more modern times, signifies business transactions that take place over the Internet and some other telecommunication networks (Cashman, Shelly & Vermaat, 2001; Carey, 2000; Djolete, 2008a; Djoleto, 2008c; Djoleto, 2009; Hecker, 2001; Kalakota & Robinson, 2000; Periasamy, Ho, Poh & Bok, 2002; Price Waterhouse, 1997; Schneider, 2003). As mentioned in the introduction section, the manifestation of the rapid development in Information Technology seen in the late twentieth and the commencing of the twenty-first century enhanced the proliferation of e-business in most to all components of business organisations (Burke et al. 2000; Gil-Garcia, 2005; Karake-Shalboub, 2002; Technology Forecast 2002-2005). As Aird (2001), Djoleto (2008a), Goral (2003a), Olsen (2000) and Periasamy, Ho, Poh and Bok (2002) stipulated to note that a significant number of business transactions are done through the internet and other telecommunication channels as e-business now typifies the way business organisations and institutions be it public or private, engage in their day-to-day business operations. The shift from the days when e-business processes were limited to very few large business organisations and institutions is credited to the invention of the World Wide Web (www), Enterprise e-business is now very popular (Djoleto, 2009; Djoleto-Okunbor, 2008a).

While there several variants of Enterprise e-business definitions exist, in this chapter enterprise e-business will be used to encompass all business-to-business and business-to-customer transactions that involve the buying and selling of goods and services and the transfer of funds through digital communications (Cashman et al.; 2001; Djoleto, 2008a; Kalakota & Robinson, 2000; Karake-Shalhoub, 2002; Price Waterhouse, 1997; Quaddus & Achjari, 2005; Schneider, 2003). Enterprise e-business finds its roots in technologies such as the Electronic Data Interchange (EDI) and Electronic Funds Transfers (EFT) from the 1960s and 1970s respectively. The former is the mechanism for the electronic interchange of funds between financial institutions such as banks (Djoleto-Okunbor, 2008c; Victor & Wilburn, 2006). The latter is the computer-to-computer interchange of strictly formatted messages that

Figure 1. Depiction of enterprise e-business system efficacy architecture

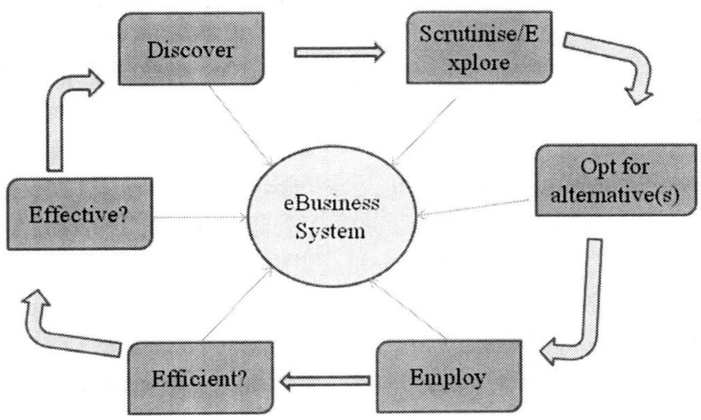

represent documents besides monetary instruments between parties (Djoleto, 2008; NIST EDI Standard, 1996).

Enterprise e-business models have evolved over the years, mainly, they are Business-to-Business (B2B or B to B); Business-to-Consumer (B2C or B to C); Business-to-Government (B2G or B to G); and Consumer-to-Consumer (C2C or C to C) (Cashman et al. 2001; Djoleto, 2008b; Schneider, 2003; Slyke & Belanger, 2003).

Business-to-Business electronic business involves one business transacting business operations with another business over the Internet. The Business-to-Consumer electronic business model occurs when consumers shop for products and services offered by business organisations on the Internet (Djoleto, 2008; Schneider, 2003). For the Business-to-Government model, just like Business-to-Business model, business transactions occur between businesses and government agencies over the Internet. The Business-to-Consumer e-business model and the Consumer-to-Consumer e-business model are similar if consumers selling products and services over the Internet are considered as operating business organisations (Djoleto, 2008a; Schneider, 2003). We note that for the purpose of this chapter, our focus will be on B to C.

E-Business Etiquettes (E-Etiquette) as a Function of Decision Making: Efficacious Consequences

It has been the pride of some organisations and institutions that have created strong presence in this *eSociety*. Achieving this business organisational or institutional presence does not come by inadvertently. It involves critical decision-making; and etiquette-savvyness plays a major role in this challenge. For organisations and institutions and their e-business, the attainment of presence dwells mainly on the creation of a strong Web presence. They want to get across the Internet society and the e-business community their reputable image. As part of the decision making, organisations and institutions of the Internet society must develop an eCulture of distinction and excellence, compass of eEthics and eLegalities and acknowledge the cultural diversities and their effects on e-business as these add to the repertoire of e-business etiquette-savvyness. Success to organisations' and institutions' e-business dwells in major part on how societal members transcend eEtiquette without any etiquette catastrophe.

Schneider (2003) noted that traditional businesses create their presence in the physical world–building office buildings, stores, warehouses and factories; and the presence in turn convey to the public the businesses' public image. Now, to even take it a step further in this chapter, organisations and institutions in addition to creating their presence through their physical properties such as the buildings just mentioned, must more importantly, put their clientele first – they [clientele] need to be assured that their business transactions with these organisations and institutions are valued. Indisputably, a beautifully built business building could only wind up being a "white elephant" or an albatross if business conducted in the building is inefficient and ineffective and this applies to e-business. This must somehow be entrenched in their Web presence.

A major business strategy is for an organisation or an institution to recognise the essence of its clientele as societal members. And with the evolution of the Internet society, eEtiquette involves eTrust, eEthical, eLegal, eSecurity and eSocial issues. The advent of the Internet and e-business has given way to global business and the meshing of various cultural and social issues. For organisations and institutions to be able to thrive in this new forged society they must pay critical attention to eEtiquette – have respect and appreciation for all stakeholders and societal members regardless of the cultural and social backgrounds and combat prejudice and ethnocentrism.

As part of the systemic rectitude of Enterprise e-business conduct in the global culture or Internet society, organisations need to put in enormous efforts to be hospitable; hospitality is something that appeals to societal members when they visit organisations' and institutions' Websites. While this serves as a critical way of creating Web presence, it also runs hand in hand with e-business etiquette.

Previously mentioned, in every society or culture, societal or cultural members need to feel safe and comfortable in their environs; this provides societal members a sense of security. The same is true for this Internet society or global culture. It should therefore, be vital to organisations and institutions to invest much so in their endeavour to forge and establish trust among members of this global culture. It is only fair to state that institutions and organisations in this regard must be able to foresee how societal members will respond to their [organisations and institutions] actions, keeping in mind eSocial, eCultural, eEthical, eTrust, eSecurity and eLegal issues; efforts in this regard will contribute to organisations' and institutiuons' Web presence and also help to skirmish any potential mar.

Fortifying their Web security must also be an ongoing challenge for organisations and institutions. For in this global culture or Internet society, it remains no secret that, that no sooner did new advances in eTechnology on Web security evolve, than security risks, which are referred to as counter operations in this chapter, perpetrated by dissidents or devious members of this eSociety also emerged. Hence, it cannot not be overemphasised, that the Internet society security remains a predicament and an encumbrance that must be tackled with much vigilance as its deterioration can deter societal members in many aspects, ranging from business transactions to other eSocial endeavours. Organisations and institutions have to make it paramount to make certain of transaction integrity in the Internet society.

In order to implement Enterprise e-business security which has been shown to be a critical factor in the rectitude of doing business in the global culture, organisations and institutions must on a regular basis review how strongly their encryption and cipher algorithms thrive. This is a critical decision making process that incorporates the eEtiquette component as it plays a major role in the success of efficacies of institutions and organisations in the global culture or eSociety.

Just for the purpose of enlightenment, examples on eGovernment will be cited in this paragraph and the next. Noted by Gil-Garcia (2005) and Wescott (2006), the expansive nature of Enterprise e-business has resulted in a number of different categories. For instance, eGovernment is an electronic business specifically designed for the conduct of business operations within government establishments. Government procurements, contract awards, organisational meetings and conferences are conducted using a variety of e-business models (Gil-Garcia, 2005; Gil-Garcia, 2006). Now, given the importance and the sensitive nature of data and transactions involved their processes, these government establishments must be critical of their e-business etiquette.

Wescott (2006) stressed that countries adopt E-Government in order to progress toward regional and global good practices and to reinforce traditional processes, buttressing good governance practices including management, accountability, information freedom and transparency, rule of law and remedy for corruption. Just as eGovernment is believed to foster good governance, in this chapter, we show that e-business may be an agent of change for organisational leadership at institutions and organisations. Furthermore, in an extensive literature review on the impact of IT on organisational structure, Karake (1992) revealed that IT is affecting organisational structure that makes organisations more efficient and productive.

REAL LIFE CASES

This section will evaluate enterprise e-business etiquette issues – the main focus of the chapter and present some additional controversies and issues that surround e-business etiquette. Additionally, real life cases presented will provide insight into the eEtiquette practices of some organisations and institutions. Four telephone interviews with higher and mid-level administrators from business organisations and HEIs were conducted to collect the data presented in this chapter (see Table 1). It is to be noted that the identities of the organisations and institutions presented in this chapter are held confidential. As previously mentioned, e-business etiquettes that govern this phenomenon are entwined with eEthical, eCultural, eLegal, eTrust, eSecurity and eSocial issues; these issues will be presented accordingly in this portion of the chapter. It should be clear that, stakeholders or members of this Internet society need some assurance of safety – legalities; as providers of services and products, organisations and institutions owe these assurances to societal members as their [organisations' and institutions'] ethical and perhaps legal compulsions, while stakeholders and members of this Internet society owe it to themselves to verify and validate these assurances; this continual judicious practice enable societal members to indulge and form an amicable Internet or Enterprise e-business culture and societal members and stakeholders engage socially in a trustful manner.

Before we look at the telephone interviews, let us consider some of the events presented in the literature.

Noted in the existing literature and in the preceding section, factors affecting Enterprise e-business solutions and systems will not be complete without examining legal, ethical, trust, social, security and cultural attributes (Johnson, 2001; Kleiner & Maury, 2002; Okunbor & Okunbor, 2005; Robertson & Sarathy, 2003; Schneider, 2003). These legal, ethical, cultural, trust, security and social issues aspects of e-business exist with limited research. Many cases proliferate on legal, ethical, social, trust, security and cultural problems with e-business solutions in industry (Hartzel & Peace, 2002; Johnson, 2001). It is imperative that organisations and institutions understand legislation governing Enterprise e-business solutions and systems, particularly legislation relating to privacy and property rights (Kleiner & Maury; Robertson & Sarathy; Schneider).

Matlay and Addis (2003), in an endeavour to provide elucidation on how HEIs have taken advantage of e-business solutions in respect of diversifying business operations, stated that HEIs that adopt e-business tend to offer more consultancy services to small businesses of the British economy. Small businesses seek services from HEI-based consultancy because of the available subsidies that come with their services. Their success in this challenge may in some measure be contingent on their eEtiquette.

In an attempt to persuade the B2C consumers, Moores (2005) revealed that the e-business

Table 1. Descriptive information on organisations and institutions in telephone interview

Type of Establishment	Ownership	Establishment Size	Services Offered
Organisation	Public′	Approximately 80,000	Telecommunication Solutions
Institution	Private	Approximately 2,000	Education & IT Services
Organisation	Private	Approximately 1,500	Policy Think Tank
Institution	Public*	Approximately 40,000	Education

N.B.:- * State owned, government assisted institution,
′ Publicly traded organisation

industry has developed Privacy (or Web assurance) seals such as TRUSTe, CPA Web Trust and BBBQOnline to ensure that websites are trusted. Trust was further defined by Moores with social connotations in terms of reliance on the integrity, veracity or reliability of a person (or thing) to fulfil a promise or complete a task. TRUSTe is a program which was released in June 1997 by a consortium of CommerceNet, the Electronic Frontier Foundation and the Boston Consulting Group. CPA WebTrust was released in September of 1997 by the American Institute of Certified Public Accountants (AICPA), with Version 3.0 released in November of 2000. BBBOnline was released in March of 1999 by the Better Business Bureau.

Divergent to the above mentioned in the preceding paragraph, there has nonetheless, been evidence that neither online consumers nor e-business sites are interested in dealing with the complicated issues surrounding trust unless there is an incentive for profit (Moores, 2005; Srinivasan, 2003). In his example Moores, cited a recent survey of over 2000 United States residents that was conducted by Harris Interactive; in the survey, 64% of respondents admitted they rarely or never read websites' privacy policies. Of the 64%, 40% cited lack of time and interest. Moores attributed the findings to the fact that on an overall worldwide study of websites by consumers' advocacy group, Consumer International revealed that only 58% of websites display a privacy policy.

Consequently, Enterprise e-business is characterised by powerful posse of instruments to capture information about consumers and this is in order to proffer consumers personalized service, (Jamal, Maier & Sunder 2003; Slyke & Belanger, 2003). The instruments and their use in business, Jamal et al. state further, also encroach on the privacy of individuals or to put it, Internet societal members in unprecedented ways. Thus, e-business has the potential to alter the concept of risk, control in organisations and institutions and demand for assurance services. The advent of Enterprise e-business as Jamal, et al. indicate, provides an opportunity to observe customer demand for disclosure, standards and auditing in the absence of regulation. Jamal, et al.(ibid) stipulate further that, since December 2001, there have been practically no government regulations on privacy policies or their disclosure in the United States of America.

Johnson (2001) and Schneider (2003) pointed out that issues of trust, culture, language, infrastructure, borders and jurisdiction, ethics, and privacy rights are critical and should be considered in all e-business installations. Because the efficacious consequences on Enterprise e-business are a reflection on decision making, appropriate leadership in these aspects of e-business is of extreme interest for e-business implementers. The eProcurement, for instance, will require proper access authentications; preservation of client's sensitive data in a database is undeniably a requirement; and roles of all leaders involved in the decision making and the managing of the system must be made clear without any ambiguity (Schneider).

In addition, it is vital to take into account how these organisations and institutions ensure consistency, integrity, effeciency and effectiveness in the application of e-business solutions (Djoleto, 2008a, Djoleto, 2009; Kleiner & Maury, 2002; Kvavik, 2002; Robertson and Sarathy, 2003; Schneider, 2003).

TELEPHONE INTERVIEW CASES

Table 1 provides demographic information on the organisations and institutions presented in the chapter.

Telephone Interview Case 1: Establishment: Silicon Valley Organisation

A publicly traded organisation located in the Silicon Valley, this first organisation presented in the table boasts of over eighty thousand (80,000) employees in several countries. This very large organisation that proffers high technology services in conjunction with enterprises, governments and service providers evidently has a large number of Enterprise e-business solutions in place. However, for the purpose of the core this chapter, the data collected via telephone interview focuses on five e-business solutions that catalyses e-business etiquette. The specific e-business installations of interest in this case are eConferencing, Supply Chain Management System (SCM), Customer Reporting Management (CRM), ePayroll and eHealth. With the well thought-out Enterprise e-business solutions and the forging and ensuring of eEtiquettes, this company boasts of tens of billions in dollars in revenue.

Leader in the high technology business, stern endeavour in ensuring proper and polite e-business environ for societal members is crucial to this Silicon Valley organisation and thus, has its clientele's needs at heart while at the same time maintaining cordiality among its own organisation members. Its clientele comes first. It is obvious that this organisation considers its e-business etiquettes to be critical and mediocrity is of zero tolerance; one of the core values of the organisation is respect of all stakeholders or societal members. Via the telephone interview, it was revealed that the organisation maintains an "extremely secure" level of Internet and for that matter e-business security. The revelation also showed that as a result of this assurance of security provided by this organisation, its clientele has expressed full satisfaction with the conduct of the organisation's e-business; undoubtedly, there has been minuscule to no complaints made by the organisation's clientele and by intra-members about the security of the organisation's e-business solutions.

The Customer Reporting Management for example, has proven to be effective. This channel has led to swift customer responses. Because of its efficiency and effectiveness, the Supply Chain Management System that channels procurement tasks has impacted greatly on this organisation in terms of revenue generation and positive image. ePayroll, eHealth and eConferencing have also helped this organisation to conduct business efficiently and effectively. The Customer Reporting Management, Supply Chain Management System and the eConferencing are examples of e-business solutions vehicles used not only for the enforcement of eEtiquette throughout the organisation, but also with clientele or societal members. As mentioned, these e-business solutions have had extreme positive impacts on this Silicon Valley organisation. Consequently, the image and reputation of this organisation has boosted up positively and the merit goes to the organisation's e-business solutions "extremely usefulness" and its judiciousness in ensuring a polite and proper societal behaviour.

While e-business solutions have greatly enhanced the promoting of efficient organisation member performance evaluation at this organisation, the overall organisational performance evaluation has also been impacted "greatly" and "positively" by e-business. Tasks at this organisation are executed in very effective, efficient and convivial manner. According to the data collected, leadership efficiency and effectiveness at this organisation has been impacted "greatly." eSocietal member encounters seem to be satisfactory.

Telephone Interview Case 2: Establishment: A Traditionally White Institution (TWI)

The second case presented is an HEI in the United States of America with private ownership and it is also Traditionally White Institution (see Introduction and United States Higher Educational System sections). Comparatively, this is a small institution (approximately 2000 in size) and in addition to offering quality education, it also offers information technology services to its clientele (members of eSociety). While its e-business facilities are limited, this institution ranks its eEtiquette as of high importance; and this is the main concern of this chapter. Noted from the interview, this institution by choice, does not offer online education.

The eConferencing, ePayroll and eProcurement systems have all had greatest positive impact on this institution as they pertain to the enforcement of eEtiquette. However, the telephone interview also revealed that the level of e-business security is moderate; there have been some mishaps – from loss of sensitive data to rampant system breakdowns and these incidents have called for some concerns of Internet societal members and complaints by clientele have occurred often. As a result, the overall clientele satisfaction of this institution's e-business security has been neutral. It is no surprise then, as the data revealed that e-business solutions have been "useful" and not "extremely useful" in sustaining the institution's positive image and reputation.

According to the data, enforcing eEtiquette at this institution is of the essence, as such, it is instilled in students. Students are in turn expected to behave politely in the Internet society – email is an example. This has led to a cordial place of business. E-business has also with almost greatest impact effected conflict resolutions among intra-members. E-business solutions have impacted promoting efficient member performance evaluation with almost "greatest impact," alternatively, they have impacted promoting overall institutional performance evaluation with "moderate impact." Efficient and effective leadership has also been impacted moderately.

Unlike the Silicon Valley organisation case, this Traditionally White Institution needs more work done, especially on its security issues. Perhaps, the disparity arises from the difference in services offered by these establishments.

Telephone Interview Case 3: Establishment: Leading Think Tank Organisation

As a private establishment, this third case presented is a policy think tank organisation. The establishment's employee size is over 1,500 with locations spanning mainly around the United States and some parts of Europe. We consider this organisation's email and eCommunication systems that serve as channels through which its e-business etiquette is enforced. Based on the interview the email and eCommunication systems are ranked as having "greatest impact" on this organisation's e-business etiquette.

With the core objective of this think tank organisation to aid business decision makers with superior research and analysis in their onerous decision making process that calls for the handling of sensitive organisational data, this leading think tank organisation ensures "extremely secure" e-business security. Till date, the clientele of this organisation shows high satisfaction of its eSecurity. E-business solutions have proven "very useful" in the up keep of the organisation's image and reputation. Complaints by its clientele have been "very rare." This organisation maintains this positive image and reputation that couples with impressive revenue of over two hundred million dollars because of the judicious practice of e-business etiquette by societal members. Rules and regulations pertaining to e-business etiquette, such how issues are handled, are in strict enforcement. eEtiquette enforcement has translated into the organisation's cordial eSociety. Member and organisational performance evaluations have been between "moderate" to "greatest" due to e-business etiquette. Leadership effectiveness and efficiency at this organisation has also been "moderately" enhanced.

Again, with this profile, this organisation's success somewhat lies in the serious endeavour in enforcing e-business etiquette. Its clientele or members of the eSociety have gathered confidence and *eTrust* in this organisation and transacting business successfully with this organisation has been pleasant.

Telephone Interview Case 4: Establishment: Another Traditionally White Institution (TWI)

Lastly, case four is a public or state owned, Traditionally White Institution and is considered one of the very large institutions in the United States - approximately forty thousand (40,000) in size. Services offered by this institution are limited to mainly providing education. eConferencing, email and eTeaching are vehicles that this institution utilises to enforce eEtiquette. While this institution is much bigger than the one presented in *Case 2* the level of e-business security revealed by the interview is also moderate. Because clientele complaints about the security of the institution's e-business solutions have been "seldom," the institution has managed to uphold high image and reputation in terms of eSecurity (this is part of the e-business etiquette repertoire). However, according to the interview data, the institution lacks established stringent e-business etiquette in place for societal members and intra-members and for that reason the level of enforcement of e-business etiquette has been of little impact.

With little e-business etiquette in practice, reliance on eEtiquette for convivial eSociety for this institution is also of "little impact." On the other hand, e-business has enhanced promoting efficient and effective intra-member and institutional performance evaluation at this Traditionally White Institution. The institution's efficient and effective leadership has also been impacted moderately.

Efficacious Consequences: Solutions and Recommendations

To effectively realise full impact of eEtiquette at organisations and institutions, these establishments need to critically identify not only their institutional needs, but also the needs of the Internet societal members in order to prevent the eSociety from being fermented. Technology or e-business experts at organisations and institutions must conduct research on existing Enterprise e-business solutions and recommend suitable e-business solutions for their organisations' and institutions' fabric. E-business etiquette must be included in institutions' and organisations' strategic decision making process.

Institutions must choose the best possible e-business solutions that will enhance the fulfilment of their organisational goals with the Internet environ in mind. It is advised that institutions refrain from partial or inadequate e-business solutions installations as this will defeat the purpose of e-business solutions. Core factors that were mentioned such as eSocial, eEthical, eCultural, eTrust, eSecurity and eLegal issues must be tackled accordingly.

Evaluations must be conducted to inform leadership and decision makers on the enhancement of e-business solutions. Institutions should conduct surveys on their eEtiquette practices; this will keep administrators abreast with the impact of their e-business solutions and eEtiquette.

Investment made by Institutions and organisations towards e-business solutions must be extended to adequate personnel training on e-business etiquette.

Organisations and institutions must keep regular documentations of their clientele satisfaction and dissatisfaction, as this will enhance their e-business etiquette improvement.

FUTURE RESEARCH DIRECTIONS

A future study will be conducted on a larger scale that will enable comprehensive analyses on organisations and institutions. This will allow for comparison to be made and will expose the disparities between the extent of systemic rectitude of e-business etiquette based on the level of the enforcement of Enterprise e-business solutions etiquette at these organisations and institutions. An international-wide scale study will be conducted.

CONCLUSION

The question lies in where all these myriad e-business solutions lead organisations and institutions from both customer and business perspectives as Internet societal members and how successful and effective these organisations and institutions e-business solutions are, as function of efficacious consequences of Enterprise e-business. The underlying issue is how effective the rectitude of e-business is at organisations and institutions and what the views of all stakeholders on Enterprise e-business at organisations and institutions within the eSociety are. As illuminated and for the context of this chapter, eEtiquette entails pertinent factors such as eEthical, eCultural, eSocial, eTrust, eSecurity and eLegal issues that should be taken extremely seriously in order to endure a peaceful global culture.

Elucidations provided in this chapter clearly show the important direct relationships between e-business etiquette, decision-making and the success of organisations and institutions that would otherwise ferment. Practicing eEtiquette judiciously is pertinent to yielding positive impact of e-business solutions at institutions and organisations. According to the data collected, decision makers at organisations and institutions must consider the essence of the entire Internet society or environs and societal members thereof, accommodate the cultural and social diversities and be vigilant to the legal, security, trust and ethical ramifications that govern the eSociety in order to see full positive impact on their investments in Enterprise e-business solutions. The results have led to the conclusions that organisations and institutions as societal members will be successful if they are stern on their e-business etiquette, because this practice will mitigate the societal chaos. E-business etiquette has been shown to positively impact overall organisational leadership efficiency and effectiveness and is also good for organisations' and institutions' fabric and the Internet society as a whole. When this condition is satisfied, full impact of e-business solutions at these institutions will be realised.

In the case of the Silicon Valley organisation, for instance, clearly the large size of this organisation and the services offered by it actually warrant their strict enforcement of e-business etiquette and the use of the Customer Reporting Management, the Supply Chain Management System, ePayroll, eHealth and eConferencing all prove to enhance efficiency and effectiveness. These channels have led to swift customer responses, increased revenue generation and overall, promoted a cordial Internet society or environ. The image and reputation of this organisation has boosted up positively and the merit goes to the organisation's e-business solutions "extremely usefulness" and its judiciousness in ensuring a polite and proper societal behaviour.

Considering the first Traditionally White Institution however, a relatively small institution with limited e-business solutions and facilities, while it has encountered some e-business security mishaps, its eEtiquette has always been of high importance. As such, even though its e-business suffers quite some, its overall image is not totally damaged. This is due to the fact that members enforce eEtiquette. The eConferencing, ePayroll and eProcurement systems have all had greatest positive impact on this institution as they pertain to the enforcement of e-business etiquette.

REFERENCES

Aird, A. (2001). E-commerce in higher education: Can we afford to do nothing? *Aridne, 2.*

Allen, W. R., Epps, E. G., & Haniff, N. Z. (Eds.). (1991). *College in black and white: African american students in predominantly white and in historically black public universities*. Albany, NY: State University of New York Press.

American Council on Education. (2001). *A brief guide to U.S. higher education*. Washington, DC: American Council on Education.

Armstrong, C. P., & Sambamurthy, V. (1999). Information Technology assimilation in firms: The influence of senior leadership and IT infrastructures. *Information Systems Research*, *10*(4), 304–327. doi:10.1287/isre.10.4.304

Cherian, E. J. (1999). *Electronic commerce changes organizational structures*. Working paper.

Djoleto, W. (2008a). E-commerce solutions and e-leadership: Impact analyses on historically black colleges and universities and their organizational leadership. *Proceedings of the WSEAS Applied Computer Science, Venice Italy*.

Djoleto, W. (2008b). Effect of e-business solutions and e-leadership: Impact analyses on higher educational institutions and their organizational leadership. *International Journal of Education and Information Technologies.*, *3*(2), 202–211.

Djoleto, W. (2009). E-commerce solutions: An impact analysis on historically black colleges and universities. *Proceedings of the WDSI 38th Annual Meeting, Kauai, Hawaii.*

Djoleto-Okunbor, W. (2008a). E-business solutions at historically black colleges and universities: Impact analyses. *Global Digital Business Review*, *3*(1), 7–12.

Djoleto-Okunbor, W. (2008b). Impact analysis of e-commerce applications at historically black colleges and universities. *MTMI Journal and Review of Business and Technology Research*, *1*(1), 1–6.

Djoleto-Okunbor, W. (2008c). The effect of e-business/e-commerce applications at higher educational institutions. *Journal of Decision Science and Information Technology*, *1*(1).

Djoleto-Okunbor, W. (2009). E-commerce solutions at historically black colleges and universities: An impact analysis. *International Journal of Effective Management*, *6*(1), 19–28.

Fraumeni, B. M. (2001). E-commerce: Measurement and measurement issues. *The American Economic Review*, *91*(2), 318–322. doi:10.1257/aer.91.2.318

Gil-Garcia, J. M. (2005). Exploring the success factors of state website functionality: An empirical investigation. *Proceedings of the National Conference on Digital Government Research, Atlanta, Georgia. Digital Government Research Center. Also ACM International Conference Proceedings, 87,* 121-130.

Gil-Garcia, J. R. (2006). Enacting websites: A mixed method study exploring e-government success in multi-organizational settings. *IEEE Proceedings of the 39th Hawaii International Conference on System Sciences.*

Goral, T. (2003a). E-commerce: Your business solution for solving problems across multiple areas of campus business, e-commerce is unbeatable. *University Business,* 40-41 C. Gupta & D.P. Chanttopadhyaya (Eds.). (1998). *Cultural otherness and beyond.* Leiden, The Netherlands: Koninklijke Brill.

Hartzel, K. S., & Peace, A. G. (2002). Ethical concerns raised by the use of the Internet in academia. *Journal of Information Ethics*, 17–31.

Hecker, D. E. (2001). Employment impact of electronic business. *Monthly Labor Review*, *124*(5), 3–16.

Henderson, J. C., & Venkatraman, N. (1999). Strategic alignment: Leveraging Information Technology for transforming organizations. *IBM Systems Journal*, *32*(1), 472–484. doi:10.1147/sj.382.0472

Jamal, K., Maier, M., & Sunder, S. (2003). Privacy in e-commerce: Development of reporting standards, disclosure and assurance services in an unregulated market. *Journal of Accounting Research*, *41*(2), 285. doi:10.1111/1475-679X.00104

Johnson, C. E. (2001). *Meeting the ethical challengers of leadership: Casting light or shadow.* London: Sage Publications.

Kalakota, R., & Rohinson, M. (2000). Electronic commerce. In *Encyclopedia of Computer Science* (4th ed., pp. 628–634). London: Nature Publishing Group.

Karake, S. (1992). *Information Technology and management control: An agency theory perspective*. Westport, CT: Praeger Publishers.

Karake-Shalboub, Z. (2002). *Trust and loyalty in electronic commerce: An agency theory perspective*. Westport, CT: Quorum Books.

Kennedy, M. (2000). The road to e-commerce. *American School & University*, *72*(11), 14–18.

Kopp, K. (2000). Nonprofits and electronic commerce. *Digital Beat Extra.* Retrieved on September 1, 2004, from www.benton.org/News/Extra/ec062000.html

Kraemer, K. L., & King, J. L. (2003). Information Technology and administrative reform: Will the time after e-government be different? [from http://jlking.people.si.umich.edu/IJEGR-Final.pdf]. *International Journal of Electronic Government Research*, 1–18. Retrieved December 15, 2006.

Kvavik, R. B. (2002). E-business in higher education. In Katz, R. N. (Ed.), *Web portals and higher education: Technologies to make IT personal.* Indianapolis: Jossey-Bass Inc.

Lee, D., Park, J., & Ahn, J. (2000). *On the explanation of factors affecting e-commerce adoption.* Working paper.

Matlay, H., & Addis, M. (2003). Adoption of ICT and e-commerce in small businesses: An HEI-based consultancy perspective. *Journal of Small Business and Enterprise Perspective.*, *10*(3), 321–335. doi:10.1108/14626000310489790

Molla, A., & Licker, P. S. (2001). E-commerce systems success: An attempt to extend and respecify the Delone and Maclean model of IS success. *Journal of Electronic Commerce Research*, *2*(4), 131–141.

Moores, T. (2005). Do consumers understand the role of privacy seals in e-commerce? *Communications of the ACM*, *48*(3), 86–91. doi:10.1145/1047671.1047674

NIST EDI Standards. (1996). Electronic data interchange (EDI). *Federal Information Processing Standards Publications Series of the National Institute of Standards and Technology.* Retrieved on April 30, 2006, from http://www.itl.nist.gov/fipspubs/fip161-2.htm

Okunbor W. & Okunbor D. (2005). Internet and e-business cultural dynamics.

Olsen, F. (2000). E-commerce may help colleges cut costs and paperwork. *The Chronicle of Higher Education*, *46*(33), A45–A46.

(2000). *Oxford advanced learner's dictionary* (6th ed.). Oxford: Oxford University Press.

Periasamy, K. P., Ho, W. M., & Bok, C. W. (2002). Organization structure for electronic commerce: A qualitative investigation. *Proceedings of 10th European Conference on Information Systems, Gdańsk, Poland,* (pp. 596-610).

Price Waterhouse. (1997). *Technology forecast.* Menlo Park, CA: Price Waterhouse World Technology Centre.

Proceedings of the WDSI 34th Annual Meeting, Vancouver, British Columbia, Canada, 2005, (pp. 258-261).

Quaddus & Archjari. (2005). A model for electronic commerce success. *Telecommunications Policy*, *20*(2-3), 127–152.

Ralston, G., Burke, K., Nakayama, D., & Tolani, A. (2000). Electronic mail. In *Encyclopedia of Computer Science* (4th ed., pp. 637–642). London: Nature Publishing Group.

Sarathy, R., & Robertson, C. J. (2003). Strategic and ethical considerations in managing digital privacy. *Journal of Business Ethics*, *46*(2), 111–126. doi:10.1023/A:1025001627419

Schneider, G. P. (2003). *Electronic commerce* (4th ed.). Boston: Thomson Learning Inc.

Shelly, G.B., Cashman, T.J. & Vermaat (2001). *Discovering computer concepts for a digital world.* Boston: Course Technology, Thompson Learning.

Srinivasan, S. (2003). On piracy and privacy. *Computer - IEEE, (36)*7, 36-38.

Technology Forecast. (2003). *Enterprise applications: Enterprise suites.* Menlo Park, CA: PriceWaterHouseCoopers Technology Center.

Turban, E., Aronson, J. E., Liang, T., & Sharda, R. (2007). *Decision support and business intelligence systems* (5th ed.). Upper Saddle River, NJ: Pearson Education, Inc.

van Slyke, C., & Belanger, F. (2003). *E-business technologies supporting the net-enhanced organization.* Danvers, MA: John Wiley & Sons, Inc.

Victor, C. R., & Wilburn, R. L. (2006). Electronic data interchange and electronic funds transfer. Retrieved on April 29, 2006, from http://www.referenceforbusiness.com/management/De-Ele/Electronic-Data-Interchange-and-Electronic-Funds-Transfer.html

Wescott, C. G. (In press). E-commerce and the application of technology to government services. In Rondinelli, D. A., & Heffron, J. M. (Eds.), *Globalization in transition: Forces of adjustment in the Asia Pacific region.*

ADDITIONAL READING

N.B.: Research on this specific subject is scant, as such related materials are recommended for further reading.

Ahkter, F., Maamar, Z., & Hobbs, D. (2006). A fuzzy logic-based approach for supporting decision-making process in B2C electronic commerce transaction. *International Journal of E-Business Research*, *2*(2), 54–67.

Aldiri, K., Hobbs, D., & Qawhaji, R. (2008). The human face of e-business: Engendering consumer initial trust through the use of images of sales personnel on e-commerce web sites. *International Journal of E-Business Research*, *4*(4), 58–78.

Allagui, A., & Azza, P.Testing an e-loyalty conceptual framework. *The Journal of Business*, *4*(1), 69–64.

Alwaked, A. M. (2008). Encouraging e-business investments in the developing nations and the ambiguous road. A proposed vision. *The Journal of Business*, *3*(1), 11–20.

Balsmeier, P., Bergiel, B. J., & Viosca, R. C. (2004). Internet fraud: A global perspective. *The Journal of Business*, *4*(1), 29–41.

Djoleto, W. (2009). E-Commerce Solutions: An impact analysis on historically black colleges and universities, *Proceedings of the WDSI 38th Annual Meeting, Kauai, Hawaii.*

Djoleto-Okunbor, W. (2008b). Impact analysis of e-commerce applications at historically black colleges and universities. *MTMI Journal and Review of Business and Technology Research*, *1*(1), 1–6.

Fabrizio Noboa, F. (2004). E-Business: A Framework For Examining Creation and Appropriation of Value. *The Journal of Business*, *4*(1), 1–15.

Gavish, B. (Ed.), *Electronic commerce res. Electronic Commerce Research*. Springer.

Koskela, K. (2008). Theoretical model for lead user identification in the context of ubiquitous services. *The Journal of Business*, *3*(1), 21–25.

Lewis, W. F., & Dickey, I. (2003). Important considerations for online branding strategies. *The Journal of Business*, *3*(2), 57–74.

Little, L., & Briggs, P. (2009). Privacy factors for successful ubiquitous computing. *International Journal of E-Business Research*, *5*(2), 1–20.

Marx, E. (1999). *Breaking through culture shock: What you need to succeed in international business*. London: Nicholas Brealey.

Mathilayakan, S. (2003). A longitudinal examination of web technology adoption and implementation in small and micro sized businesses. *The Journal of Business*, *3*(2), 1–23.

Nguyen, Q. (2007). The network of banks and the private regulation of e-commerce. *International Journal of e-Business Management. 1*(1). 4-12.

Palaniswami, S., Lopez, D., Sprague, D., Okonokwo, P., & Odeydele, S. (2004). Impact of information technology and e-commerce on supply chain management: Survey evidence from manufacturing companies in michigan. *The Journal of Business*, *4*(1), 16–28.

Pearson, S., & Allison, D. (2009). A model-based privacy compliance checker. *International Journal of E-Business Research*, *5*(2), 63–83.

Penny, C., & Fox, C. (2008). *Bridging the culture gap [electronic resource]: A practical guide to international business communication* (2nd ed). London, UK: Kogan Page.

Rhee, J. H. (2004). Information technology and e-commerce strategy of entrepreneurial ventures: A contingency approach based on information-processing theory. *The Journal of Business*, *4*(1), 42–58.

Sabbath, A. (2002). *International business etiquette. asia & pacific rim: What you need to know to conduct business abroad with charm and savvy*. New York, New York: ASJA Press.

Sarkar, S. (2003). Economic and tax issues of electronic commerce. *The Journal of Business*, *3*(2), 38–56.

Sethi, S. P. (2003). *Setting global standards: Guidelines for setting code of conducts in multinational corporations*. Hoboken, New Jersey: J. Wiley.

Shneor, R., & Flaten, B. (2008). The internet-enable internationalization process: A focus of stages and sequences. *The Journal of Business*, *3*(1), 45–52.

Shumanov, M., & Ewing, M. (2007). Developing a global CRM strategy. *International Journal of E-Business Research*, *3*(2), 70–82.

Spinelli, R. (2008). The digital economy: A conceptual framework for a new economic paradigm. *The Journal of Business*, *3*(1), 53–60.

Ultan, S., Murray, S., & Thomas, A. (2010). The influence of quality on e-commerce success: An empirical application of the delone and mclean IS success model. *International Journal of E-Business Research*, *9*(1), 68–84.

Volkof, M. (2003). What do we know about why internet users complain. An exploratory study. *The Journal of Business*, *3*(2), 75–88.

Yee, G., & Korba, L. (2005). Semiautomatic derivation and use of personal privacy policies in e-business. *International Journal of E-Business Research*, *1*(1), 54–69.

KEY TERMS AND DEFINITIONS

Counter Operations: Dissidents' attacks or operations considered ill or heinous on eSociety members.

eCulture: Culture that transcends the Internet society or global culture.

eEthical: Of what is considered rectitude in the Internet society.

eEtiquette: This is the proper, recognised or formal rules of rectitude or polite conduct or behaviour in the Internet or Web society or among its members.

eLegal: The permissible and lawful issues that govern the Internet society.

eSecurity: The various safety measures established to ensure the protection members of the Internet society or the global culture.

eSocial: Network activities of various members of the Internet society from all walks of life.

eSociety: The Internet or virtual society.

eTrust: The belief or confidence engendered in the Internet societal members.

Intra-Members: members of the organisation.

Traditionally White Institutions: Higher educational institutions established in the United States of America originally open to the white race only during America's segregation era. Currently, such institutions are open to people of any colour and any race due to a 1954 United States Supreme Court ruling of segregation as being unconstitutional.

Compilation of References

Aarts, J., Ash, J., & Berg, M. (2007). Extending the understanding of computerized physician order entry: Implications for professional collaboration, workflow and quality of care. *International Journal of Medical Informatics*, *76*, S4–S13. doi:10.1016/j.ijmedinf.2006.05.009

Accessiweb. (2008). Du référentiel AccessiWeb créé par le consortium W3C. Retrieved from http://www.accessiweb.org/

Adam, O., Hofer, A., Zang, S., Hammer, C., Jerrentrup, M., & Leinenbach, S. (2005). A collaboration framework for cross-enterprise business process management. *Pre-proceedings of the First International Conference on Interoperability of Enterprise Software and Application*. Geneva, Switzerland.

Adamson, I., Chan, K., & Handford, D. (2003). Relationship marketing: Customer commitment and trust as a strategy for the smaller Hong Kong corporate banking sector. *International Journal of Bank Marketing*, *21*(6/7). doi:10.1108/02652320310498492

Agarwal, A., & Hedge, A. (2008). The impact of Web page usability guideline implementation on aesthetics and perceptions of the e-retailer. *Proceedings of the Human Factors and Ergonomics Society*, (pp. 528-532).

Agency for Healthcare Research and Quality. (2009). *National Resource Center for Health Information Technology*. Retrieved November 13, 2009, from http://healthit.ahrq.gov

Agency for Healthcare Research and Quality. (2009). *PSNet Primer on CPOE*. Retrieved November 13, 2009, from http://healthit.ahrq.gov

Aird, A. (2001). E-commerce in higher education: Can we afford to do nothing? *Aridne, 2*.

Aissi, S., Malu, P., & Srinivasan, K. (2002). E-business process modeling: The next big step. *Computer*, *35*(5), 55–62. doi:10.1109/MC.2002.999776

Akinci, S., Aksoy, S., & Atilgan, E. (2004). Adoption of Internet banking among sophisticated consumer segments in an advanced developing country. *International Journal of Bank Marketing*, *22*(2/3), 212–219. doi:10.1108/02652320410530322

Al-Hakim, L. (2005). Modelling interdependencies for electronic supply chain management. In *proceedings from 2005 Information Resources Management Association (IRMA) International Conference: Managing Modern Organisations with Information Technology*, 15-18 May 2005, San Diego.

Alkhateeba, F. M., Khanfarb, N. M., Doucettec, W. R., & Loudond, D. (2009). Characteristics of physicians targeted by the pharmaceutical industry to participate in e-detailing. *Health Marketing Quarterly*, *26*(2), 98–116. doi:10.1080/07359680802619792

Allen, W. R., Epps, E. G., & Haniff, N. Z. (Eds.). (1991). *College in black and white: African american students in predominantly white and in historically black public universities*. Albany, NY: State University of New York Press.

Alliance for Healthcare Reform. (2009). *Definition of healthcare information technology*. Retrieved November 13, 2009, from http://www.allhealth.org

Alter, S. (1999). *Information Systems–a management perspective*. Addison Wesley Longman.

Aman, A., & Nicholson, B. (2003). The process of offshore development: Preliminary studies of UK companies in Malaysia. In Korpela, M., Montealegre, R., & Poulymenakou, A. (Eds.), *Organizational Information Systems in the context of globalization* (pp. 201–216). Boston: Kluwer Academic Publishers.

American Council on Education. (2001). *A brief guide to U.S. higher education*. Washington, DC: American Council on Education.

Anderson, B., Hansen, J., Lowry, P., & Summers, S. (2005). Model checking for design and assurance of e-business processes. *Decision Support Systems, 39*, 333–344. doi:10.1016/j.dss.2003.12.001

Anderson, D., & Lee, H. (2000). The Internet-enabled supply chain: From the first click to the last mile. *ASCET, 2*. Retrieved on April 20, 2010, from http://web.efzg.hr/dok/INF/bpm/BPM-priprema-03-SCM.pdf

Anderson, D., & Lee, H. (2001). New supply chain business models–the opportunities and challenges. *ASCET, 3*. Retrieved on April 20, 2010, from http://jobfunctions.bnet.com/abstract.aspx?docid=58793

Ando, M., & Ueno, M. (2008). Effect of pointer presentation on multimedia e-learning materials. In *Proceedings of World Conference on Educational Multimedia, Hypermedia and Telecommunications*, (pp. 5549-5559). Chesapeake, VA: AACE.

Andresen, J., Baldwin, A., Betts, M., Carter, C., Hamilton, A., & Stokes, E. (2000). A framework for measuring IT innovation benefits. *Electronic Journal of Information Technology in Construction, 5*, 57–72.

Andreu, L., Aldas, J., Bigne, J. & Mattila, A. (2009). An analysis of e-business adoption and its impact on relational quality in travel agency–supplier relationships. *Tourism Management*.

Andrew, J. P., Blackburn, A., & Sirkin, H. L. (2000). *The B2B opportunity: Creating advantage through e-marketplaces*. Boston: The Boston Consulting Group.

Androutsellis-Theotokis, S., Spinellis, D., & Karakoidas, V. (2004). Performing peer-to-peer e-business transactions: A requirements analysis and preliminary design proposal. *Preceedings of the IADIS, International e-Commerce 2004 Conference*, (pp. 399-404).

Anwar, M., & Greer, J. (2008). *Enabling reputation-based trust in privacy-enhanced learning systems*. (5091 LNCS), (pp. 681-683).

Anyiwo, D. (2006). *Perspectives on Semantic Web Services security.* Paper presented at the 1st Annual Web Services Security Conference, College Park, Maryland.

APHIS. (2010). Questions and answers: New animal disease traceability framework. Retrieved on February 7, 2010, from http://www.aphis.usda.gov/publications/animal_health/content/printable_version/faq_traceability.pdf.

Applegate, L. (2002). *E-business handbook*. The St. Lucie Press.

AQU. (2003). *Procés d'avaluació de la Qualitat del Sistema Universitari Catalan*. Barcelona: Agencia per a la Qualitat del Sistema Universitari en Catalunya.

Armellini, A., & Jones, S. (2008). Institutional embedding of e-learning: Findings from the Adelie Pathfinder Project. In Walker, S., Ryan, M., & Teed, R. (Eds.), *Designing for Learning – the e-learning@greenwich/conference*. University of Greenwich.

Armstrong, C. P., & Sambamurthy, V. (1999). Information Technology assimilation in firms: The influence of senior leadership and IT infrastructures. *Information Systems Research, 10*(4), 304–327. doi:10.1287/isre.10.4.304

Ash, J. S., Stavri, Z., & Kuperman, G. J. (2003). A consensus statement on considerations for a successful CPOE implementation. *Journal of the American Medical Informatics Association, 10*(3), 229–234. doi:10.1197/jamia.M1204

Ashford, W. (2009). Cyber-criminals to favour financial scams in 2009, says McAfee. *Computer Weekly*. Retrieved February 26, 2010, from http://www.computerweekly.com/Articles/2009/01/20/234325/cyber-criminals-to-favour-financial-scams-in-2009-says.htm

Aspillaga, M. (1991). Screen design: Location of information and its effects on learning. *Journal of Computer-Based Instruction, 18*(3), 89–92.

Association of Business Schools. (2009/10). *Pillars of the sustainable economy: sustainability and current market*. Retrieved on May 11, 2010, from www.the-abs.org.uk/files/ABS%20Pillars_Annual%20Report%2009.pdf

Atlantic-Consulting. (1998). *LCA study of the product group personal computers in the EU Ecolabel scheme.* Retrieved from http://ec.europa.eu/environment/ecolabel/pdf/personal_computers/lcastudy_pc_1998.pdf

Australian Bankers Association. (2000). *The banking industry and overview.*

Avgerou, C., & Madon, S. (2005). Information society and the digital divide problem in developing countries. In Berleur, J., & Avgerou, C. (Eds.), *Perspectives and policies on ICT in society* (pp. 205–218). New York: Springer. doi:10.1007/0-387-25588-5_15

Avgerou, C. (2003). The link between ICT and economic growth in the discourse of development. In Korpela, M., Montealegre, R., & Poulymenakou, A. (Eds.), *Organizational Information Systems in the context of globalization* (pp. 373–386). Boston: Kluwer Academic Publishers.

Babcock, C. (2010). Cloud security alliance to tackle cloud standards. *Information Week.* Retrieved March 2, 2010, from http://www.informationweek.com/news/software/hosted/showArticle.jhtml?articleID=223101102&cid=RSSfeed_IWK_All

Balkenhol, B. (Ed.). (2007). *Microfinance and public policy: Outreach, performance and efficiency.* Palgrave Macmillan.

Barber, B. (1983). *The logics and limits of trust.* New Brunswick, NJ: Rutgers University Press.

Barber, R., & Integralis, A. (2001). Hacking techniques: The tools that hackers use, and how they are evolving to become more sophisticated. *Computer Fraud & Security,* (3): 9–12. doi:10.1016/S1361-3723(01)03014-7

Barnes, S. J., & Corbitt, B. (2003). Mobile banking: Concept and potential. *International Journal of Mobile Communications, 1*(3), 273–288. doi:10.1504/IJMC.2003.003494

Barnes, J. (1997). *Exploring the importance of closeness in customer relationships.* (pp. 227-40). Paper presented at American Marketing Association Conference, Dublin.

Barnett, R., & Coate, K. (2005). *Engaging the curriculum in Higher Education. Berkshire.* McGraw-Hill.

Bates, A. (2000). *Managing technological change. Strategies for college and university leaders.* San Francisco: Jossey-Bass Inc.

Bautista, G. (2006). *Didáctica universitaria en entornos virtuales de enseñanza-aprendizaje.* Madrid: Narcea Ediciones.

Beachboard, J., et al. (2008). Improving Information Security risk analysis practices for small- and medium-sized enterprises: A research agenda. *Issues in Informing Science and Information Technology, 5.*

Beamon, B. M., & Bermudo, J. M. (2000). A hybrid push/pull control algorithm for multistage, multiline production systems. *Production Planning and Control, 1*(4), 349–356. doi:10.1080/095372800232072

Bec, T. A. *(2008).* Harnessing technology: Next generation learning. *Retrieved onJanuary18, 2010, from*http://publications.becta.org.uk/display.cfm?resID=37348&page=1835

Beetham, H., & Sharpe, R. (Eds.). (2007). *Rethinking pedagogy for a digital age.* Routledge.

Bennet, S., Maton, K., & Kervin, L. (2008). The digital natives debate: A critical review of the evidence. *British Journal of Educational Technology, 39*(5), 775–786. doi:10.1111/j.1467-8535.2007.00793.x

Bennett, R. (2002). Lecturers' attitudes to new teaching methods. *International Journal of Management Education, 2*(1), 42–58. doi:10.3794/ijme.21.36

Bennett, R. (1996). Relationship formation and governance in consumer markets: Transactional analysis versus the behaviourist approach. *Journal of Marketing Management, 12*(12), 417–436. doi:10.1080/0267257X.1996.9964425

Berry, L. (1983). *Relationship marketing.* Chicago: AMA.

Berry, L. (2002). Relationship marketing of services–perspectives from 1983 and 2000. *Journal of Relationship Marketing,* 1.

Berry, L., & Parasuraman, A. (1991). *Marketing services—competing through quality.* New York: Free Press.

Besley, T., & Coate, S. (1995). Group lending, repayment incentives and social collateral. *Journal of Development Economics*, *46*(1), 1–18. doi:10.1016/0304-3878(94)00045-E

Bhakoo, V., & Sohal, A. (2008). An assessment into the drivers of e-business adoption within the Australian pharmaceutical supply chain. In *proceedings of the 39th Annual Meeting of the Decision Sciences Institute at Baltimore*, Maryland, 22-25 November.

Bhaskaran, S. (2010). *Using the Internet to enhance small to medium sized food businesses. Rural Industries R&D Report No: 10/004*. Canberra, Australia: RIRDC.

Bishop, M. (2005). *Introduction to computer security*. Addison-Wesley.

Bitner, M. (2001). Service and Technology: Opportunities and paradoxes. *Managing Service Quality*, *11*(6). doi:10.1108/09604520110410584

Bitner, M., Ostrom, A., & Meuter, M. (2002). *Implementing successful self-service technologies*. Paper presented at Academy of Management Executive, (pp. 96-119).

Blackmore, B. S., Stout, W., Wang, M., & Runov, B. (2005). Robotic agriculture–the future of agricultural mechanisation? In J. Stafford (Ed.), *Proceedings of the 5th European Conference on Precision Agriculture*. (pp. 621-628). The Netherlands: Wageningen Academic Publishers.

Blau, P. (1964). *Exchange and power in social life*. New York: John Wiley and Sons.

Blumenthal, D. (2010). *Homepage information*. Retrieved March 28, 2010, from http://healthit.hhs.gov

Blunkett, D. (2000). *Secretary of State for education speech given at the University of Greenwich*.

Bocij, P., Chaffey, D., Greasley, A., & Hickie, S. (2003). *Business Information Systems* (2nd ed.). Financial Times, Prentice Hall.

Bodin, L., Gordon, L., & Loeb, M. (2005). Evaluating information security investments using the analytic hierarchy process. *Communications of the ACM*, *48*(2), 78–83. doi:10.1145/1042091.1042094

Boehlje, M. (1999). Structural changes in the agricultural industries: How do we measure, analyse and understand them? *American Journal of Agricultural Economics*, *81*(5), 10–28. doi:10.2307/1244080

Bogomolova, S., & Romaniuk, J. (2005). *Why do they leave? An examination of the reasons for customer defection in the business banking industry*. Paper presented at ANZAMC 2005.

Boni, W., & Kovacich, G. L. (2000). *Netspionage-the global threat to information*. Butterworth-Heinemann.

Boss, S., McGranahan, D., & Mehta, A. (2000). Will the banks control online banking? *The McKinsey Quarterly*, 70–77.

Bourdieu, P. (1983). *The forms of capital* (pp. 183–198).

Boyes, G. (2003). E-Business opportunities in financial services. *Journal of Financial Services Marketing, 8*.

Brown, G., Bull, J., & Pendlebury, M. (1997). *Assessing student learning in higher education*. London: Routledge.

Brown, A., Remenyi, D., & Bajomo, A. (2004). E-banking for SMEs: A case study of Fortis Bank UK. *International Journal of Electronic Business*, *2*(4), 366–382. doi:10.1504/IJEB.2004.005639

Brown, M., Fry, H., & Marshall, S. (2007). Reflective practice. In Fry, H., Ketteridge, S., & Marshall, S. (Eds.), *A handbook for teaching and learning in higher education* (2nd ed., pp. 215–225). Oxford: Routledge-Falmer.

Bruns, A., Cobcroft, R., Smith, J., & Towers, S. (2007). *Mobile learning technologies and the move towards 'user-led education'*. Paper presented at the Mobile Media 2007 conference, Sydney.

Bryceson, K. P. (2006). *E-issues in agribusiness–the what, why, how*. Wallingford, UK: CABI International. doi:10.1079/9781845930714.0000

Bryceson, K. P., & Slaughter, G. (2010). Alignment of performance metrics in a multi-enterprise agribusiness: A case of integrated autonomy? *International Journal of Productivity and Performance Management*, *59*(4), 325–250. doi:10.1108/17410401011038892

Bryceson, K. P., & Smith, C. (2008). Abstraction and modeling of agri-food chains as complex decision making systems. In Fritz, M., Rickert, U., & Schiefer, G. (Eds.), *EAAA Series on Systems Dynamics and Innovation in Food Networks* (*Vol. 2*, pp. 147–159).

Bryceson, K. P., & Kandampully, J. (2004). The balancing act: E issues in the Australian agri-industry sector. In *Proceedings of the McMaster World Congress on the Management of Electronic Business*, Hamilton, Ontario, Canada, January 14-16.

Bryceson, K. P., & Marvenek, S. (1998). The role of satellite data and precision farming technologies for on-farm monitoring in Australia. In *Proceedings of the First International Conference on Geospatial Information in Agriculture & Forestry*, 1-3 June, Florida.

Bryceson, K. P., & Pritchard, A. (2003). *The impact of e-business technologies on the grains value chain.* Final Report to the Grains R&D Corporation, GRDC, Canberra, Australia.

Brynjolfsson, E., & Hitt, L. M. (2003). Computing productivity: Firm-level evidence. *The Review of Economics and Statistics*, *85*(4), 793–808. doi:10.1162/003465303772815736

BT. (2003). *E-business and the environment. Better world–our commitment to society*. Report retrieved from www.groupbt.com/betterworld

Buhalis, D., & Deimezi, O. (2003). Information Technology penetration and e-commerce developments in Greece, with a focus on small to medium-sized enterprises. *Electronic Markets*, *13*(4), 309–324. doi:10.1080/1019678032000135563

Burgers, W. P., Hill, C. W. L., & Kim, W. C. (1993). A theory of global strategic alliances: The case of the global auto industry. *Strategic Management Journal*, *14*(6), 419–432. doi:10.1002/smj.4250140603

Burnaby, P., & Hass, S. (2009). Ten steps to enterprise-wide risk management. *Corporate Governance*, *9*(5), 539–550. doi:10.1108/14720700910998111

Burns, R. B. (1994). *Introduction to research methods.* London: Longman Cheshire Pty Ltd.

Butcher, C., Davies, C., & Highton, M. (2006). *Designing learning–from module outline to effective teaching, key guides for effective teaching in HE*. New York: Routledge.

Butler, J. K. (1991). Toward understanding and measuring conditions of trust: Evolution of the conditions of trust inventory. *Journal of Management*, *17*, 643–663. doi:10.1177/014920639101700307

Butterfield, S. (2010). Remote monitoring: Out of sight, right in line. *ACP Internist*, 1.

Buzhardt, J., Abbott, M., Greenwood, C., & Tapia, Y. (2005). Usability testing of the class-wide peer tutoring-learning management system. *Journal of Special Education Technology*, *20*(1), 19–29.

Byrne, J. (1993). The futurist who fathered the ideas. *Business Week.*

Campbell, E. M., Guappone, K. P., Sittig, D. F., Dykstra, R. H., & Ash, J. S. (2008). Computerized provider order entry adoption: Implications for clinical workflow. *Journal of General Internal Medicine*, *24*(1), 21–26. doi:10.1007/s11606-008-0857-9

Cantoni, V., Cellario, M., & Porta, M. (2004). Perspectives and challenges in e-learning: Towards natural interaction paradigms. *Journal of Visual Languages and Computing*, *15*, 333–345. doi:10.1016/j.jvlc.2003.10.002

Capoor, K., & Ambrosi, P. (2006). *State And trends of the carbon market, 2006.* Washington, DC: IETA and the World Bank.

Carayannis, E. G., Alexander, J., & Geraghty, J. (2001). Service sector productivity: B2B electronic commerce as a strategic driver. *The Journal of Technology Transfer*, *26*(4), 337–350. doi:10.1023/A:1011130802700

Carson, D., Gilmore, A., & Walsh, S. (2004). Balancing transaction and relationship marketing in retail banking. *Journal of Marketing Management*, *20*(3/4), 431–455. doi:10.1362/026725704323080489

Caswell, J. A. (1998). How attributes of safety and process attributes affect markets for food. *Agricultural and Resource Economics Review*, *27*(2), 151–158.

CAW. (2004). *E-waste background. Californians Against Waste.* Retrieved from www.cawrecycles.org/Ewaste/background%20e-waste.html

Cebrian de la Serna, M., & Vain, P. D. (2008). Una mirada acerca del rol docente universitario, desde las prácticas de la enseñanza en entornos no presenciales. *Pixel-Bit: Revista de Medios y Educación, 32*, 117–129.

Cerullo, V., & Cerullo, M. (2004). Business continuity planning: A comprehensive approach. *Information Systems Management, 21*(3), 70–80. doi:10.1201/1078/44432.21.3.20040601/82480.11

Chaffey, D. (2002). *E-business and e-commerce management. Financial Times*. Prentice Hall.

Chakrabarti, G., & Manimaran, A. (2002). Internet infrastructure security: A taxonomy. *Network, 16*(6), 13–21.

Chang, J. E., Simpson, T. W., Rangaswamy, A., & Tekchadaney, J. R. (2002). *A good website can convey the wrong brand image! A preliminary report*. Working Paper, E-Business Research Center (EBRC), Université de Pennsylvanie.

Changchit, C., Joshi, K. D., & Lederer, A. L. (1998). Process and reality in Information Systems benefit analysis. *Information Systems Journal, 8*, 145–162. doi:10.1046/j.1365-2575.1998.00031.x

Chen, I. J., & Popovich, K. (2003). Understanding Customer Relationship Management (CRM): People, process and technology. *Business Process Management Journal, 9*(5), 672–688. doi:10.1108/14637150310496758

Cherian, E. J. (1999). *Electronic commerce changes organizational structures*. Working paper.

Chew, D. H., & Gillan, S. L. (2005). *Corporate governance at the crossroads, a book of readings*. New York: McGraw-Hill.

Chiasson, M., Reddy, M., Kaplan, B., & Davidson, E. (2007). Expanding multi-disciplinary approaches to healthcare Information Technologies: What does Information Systems offer medical informatics? *International Journal of Medical Informatics, 76*, s89–s97. doi:10.1016/j.ijmedinf.2006.05.010

Chinagateway. (2004). *China's electronics industry*. PRC Ministry of Posts and Telecommunications. Retrieved from http://en.chinagate.com.cn/english/1474.htm

Chou, C. (2003). Interactivity and interactive functions in Web-based learning systems: A technical framework for designers. *British Journal of Educational Technology, 34*(3), 265–279. doi:10.1111/1467-8535.00326

Chu, L. F., & Chan, B. K. (1998). Evolution of website design: Implications for medical education on the Internet. *Computers in Biology and Medicine, 28*, 460–472. doi:10.1016/S0010-4825(98)00027-4

Chua, C., Straub, D., Khoo, H., Kadiyala, S., & Kuechler, D. (2005). The evolution of e-commerce research: A stakeholder perspective. *Journal of Electronic Commerce Research, 6*(4).

Clark, J. M., & Paivio, A. (1991). Dual coding theory and education. *Educational Psychology Review, 3*, 149–210. doi:10.1007/BF01320076

Clay, K., & Strauss, R. P. (2002). Institutional barriers to electronic commerce: An historical perspective. *Advances in Strategic Management, 19*, 247–273. doi:10.1016/S0742-3322(02)19008-5

Cohen, A. (2000). Catching the green express from euston. *Financial Times*.

Coleman, J. S. (1990). *Foundations of social theory*. Cambridge, MA: Harvard University Press.

Coles, J. (2007). *New job titles are loopy labels*. Retrieved January 15, 2010, from http://www.thesun.co.uk/sol/homepage/news/article24172.ece

Colgate, M., & Alexander, N. (1998). Banks, retailers and their customers: A relationship marketing perspective. *International Journal of Bank Marketing, 16*(4). doi:10.1108/02652329810220693

Colgate, M., & Hedge, R. (2001). An investigation into the switching process in retail banking services. *International Journal of Bank Marketing, 19*(5). doi:10.1108/02652320110400888

Colgate, M., Stewart, K., & Kinsella, R. (1996). Customer defection: A study of the student market in Ireland. *International Journal of Bank Marketing, 14*(3). doi:10.1108/02652329610113144

Collins, L. (2008). *Generation Y in cyberspace: How do digital natives manage social stigma on MySpace*. Paper presented at the Australian Sociology Association conference.

Collis, B., & Moonen, J. (2002). *Flexible learning in a digital world: Experiences and expectations*. London: Kogan Page.

Collis, B., & Van der Wende, M. (2002). *Models of technology and change in higher education: An international comparative survey on the current & future use of ICT in higher education*. University of Twente, Enschede: Centre for Higher Education Policy Studies. Accessed on October 26, 2005, from http://www.utwente.nl/cheps/The%20CHEPS%20Team/Staff/Chairs/wende.doc/

Comer, J. M., Mehta, R., & Holmes, T. L. (1998). Information Technology: Retail users versus nonusers. *Journal of Interactive Marketing*, *12*(2), 49–62. doi:10.1002/(SICI)1520-6653(199821)12:2<49::AID-DIR5>3.0.CO;2-D

Commere, B. (2003). *Controlling the cold chain to ensure food hygiene and quality*. Bulletin of the International Institute of Refrigeration (IIR) No: 2003(2).

Conning, J. (2000). *Group lending, moral hazard and the creation of social collateral*. IRIS Working Paper no. 165. University of Maryland.

Conole, G., White, S., & Oliver, M. (2007). The impact of e-learning on organisational roles and structures. In Conole, G., & Oliver, M. (Eds.), *Contemporary perspectives in e-learning research: Themes, methods and impact on practice. Part of the Open and Distance Learning Series*. London: RoutledgeFalmer.

Conole, G. (2010). Personalisation through technology enhanced learning. In O'Donoghue, J. (Ed.), *Technology-supported environments for personalized learning method and case studies* (pp. 1–14). Hershey, PA: IGI Global.

Cook, J., & Light, A. (2006). New patterns of power and participation? Designing ICT for informal and community learning. *E-learning*, *4*(1), 51–61. doi:10.2304/elea.2006.3.1.51

Cook, J., Holley, D., Smith, C., Bradley, C., & Haynes, R. (2006). *A blended m-learning design for supporting teamwork in formal and informal settings*. Mlearn Conference, Banff, Alberta, Canada.

COSO. (2008). *The Committee of Sponsoring Organizations*. Retrieved December 25, 2009, from http://www.coso.org/default.htm

Costanzo, C. (2000). *Pioneer Internet-only bank has a new strategy*. American Banker.

Coursaris, C., Swierenga, S., & Watrall, E. (2008). An empirical investigation of color, temperature, and gender effects on Web aesthetics. *Journal of Usability Studies*, *3*(3), 103–117.

Cronin, M. (2000). *Unchained value: The new logic of digital business*. HBSWK Publications.

Crowther, M. S., Keller, C. C., & Waddoups, G. L. (2004). Improving the quality and effective of computer-mediated instruction through usability evaluations. *British Journal of Educational Technology*, *35*(3), 289–303. doi:10.1111/j.0007-1013.2004.00390.x

CRUE. (2004). *Información académica, productiva y financiera de las Universidades Españolas*. Madrid: Confederación de Rectores de Universidades Españolas.

Cull, R., Demirguç-Kunt, A., & Morduch, J. (2007). Financial performance and outreach: A global analysis of leading microbanks. *The Economic Journal*, *117*(517), F107–F133. doi:10.1111/j.1468-0297.2007.02017.x

Dabholkar, P. (1994). Incorporating choice into an attitudinal framework: Analysing model of mental comparison process. *The Journal of Consumer Research*, *21*, 100–118. doi:10.1086/209385

Dahlbom, B. (2001). Postface: From infrastructure to networking, from control to drift. In Ciborra, C. A. (Ed.), *From control to drift-The Dynamics of Corporate Information Infrastructure*. Oxford: Oxford University Press.

Dailey, L. (2004). Navigational Web atmospherics explaining the influence of restrictive navigation cues. *Journal of Business Research*, *57*, 795–803. doi:10.1016/S0148-2963(02)00364-8

Damanpour, F. (2001). E-business e-commerce evolution: Perspective and strategy. *Managerial Finance*, *27*(7), 16–33. doi:10.1108/03074350110767268

Daniel, J. S. (1996). *Mega-universities and knowledge media: Technology strategies for higher education*. London: Kogan Page.

Daniel, E., Wilson, H., & Myers, A. (2002). Adoption of e-commerce by SMEs in the UK. *International Small Business Journal*, *20*(3), 253–269. doi:10.1177/0266242602203002

Dann, S., & Dann, S. (2001). *Strategic Internet marketing*. Milton, Australia: John Wiley and Sons.

D'Aprix, R., Clemons, J. G., Windsor-Lewis, S., Likely, F., Baron, A. L., & Moorcroft, D. (2006). *Fast forward: Future trends in corporate communication*. UK: Strategic Communication Management. (Original work published 2005)

Davenport, T. H., & Harris, J. G. (2007). *Competing on analytics*. Cambridge, MA: Harvard Business School Press.

De Groot, A. (1965). *Thought and choice in chess*. The Hague, Netherlands: Mouton.

De Win, B., Van den Bergh, J., Matthijs, F., De Decker, B. & Joosen, W. (2000). Security architecture for electronic commerce applications. *SEC*, 491-500.

DEFRA. (2000). *Environmental reporting–guidelines for company reporting on greenhouse gas emissions*. Retrieved from http://www.defra.gov.uk/environment/business/reporting/pdf/envrpgas-annexes.pdf

DeLone, W. H., & McLean, E. R. (2004). Measuring e-commerce success: Applying the DeLone & McLean Information Systems success model. *International Journal of Electronic Commerce*, *9*(1), 31–48.

Denning, D. (2006). *Information warfare and security* (12th ed.). Addison-Wesley.

Department of Health and Human Services, Office of the Secretary. (2010). *Health Information Technology: Initial set of standards, implementation specifications, and certification criteria for Electronic Health Record technology*. Federal register: Rules and regulations.

Deutsch, M. (1960). The effect of motivational orientation upon trust and suspicion. *Human Relations*, *13*, 123–139. doi:10.1177/001872676001300202

Devaraj, S., & Kohli, R. (2003). Performance impacts of Information Technology: Is actual usage the missing link? *Management Science*, *49*(3), 273–289. doi:10.1287/mnsc.49.3.273.12736

Dibb, S., & Meadows, M. (2004). Relationship marketing and CRM: A financial services case study. *Journal of Strategic Marketing*, *12*, 111–125. doi:10.1080/0965254042000215177

Dimitrakos, T., et al. (2002). CORAS-a framework for risk analysis of security critical systems. *ERCIM News, 49*.

Divard, R., & Urien, B. (2001). *Le consumer vit dans un monde en couleurs* (pp. 3–24).

Djoleto, W. (2009). E-commerce solutions: An impact analysis on historically black colleges and universities. *Proceedings of the WDSI 38th Annual Meeting, Kauai, Hawaii*.

Djoleto-Okunbor, W. (2009). E-commerce solutions at historically black colleges and universities: An impact analysis. *International Journal of Effective Management*, *6*(1), 19–28.

Dobermann, A., Blackmore, S., Cook, S. E., & Adamchuk, V. I. (2004). Precision farming: Challenges and future directions. In *New directions for a diverse planet, Proceedings of the 4th International Crop Science Congress*. (pp 1-19). 26 Sep – 1 Oct 2004, Brisbane, Australia.

Downey, D., Etzioni, O., & Soderland, S. (2005). *A probabilistic model of redundancy in information extraction*. Seattle: University of Washington.

Dragstedt, C. A. (1956). Personal health log: Guest editorial. *Journal of the American Medical Association*, *160*(15), 1320.

Dreher & Dougherty. (2001). *Human resource strategy: A behavioral perspective for the general manager*. New York: McGraw-Hill.

Drucker, P. F. (2002). *Managing in the next society* (p. 118). New York: St. Martin Press.

Duart, J. M., & Sangrà, A. (2000). *Aprender en la virtualidad.* Barcelona: Gedisa.

Dunne, A. J. (2001). Supply chain management: Fad, panacea or opportunity? *Agribusiness Perspectives, 48.* Melbourne University.

Dwyer, F. M., & Lamberski, R. J. (1982-83). A review of the research on the effects of the use of colour in the teaching-learning process. *International Journal of Instructional Media, 10*(4), 303–328.

Dwyer, F., & Tanner, J. (2002). *Business marketing: Connecting strategies. Relationship and learning.* Boston: McGraw Hill.

Easingwood, C., & Storey, C. (1996). The value of multi-distribution systems in the financial services sector. *Service Industries Journal, 16*(2), 223–241. doi:10.1080/02642069600000023

Ebling, T. (2001). The economics of online banking. *Target Marketing*, 67.

Ehlers, U. D., & Carneiro, P. (2008). Entornos de aprendizaje personales. *Revista E-learning Papers, 9*, 2.

Ehrmann, S. C. (1999). Access and/or quality? Defining choices in the third revolution. *The Educom Review, 34*(5). Retrieved on October 25, 2005, from http://www.educause.edu/ir/library/html/erm/erm99/erm9956.html.

Eisenhardt, K. M. (1989). Building theories from case study research. *Academy of Management Review, 14*, 532–550. doi:10.2307/258557

Eisenhardt, K. M. (1989). Building theories from case study research. *Academy of Management Review, 14*(4), 532–550. doi:10.2307/258557

Ellis, K., & Blashki, K. (2004). Toddler techies: A study of young children's interaction with computers. *Information Technology in Childhood Education Annual, 2004*, 77–96.

Elton, L. (1988). Student motivation and achievement. *Studies in Higher Education, 13*(2), 215–221. doi:10.1080/03075078812331377886

Elton, L. (2001). Research and teaching: Conditions for a positive link. *Teaching in Higher Education, 6*(1), 43–56. doi:10.1080/13562510020029590

Enterprise Ireland. (2004). *IT/e-business status and issues of small and medium sized Irish SMEs.* BSM Ltd Report. Retrieved from http://www.enterprise-ireland.com/ebusiness/eBIT_ICTissues.htm

Entwhistle, N. (2000). Approaches to studying and levels of understanding: The influences of teaching and assessment. In J. Forest (Ed.), *Higher education: Handbook of theory and research.* (p. 15). New York: Agathon Press.

Ericksen, P. J. (2007). Conceptualizing food systems for global environmental change research. Global environmental change. Retrieved Februrary 9, 2007, from http://www.eci.ox.ac.uk/publications/downloads/ericksen07-foodsystems.pdf

Ericksen, P. J. (2008). What is the vulnerability of a food system to global environmental change? *Ecology and Society, 13*(2), 14. Retrieved from http://www.ecologyandsociety.org/vol13/iss2/art14/

Eriksson, K., & Marquardt, R. (2001) *Is relationship marketing theory applicable to Internet bank relationships?* Paper presented at the 17th IMP-conference in Oslo, Norway.

Ettinger, A., Holton, V., & Blass, E. (2006). E-learner experiences: What is the future for e-learning? *Industrial and Commercial Training, 38*(4), 208–212. doi:10.1108/00197850610671991

Etzioni, O., Cafarella, M., Downey, D., Popescu, A., Shaked, T., & Soderland, S. (2004). *Unsupervised named-entity extraction from the Web: An experimental study.* Seattle: University of Washington.

EUF. (2004). Information *Technology and resource use, energy end-use forecasting at Lawrence Berkeley National Laboratory.* Retrieved from http://enduse.lbl.gov/Projects/InfoTech.html

European-Commission. (2000). *Directive on waste electrical and electronic equipment and directive on the restriction of the use of certain hazardous substances in electrical and electronic equipment.* Explanatory memorandum. Brussels.

Evans, M. (2004). *Killing thinking: The death of the universities.* London: Continuum.

FAO. (2009). *The state of food insecurity in the world.* Retrieved on March 12, 2010, from http://www.fao.org/publications/en/

Farrell, L., Sampat, P., Sarin, R., & Slack, K. (2004). *Dirty metals-mining, communities and the environment.* Earthworks and Oxfam America.

Feinman, T., Goldman, D., Wang, R., & Cooper, N. (1999). Security basics: A Whitepaper. Retrieved from http:/ /www. pwc.com

Fink, D. (1998). Guidelines for successful adoption of Information Technology in small and medium enterprises. *International Journal of Information Management, 18*(4), 243–253. doi:10.1016/S0268-4012(98)00013-9

Firestorm. (2008). *Public law 110-53, Title IX.* Retrieved March 2nd, 2010, from http://www.publiclaw110.com/?p=3

Firpo, J. (2009). Banking the unbanked: Issues in designing technology to deliver financial services to the poor. In Matthäus-Maier, I., & von Pischke, J. D. (Eds.), *New partnerships for innovation in microfinance.* Springer Publishers. doi:10.1007/978-3-540-76641-4_11

Fleiss, J. L. (1981). *Statistical methods for rates and proportions* (2nd ed., pp. 38–46). New York: John Wiley.

Fletcher, O. (2010). China working to toughen up hacking laws. *Infoworld.* Retrieved February 2, 2010, from http://www.infoworld.com/d/security-central/china-working-toughen-hacking-laws-681

Ford, D. (1990). *Understanding business markets: Interaction, relationships and networks. The IMP Group.* Academic Press Limited.

Forsyth, A. & Furlong, A. (2000). *Socioeconomic disadvantage and access to higher education.*

Frasconi, P., Gori, M., & Sperduti, A. (1997). *A general frame work for adaptive processing of data structures.* Italy: Universita degli Studi di Firenze.

Fraumeni, B. M. (2001). E-commerce: Measurement and measurement issues. *The American Economic Review, 91*(2), 318–322. doi:10.1257/aer.91.2.318

Frederick, L. (2009). Information Technology innovations that extend rural microfinance outreach. In Matthäus-Maier, I., & von Pischke, J. D. (Eds.), *New partnerships for innovation in microfinance.* Springer Publishers. doi:10.1007/978-3-540-76641-4_10

Fresco, L. (2009). Challenges for food system adaptation today and tomorrow. In *Environmental Science & Policy, Special Issue: Food Security and Environmental Change, Food Security and Environmental Change: Linking Science, Development and Policy for Adaptation, 12*(4), 378-385.

Freytag, P. V., & Clarke, A. H. (2001). Business-to-business market segmentation. *Industrial Marketing Management, 30,* 473–486. doi:10.1016/S0019-8501(99)00103-0

Fry, H., Ketteridge, S., & Marshall, S. (2003). *A handbook for teaching and learning in higher education.* London: Kogan Page.

Fukuda-Parr, S. (2003). The human development paradigm: Operationalizing Sen's ideas on capabilities. *Feminist Economics, 9*(2–3), 301–317.

Furnell, S. (2005). *Computer insecurity.* London: Springer-Verlag.

Fusilier, M., & Durlabhji, S. (2009). E-business education worldwide: On the right track? *The International Journal of Management Education, 8*(2), 23–30. doi:10.3794/ijme.82.253

Galliers, R. D., & Sutherland, A. R. (1991). Information Systems management and strategy formulation: The stages of growth model revisited. *Journal of Information Systems, 1*(1), 89–114.

Gan, Y., & Zhu, Z. (2007). A learning framework for knowledge building and collective wisdom advancement in virtual learning communities. *Journal of Educational Technology & Society, 10*(1), 206–226.

Gao, P., & Owolabi, O. (2008). Consumer adoption of Internet banking in Nigeria. *International Journal of Electronic Finance, 2*(3), 284–299. doi:10.1504/IJEF.2008.020598

Gardener, E., Howcroft, B. & Williams, J. (1999). The new retail banking revolution. *The Service Industries Journal, 19*(2).

Garrette, B., & Mitchell, W. (1998). *Acquiring partners' capabilities: Outcomes of scale and link alliances between competitors. Managing strategically in an interconnected world.* New York: John Wiley and Sons.

Garrison, D. R., & Anderson, T. (2005). *El e-learning en el siglo XXI: Investigación y práctica*. Barcelona: Octaedro.

Garrison, D.R. & Anderson, T. (2003). *E-Learning in the 21st Century*. London: Routledge Falmer.

Gauss, B., & Urbas, L. (2003). Individual differences in navigation between sharable content objects–an evaluation study of a learning module design. *British Journal of Educational Technology*, *34*(4), 499–509. doi:10.1111/1467-8535.00346

Gefen, D., Karahanna, E., & Straub, D. W. (2003). Trust and TAM in online shopping: An integrated model. *Management Information Systems Quarterly*, *27*(1), 51–90.

Gehling, B., & Stankard, D. (2005). E-commerce security. *InfoSecCD '05: Proceedings of the 2nd annual conference on Information security curriculum development* (pp. 32-37). New York: ACM Press.

Gerpott, T. J., & Kornmeier, K. (2009). Determinants of customer acceptance of mobile payment systems. *International Journal of Electronic Finance*, *3*(1), 1–30. doi:10.1504/IJEF.2009.024267

Ghatak, M., Gangopadhyay, S. & Lensink, R. (2005). Joint liability lending and the peer selection effect. *Economic Journal.*

Giaglis, G. M., Paul, R. J., & Doukidis, G. I. (1999). Dynamic modelling to assess the business value of electronic commerce. *International Journal of Electronic Commerce*, *3*(3), 35–51.

Gibbs, G. (1981). *Teaching students to learn: A student-centred approach.* Milton Keynes: The Open University Press.

Gibson, J. J. (1977). The theory of affordances. In Shaw, R., & Bransford, J. (Eds.), *Perceiving, acting and knowing*. Hillsdale, NJ: Erlbaum.

Gil-Garcia, J. M. (2005). Exploring the success factors of state website functionality: An empirical investigation. *Proceedings of the National Conference on Digital Government Research, Atlanta, Georgia. Digital Government Research Center. Also ACM International Conference Proceedings, 87,* 121-130.

Gil-Garcia, J. R. (2006). Enacting websites: A mixed method study exploring e-government success in multi-organizational settings. *IEEE Proceedings of the 39th Hawaii International Conference on System Sciences.*

Gladwell, M. (2000). *The tipping point–how little things can make a big difference.* New York: Little, Brown, and Company.

Gloor, P. (2000). *Making the e-Business transformation.* London: Springer-Verlag.

Golan, E., Krissoff, B., Kuchler, F., Nelson, K., Price, G., & Calvin, L. (2003). Traceability in the US food supply: Dead end or euperhighway? *Choices Magazine*, *18*(2), 36–40.

Golan, E., Krissoff, B., Kuchler, F., Calvin, L., Nelson, K. & Price, G. (2004). *Traceability in the US Food Supply: Economic theory and industry studies.* AER-830, USDA/ERS.

Gollman, D. (2003). *Computer security*. John Wiley & Sons.

González, M. E., Mueller, R. D., & Mack, R. W. (2008). An alternative approach in service quality: An e-banking case study. *The Quality Management Journal*, *15*(1), 41–60.

Goodhue, D. L., Wixom, B. H., & Watson, H. J. (2002). Realizing business benefits through CRM: Hitting the right target in the right way. *Management Information Systems Quarterly Executive*, *1*(2), 79–94.

Goodman, J., Alakeson, V. & Jorgensen, B. (2004). *Encouraging green telework: Forum for the future.*

Greenly, M., & Gugerty, B. (2002). How bar coding reduces medication errors. *Nursing*, *32*(5), 70.

Greenspan, R. (2003). *What they're doing down under.* Retrieved on April 15, 2004, from http://www.clickz.com/stats/big_picture/geographics/article.php/5911_3090361

Grimm, M. (1999). Ford connects: Hatches massive relationship program via Web. *Brandweek, 40*(1), 19–20.

Grönroos, C. (1994). From marketing mix to relationship marketing-towards a paradigm shift in marketing. *Management Decision, 32*(2). doi:10.1108/00251749410054774

Grover, V., Teng, J., Segar, A. H., & Fiedler, K. (1998). The influence of Information Technology diffusion and business process change on perceived productivity: The IS executive's perspective. *Information & Management, 34*(3), 141–159. doi:10.1016/S0378-7206(98)00054-8

GS1. (2004). *Hospital pharmaceutical e-commerce looks healthy*. Media Press Release, GS1 Australia.

Gugerty, B., Maranda, M., & Rook, D. (2006). The clinical Information System implementation evaluation scale. *Studies in Health Technology and Informatics, 122*, 621–625.

Gupta, S., Koulamas, C., & Kyparisis, G. J. (2009). E-business: A review of research published in production and operations management (1992–2008). *Production and Operations Management, 18*(6), 604–620. doi:10.1111/j.1937-5956.2009.01100.x

Gutiérrez, C., et al. (2004). Web Services security: Is the problem solved? *Information Security Journal: A Global Perspective, 13*(3), 22–31.

H.R. 1–111th Congress. (2009). American Recovery and Reinvestment Act of 2009. Retrieved January 12, 2010, from http://www.govtrack.us/congress/bill.xpd?bill=h111-1

Hackbarth, G., & Kettinger, W. J. (2004). Strategic aspirations for Net-enabled business. *European Journal of Information Systems, 13*(4), 273–285. doi:10.1057/palgrave.ejis.3000511

Haggis, T. (2003). Constructing images of ourselves? A critical investigation into approaches to learning research in higher education. *British Educational Research Journal, 29*(1), 89–104. doi:10.1080/0141192032000057401

Hall, R. H., & Hanna, P. (2004). The impact of Web page text-background colour combinations on readability, retention, aesthetics, and behavioral intention. *Behaviour & Information Technology, 23*(3), 183–195. doi:10.1080/01449290410001669932

Halperin, R. (2008). *The role of institutional factors in the formation of e-learning practices, advances in e-learning: Experiences and methodologies*. Hershey, PA: IGI Global.

Halperin, K. (2001). *Balancing act*. Company Business and Marketing.

Hammoudi, A., Hoffmann, R., & Surry, Y. (2009). Food safety standards and agri-food supply chains: An introductory overview. *European Review of Agriculture Economics, 36*(4), 469–478. doi:10.1093/erae/jbp044

Hampton-Sosa, W., & Koufaris, M. (2005). The effect of website perceptions on initial trust in the owner company. *International Journal of Electronic Commerce, 10*(1), 55–81.

Harker, M. (1999). Relationship marketing defined? An examination of current relationship marketing definitions. *Marketing Intelligence & Planning, 17*(1). doi:10.1108/02634509910253768

Harris, G. (2000). The big four banks: Dealing with the online challenge. *Journal of Financial Services Marketing, 4*(4), 296–305.

Harris, S. (2008). *CISSP all-in-one exam guide* (2nd ed.). McGraw-Hill/Osborne.

Hartzel, K. S., & Peace, A. G. (2002). Ethical concerns raised by the use of the Internet in academia. *Journal of Information Ethics*, 17–31.

Haselgrove, S. (1994). *The student experience*. Buckingham, UK: SRHE & Open University Press.

Hasselknippe, H. (2003). Systems for carbon trading: An overview. *Climate Policy, 3*(2), S43–S57. doi:10.1016/j.clipol.2003.09.014

Hausman, W. H. (2002). Supply chain performance metrics. In Billington, C., Harrison, T., Lee, H., & Neale, J. (Eds.), *The practice of supply chain management*. Boston: Kluwer.

Health Information Technology Standards Panel (HITSP). (2008). *IS 77: Remote monitoring interoperability specification*. Retrieved March 22, 2010, from http://www.hitsp.org

HeartQuotes. (2009). TEAM–Together Everyone Achieves More. Retrieved December 24, 2009, from http://www.heartquotes.net/teamwork-quotes.html

Hecker, D. E. (2001). Employment impact of electronic business. *Monthly Labor Review, 124*(5), 3–16.

Heeks, R. (2006). Health Information Systems: Failure, success and improvisation. *International Journal of Medical Informatics, 75*, 125–137. doi:10.1016/j.ijmedinf.2005.07.024

Henderson, J. C., & Venkatraman, N. (1999). Strategic alignment: Leveraging Information Technology for transforming organizations. *IBM Systems Journal, 32*(1), 472–484. doi:10.1147/sj.382.0472

Herbane, B., Elliot, D., & Swartz, E. (1997). Contingency and continua: Achieving excellence through business continuity planning. *Business Horizons, 40*(6). doi:10.1016/S0007-6813(97)90063-X

Hermes, N., & Lensink, R. (2007). The empirics of microfinance: What do we know? *The Economic Journal, 117*(517), F1–F10. doi:10.1111/j.1468-0297.2007.02013.x

Herrmann, G., & Pernul, G. (1999). Viewing business-process security from different perspectives. *International Journal of Electronic Commerce, 3*(3).

Higher Education Academy. (2010). *Subject centre spotlight: ESCalate*. Retrieved on May 10, 2010, from http://www.heacademy.ac.uk/resources/detail/ourwork/ipp/Issue3_Escalate

Higher Education Academy/ Equality Challenge Unit. (2008). Ethnicity, gender and degree attainment project. Retrieved from http://www.heacademy.ac.uk/assets/York/documents/ourwork/research/Ethnicity_Gender_Degree_Attainment_report_Jan08.pdf

HIMSS. (2010). *HIMSS analytics U.S. EMR adoption model*. Retrieved March 30, 2010, from http://www.himssanalytics.org

Hinostroza, J. E., & Mellar, H. (2001). Pedagogy embedded in educational software design: Report of a case study. *Computers & Education, 37*, 27–40. doi:10.1016/S0360-1315(01)00032-X

Hirschheim, R., Earl, M., Feeny, D., & Lockett, M. (1988). An exploration into the management of the information function: Key issues and an evolutionary model. *Proceedings of Information Technology Management for Productivity and Strategic Advantage, IFIP TC-8 Open Conference*, Singapore.

HL7. (2007). *HL7 continuity of care document, a healthcare IT interoperability standard, is approved by balloting process and endorsed by healthcare IT standards panel.* Retrieved March 17, 2010, from http://www.hl7.org/documentcenter/public/pressreleases/20070212.pdf

Hobbs, J. E. (2006). Liability and traceability in agri-food supply chains. In Ondersteijn, C. J. M., Wijnands, J. H. M., Huirne, R. B. M., & Van Kooten, O. (Eds.), *Quantifying the agri-food supply chain* (pp. 87–100). The Netherlands: Springer. doi:10.1007/1-4020-4693-6_7

Hobbs, J. E. (2003). *Traceability & country of origin labelling.* Paper presented at the Policy Dispute Information Consortium, 9th Agricultural and Food Policy Information Workshop, Montreal, April 25 2003. Retrieved on March 25, 2010, from http://www.farmfoundation.org/news/articlefiles/859-hobbs.pdf

Hodge, M. (2002). *Minister for lifelong learning.* Paper presented at Education Conference.

Holley, D., & Oliver, M. (2000). Pedagogy and new power relationships. *International Journal of Management Education, 1*(1), 11–21. doi:10.3794/ijme.11.b

Holley, D., & Oliver, M. (2010). Student engagement and blended learning: portraits of risk. [Elsevier Publishing.]. *Computers & Education, 54*, 693–700. doi:10.1016/j.compedu.2009.08.035

Holley, D., Sinfield, S., & Burns, T. (2006). It was horrid, very very horrid: A student perspective on coming to an inner-city university in the UK. *Social Responsibility Journal, 2*(1), 36–42. doi:10.1108/eb045820

Holley, D. (2008). Using biographic narrative to explore students' experiences of online learning. In Frame, P., & Burnett, J. (Eds.), *Using auto/biography in learning and teaching* (pp. 53–58).

Holley, D., Greaves, L., Bradley, C., & Cook, J. (2010). You can take out of it what you want: How learning objects within blended learning designs encourage personalised learning. In O'Donoghue, J. (Ed.), *Technology-supported environments for personalized learning method and case studies* (pp. 285–304). Hershey, PA: IGI Global.

Holley, D., & Oliver, M. *(2009). A private revolution–how technology is enabling students to take their work home.* Journal of Enhancing Learning in the Social Sciences, 1*(*3). *Retrieved onApril30, 2009, from*http://www.eliss.org.uk/

Holsapple, C., & Singh, M. (2000). Toward a unified view of electronic commerce, electronic business, and collaborative commerce: A knowledge management approach. *Knowledge and Process Management, 7*(3), 151–164. doi:10.1002/1099-1441(200007/09)7:3<151::AID-KPM83>3.0.CO;2-U

Höne, K., & Eloff, J. (2002). Information security policy—what do international information security standards say? *Computers & Security, 21*(5), 402–409. doi:10.1016/S0167-4048(02)00504-7

Hopkins, D. (2006) *Choice and voice in personalised learning.* (pp 1-22). retrieved January 18, 2010, from http://titania.sourceoecd.org/vl=2327409/cl=25/nw=1/rpsv/cw/vhosts/oecdthemes/99980029/v2006n3/contp1-1.htm

Hoque, F. (2000). *E-enterprise-business models, architecture, and components.* Cambridge, UK: Cambridge University Press. doi:10.1017/CBO9780511584923

Horno, M. C., & Ibarretxe, I. (2008). *Metodologías educativas y uso de las TIC en el área de lingüística.* Paper presented at the II Jornadas de Innovación docente, Tecnologías de la información y de la comunicación e investigación educativa. Universidad de Zaragoza. Retrieved June 3, 2010, from http://ice.unizar.es/uzinnova/jornadas/pdf/27.pdf

Hounsell, D. (2004). No comment? Feedback, learning and development. In Slowey, M., & Watson, D. (Eds.), *Higher education and the life course.* Buckingham, UK: Open University Press.

Howcroft, B. (1993). Branch networks and alternative distribution channels. *International Journal of Bank Marketing, 11*(6). doi:10.1108/02652329310045738

Hughes, T. (2003). Marketing challenges in e-banking standalone or integrated? *Journal of Marketing Management, 19,* 1067–1085. doi:10.1362/026725703770558321

Hughes, R., & Hughes, A. (2004). *The adoption of Internet banking services by young Australian consumers.* Paper presented at the AusWeb Conference, Gold Coast, Australia.

Hwang, H. G., Chen, R. F., & Lee, J. M. (2007). Measuring customer satisfaction with Internet banking: An exploratory study. *International Journal of Electronic Finance, 1*(3), 321–335. doi:10.1504/IJEF.2007.011502

ILO (International labour organisation). (2007). *Resolution concerning updating the international standard classification of occupations.* Geneva, Switzerland.

Institute for Educational Sciences, National Center for Educational Statistics. (2006). *The health literacy of America's adults. Results from the 2003 National Assessment of Adult Literacy.*

Internet Society. (2010). *A brief history of the Internet.*

Irani, Z. (2002). Information Systems evaluation: Navigating through the problem domain. *Information & Management, 40,* 11–24. doi:10.1016/S0378-7206(01)00128-8

Itten, J. (1970). *The elements of color.* New York: Van Nostrand Reinhold Company.

Ivatury, G. (2009). Using technology to build inclusive financial systems. In Matthäus-Maier, I., & von Pischke, J. D. (Eds.), *New partnerships for innovation in microfinance.* Springer Publishers. doi:10.1007/978-3-540-76641-4_9

Jamal, K., Maier, M., & Sunder, S. (2003). Privacy in e-commerce: Development of reporting standards, disclosure and assurance services in an unregulated market. *Journal of Accounting Research, 41*(2), 285. doi:10.1111/1475-679X.00104

Jaquith, A. (2002). *The security of applications: Not all are created equal.*

Jarmon, L., Traphagan, T., Mayrath, M., & Trivedi, A. (2009). Virtual world teaching, experiential learning, and assessment: An interdisciplinary communication course in Second Life. *Computers & Education, 53,* 169–182. doi:10.1016/j.compedu.2009.01.010

Jevons, C., & Gabbott, M. (2000). Trust, brand equity and brand reality in Internet business relationships: An interdisciplinary approach. *Journal of Marketing Management, 16*(6), 619–634. doi:10.1362/026725700785045967

Jha, A. K., Desroches, C. M., Campbell, E. G., Donelan, K., Rao, S. R., & Ferris, T. G. (2009). Use of electronic health records in US hospitals. *The New England Journal of Medicine, 360*(16), 1628–1638. doi:10.1056/NEJMsa0900592

Jisc Infonet. (2010). *Effective use of VLEs*. Retrieved on January 27, 2010, from http://www.jiscinfonet.ac.uk/InfoKits/effective-use-of-VLEs

JISC. (2004). *Effective practice with e-learning*. Retrieved from www.jisc.ac.uk/practice

Johns, R., & Perrott, B. (2008). The impact of internet banking on business-customer relationships. Are you being self-served? *International Journal of Bank Marketing, 26*(7), 465–482. doi:10.1108/02652320810913846

Johns, R., Low, D., & Blackman, D. (2009). Information Technology and relationship marketing in an inter-firm context: Implications for research. Paper presented at the ANZMAC Conference, Melbourne.

Johnson, L., Levine, A., Smith, R., & Stone, S. (2010). *The 2010 horizon report*. Austin: The New Media Consortium.

Johnson, M. (2008). Information risk of inadvertent disclosure: An analysis of file-sharing risk in the financial supply chain. *Journal of Management Information Systems, 25*(2), 97–123. doi:10.2753/MIS0742-1222250205

Johnson, C. E. (2001). *Meeting the ethical challengers of leadership: Casting light or shadow.* London: Sage Publications.

Johnson, L., Pokorny, H., Djebbour, Y., & Curran, J. (2009). *Look who's here: Engagement and attendance initiatives*. HEA Teaching and Research Development Grant. Retrieved from http://www.heacademy.ac.uk/business/projects/detail/trdg/2008-09/0809trdg_Linda_Johnson_LMU

Johnston, R. (1995). The determinants of service quality: Satisfiers and dissatisfiers. *International Journal of Service Industry Management, 6*(5), 53–71. doi:10.1108/09564239510101536

Jokinen, P., Malaska, P., & Kaivo-oja, J. (1998). The environment in an information society: A transition stage towards more sustainable development? *Futures, 30*(6), 485–498. doi:10.1016/S0016-3287(98)00054-8

Jonassen, D. H., & Wang, S. (1993). Acquiring factual knowledge from semantically structured hypertext. *Journal of Computer-Based Instruction, 20*(1), 1–8.

Jones, N. (1996). Talking pages. *Marketing Week, 19*(18), 37–40.

Jones, P., Beynon-Davies, P., & Muir, E. (2003). E-business barriers to growth within the SME sector. *Journal of Systems & Information Technology, 7*(1-2), 1–26. doi:10.1108/13287260380000771

Jones, S., Wilikens, M., Morris, P., & Nasera, M. (2000). Trust requirements in e-business. *Communications of the ACM, 43*(12), 81–87. doi:10.1145/355112.355128

Joseph, M., McClure, C., & Joseph, B. (1999). Service quality in the banking sector: The impact of technology on service delivery. *International Journal of Bank Marketing, 17*(4). doi:10.1108/02652329910278879

Joseph, R. (1998). *Customer orientation and the creation of new technology based services*. Electronic Media and Executive Education.

Joshi, J., Aref, W., Ghafoor, A., & Spafford, E. (2001). Security models for Web-based applications. *Communications of the ACM, 44*(2), 38–44. doi:10.1145/359205.359224

Jungck, P., & Shim, S. (2004). Issues in high speed Internet security. *Computer, 37*(7), 36–42. doi:10.1109/MC.2004.58

Kaefer, F., & Bendoly, E. (2004). Measuring the impact of organizational constraints on the success of business-to-business e-commerce efforts: A transactional focus. *Information & Management, 41*, 529–541. doi:10.1016/S0378-7206(03)00088-0

Kaelber, D. C., Jha, A. K., Johnston, D., Middleton, B., & Bates, D. W. (2008). A research agenda for personal health records (PHRs). *Journal of the American Medical Association, 15*, 729–736.

Kalakota, R., & Rohinson, M. (2000). Electronic commerce. In *Encyclopedia of Computer Science* (4th ed., pp. 628–634). London: Nature Publishing Group.

Kandampully, J., & Duddy, R. (1999). Competitive advantage through anticipation, innovation and relationships. *Management Decision*, *37*(1), 51–56. doi:10.1108/00251749910252021

Kang, K.-S., & Corbitt, B. (2001). Effectiveness of graphical components in website e-commerce application-a cultural perspective. *Electronic Journal on Information Systems in Developing Countries*, *7*(2), 1–6.

Kanneganti, R., & Chodavarapu, P. (2008). *SOA security*. Greenwich, CT: Mannings Publications.

Kapplinsky, R. (1999). Globalisation and unequalisation: What can be learned from value chain analysis. *The Journal of Development Studies*, *37*(2), 117–146. doi:10.1080/713600071

Karake, S. (1992). *Information Technology and management control: An agency theory perspective*. Westport, CT: Praeger Publishers.

Karake-Shalboub, Z. (2002). *Trust and loyalty in electronic commerce: An agency theory perspective*. Westport, CT: Quorum Books.

Katsikas, S., Lopez, J., & Pernul, G. (2005). *Trust, privacy and security in e-business: Requirements and solutions*. (LNCS 3746), (pp. 548-558).

Kauffman, R. J., & Walden, E. A. (2001). Economics and electronic commerce: Survey and directions for research. *International Journal of Electronic Commerce*, *5*(4), 5–16.

Kauffman, R., & Wood, C. (2007). Revolutionary research strategies for e-business: A philosophy of science view in the age of the Internet. In Kauffman, R., Wood, C., & Kauffman, R. A. (Eds.), *Economics, Information Systems, and electronic commerce research: Advanced empirical methodologies*. Armonk, NY: M. E. Sharpe.

Kauremaa, J., Nurmilaakso, J.M. & and Tanskanen, K. (2009). E-business enabled operational linkages: The role of RosettaNet in integrating the telecommunications supply chain. *International Journal of Production Economics*.

Kawamoto, K., Koomey, J. G., Nordman, B., Brown, R. E., Piette, M. A., & Ting, M. (2001). *Electricity used by office equipment and network equipment in the U.S.: Detailed report and appendices*. Orlando: Lawrence Berkeley National Laboratory.

Kaynak, E., & Harcar, T. D. (2005). Consumer value creation in mobile banking services. *International Journal of Technology Marketing*, *1*(1), 62–78. doi:10.1504/IJTMKT.2005.008125

Keaveney, S. (1995). Customer switching behaviour in service industries: An exploratory study. *Journal of Marketing*, 59.

Kelley, S., & Davis, M. (1994). Antecedents to customer expectations for service recovery. *Journal of the Academy of Marketing Science*, *22*(1), 52–61. doi:10.1177/0092070394221005

Kennedy, M. (2000). The road to e-commerce. *American School & University*, *72*(11), 14–18.

Kern, S. I. (2009). PHRs can boost efficiency, but also hold risk. *Medical Economics*, *86*(15).

Khandker, S. R. (2005). Microfinance and poverty: Evidence using panel data from Bangladesh. *The World Bank Economic Review*, *19*(2), 263–286. doi:10.1093/wber/lhi008

King, C. III. (2000). *Marketing, product differentiation, and competition in the market for antiulcer drugs*. Harvard Business School Review.

Kintsch, W. (1970). Models for free recall and recognition. In Norman, D. A. (Ed.), *Models of human memory*. New York: Academic Press.

Klein, H. K., & Myers, M. D. (1999). A set of principles for conducting and evaluating interpretive field studies in Information Systems. *Management Information Systems Quarterly*, *23*(1), 67–94. doi:10.2307/249410

Kobayashi, S. (1986). Theoretical issues concerning superiority of pictures over words and sentences in memory. *Perceptual and Motor Skills*, *63*, 783–792.

Kola, J. & Latvala, T. (2003). Impact of information on the demand for credence characteristics. *International Food and Agribusiness Management Review, 5*(2).

Kolodinsky, J. M., Hogarth, J. M., & Hilgert, M. A. (2004). The adoption of electronic banking technologies by US consumers. *International Journal of Bank Marketing*, *22*(4), 233–241. doi:10.1108/02652320410542536

Kopp, K. (2000). Nonprofits and electronic commerce. *Digital Beat Extra.* Retrieved on September 1, 2004, from www.benton.org/News/Extra/ec062000.html

Korchak, R., & Rodman, R. (2001). E-business adoption among US small manufacturers and the role of manufacturing extension. *Economic Development Review (Schiller Park, Ill.), 17*(3), 20–25.

Koskosas, I., & Paul, R. (2003). A socio-organisational approach to Information Systems security risks. [IJRAM]. *International Journal of Risk Assessment and Management, 4*(2/3). doi:10.1504/IJRAM.2003.003527

Kraemer, K. L., & King, J. L. (2003). Information Technology and administrative reform: Will the time after e-government be different? [from http://jlking.people.si.umich.edu/IJEGR-Final.pdf]. *International Journal of Electronic Government Research*, 1–18. Retrieved December 15, 2006.

Krajewski, L., & Ritzman, L. (1993). *Operations management: Strategy and analysis.* Boston: Addison Wesley.

Krishna, S., & Walsham, G. (2005). Implementing public Information Systems in developing countries: Learning from a success story. *Information Technology for Development, 11*(2), 123–140. doi:10.1002/itdj.20007

Kuperman, G. J., & Gibson, R. F. (2003). Computer physician order entry: Benefits, costs, and issues. *Annals of Internal Medicine, 139*, 31–39.

Kurose, J. F., & Ross, K. W. (2003). *Computer networking: A top-down approach featuring the Internet* (2nd ed.). Pearson Education, Inc.

Kutter, M., & Hartung, F. (2000). Introduction to watermarking techniques. In Petitcolas, S. K. (Ed.), *Information hiding techniques for steganography and digital watermarking.*

Kvavik, R. B. (2002). E-business in higher education. In Katz, R. N. (Ed.), *Web portals and higher education: Technologies to make IT personal.* Indianapolis: Jossey-Bass Inc.

Kwok, S., Yang, C., & Tam, K. (2004). Intellectual property protection for electronic commerce applications. *Journal of Electronic Commerce Research, 5*(1).

Lachance, J. R. (2000). International Symposium of the International Personnel Management Association. *Public Personnel Management, 29*(3), 305.

Laforet, S., & Li, X. Y. (2005). Consumers' attitudes towards online and mobile banking in China. *International Journal of Bank Marketing, 23*(4/5), 362–381. doi:10.1108/02652320510629250

Lafourcade, A. C., & Isern, J. Mwangi. P. & Brown, M. (2005). *Overview of the outreach and financial performance of microfinance institutions in Africa.* Microfinance Information Exchange (MIX).

Lang, B., & Colgate, M. (2003). Relationship quality, on-line banking and the Information Technology gap. *International Journal of Bank Marketing, 21*(1), 29–37. doi:10.1108/02652320310457785

Lang, R. D. (2009). Personal health records ready for prime time? *Journal of Healthcare Information Management, 23*(3).

Laskey, K. (2005). Semantic interoperability for Web Services: The needs and challenges for the demanding consumer. *W3C Workshop on Frameworks for Semantics in Web Services.*

Laudon, K., & Laudon, J. (1998). *Information Systems and the Internet* (4th ed.). Dryden Press.

Laukkanen, T., & Lauronen, J. (2005). Consumer value creation in mobile banking services. *International Journal of Mobile Communications, 3*(4), 325–338. doi:10.1504/IJMC.2005.007021

Lawson, R., Alcock, C., Cooper, J., & Burgess, L. (2003). Factors affecting adoption of electronic technologies by SMEs: An Australian study. *Journal of Small Business and Enterprise Development, 10*(3), 265–276. doi:10.1108/14626000310489727

Lay, J., Ong, C., Yang, C., & Wang, C. (2008). Assessing and managing employee readiness for embracing e-business. *Proceedings of the 2008 ACM SIGMIS Computer Personnel Research Annual Conference.* Charlottesville, VA: ACM.

Lee, S., & Koubek, R. J. (2010). The effects of usability and web design attributes on user preference for e-commerce websites. *Computers in Industry, 61*, 329–341. doi:10.1016/j.compind.2009.12.004

Lee, J., & Runge, J. (2001). Adoption of Information Technology in small business: Testing drivers of adoption for entrepreneurs. *Journal of Computer Information Systems*, *42*(1), 44–57.

Lee, D., Park, J., & Ahn, J. (2000). *On the explanation of factors affecting e-commerce adoption*. Working paper.

Lee, H.L. & Amaral J. (2002). *Continuous and sustainable improvement through supply chain performance management*. Stanford global supply chain management forum paper SGSCMF – W1.

Lefebvre, H. (1991). *The production of space (trans D.Nicholson-Smith)*. Oxford: Blackwell.

Leiva, M. (2007). *Teaching strategies for English language learners in mathematics: Implications for teaching all students*. Houghton Mifflin. Retrieved on December 2, 2007 from www.Beyond-the-Book.com

Lejk, M., & Deeks, D. (2002). *Systems analysis techniques* (2nd ed.). Addisson-Wesley, Pearson Education Limited.

Lemoine, J.-F. (2008). Atmosphère des sites web marchands et réactions des apprenants. *Revue Française de Marketing, 217*, 2/5.

Leonard-Barton, D., & Deschamps, I. (1988). Managerial influence in the implementation of new technology. *Management Science*, *34*(10), 1252–1265. doi:10.1287/mnsc.34.10.1252

Levesque, T., & McDougall, G. H. (1996). Determinants of customer satisfaction in retail banking. *International Journal of Bank Marketing*, *14*(7), 12–20. doi:10.1108/02652329610151340

Levy, M., & Powell, P. (2003). Exploring SME Internet adoption: Towards a contingent model. *Electronic Markets*, *13*(2), 173–181. doi:10.1080/1019678032000067163

Levy, A. (2005). *Making pollution pay*. Bloomberg Markets.

Lewicki, R. J., & Bunker, B. B. (1996). Developing and maintaining trust in work relationships. In Kramer, R. M., & Tyler, T. (Eds.), *Trust in organizations: Frontiers of theory and research* (pp. 114–139). Thousand Oaks, CA: Sage Publications.

Lewis, J. D., & Weigert, A. J. (1985). Trust as a social reality. *Social Forces*, *63*(4), 967–985. doi:10.2307/2578601

Lewis, P. (2003). *Control traffic farming becomes mainstream*. Retrieved on January 12, 2010, from http://www.abc.net.au/landline/stories/s908970.htm

Leymann, F., Roller, D., & Schmidt, M.-T. (2002). Web services and business process management. *IBM Systems Journal*, *41*(2), 1015–1019. doi:10.1147/sj.412.0198

Li, F. (2002). Internet banking: From new distribution channel to new business models. *International Journal of Business Performance Management*, *4*(2-4), 136–160. doi:10.1504/IJBPM.2002.000112

Liker, J. K. (2004). *The Toyota way*. New York: McGraw-Hill.

Lin, C., Huang, Y., & Burn, J. (2007). Realising B2B e-commerce benefits: The link with IT maturity, evaluation practices, and B2B e-commerce adoption readiness. *European Journal of Information Systems*, *16*(6), 806–819. doi:10.1057/palgrave.ejis.3000724

Lin, C., & Pervan, G. (2003). The practice of IS/IT benefits management in large Australian organizations. *Information & Management*, *41*(1), 13–24. doi:10.1016/S0378-7206(03)00002-8

Lin, C., Pervan, G., Lin, H.-C., & Tsao, H. (2008). An investigation into business-to-business electronic commerce organizations. *Journal of Research and Practices in Information Technology*, *40*(1), 3–18.

Lin, C., Pervan, G., & McDermid, D. (2005). IS/IT investments evaluation and benefits realization issues in Australia. *Journal of Research and Practices in Information Technology*, *37*(3), 235–251.

Lin, C., & Huang, Y. (2009). A framework for business-to-business e-commerce evaluation challenges and critical success factors. In Xu, J., & Quaddus, M. (Eds.), *E-business in the 21st Century: Realities, challenges and outlook* (pp. 233–254). World Scientific Publishing Co.

Liska, A. (2003). *The practice of network security-deployment strategies for production environment*. Prentice Hall PTR, Pearson Education Inc.

Liu, Q., Safavi-Naini, R., & Sheppard, N. (2003). Digital rights management for content distribution. *ACSW Frontiers '03: Proceedings of the Australasian information security workshop conference on ACSW frontiers,* (pp. 49-58). Darlinghurst, Australia: Australian Computer Society, Inc.

Livingston, L. A. (1991). The effect of colour on performance in an instructional gaming environment. *Journal of Research of Computing in Education, 24*(2), 246–253.

Loader, B. (Ed.). (1998). *Cyberspace divide: Equality, agency, and policy in the information society.* London, New York: Routledge. doi:10.4324/9780203169537

Locke, S., & Cave, J. (2002). Information communication technology in New Zealand SMEs. *Journal of American Academy of Business, 2*(1), 235–240.

Lorenzi, N. M., & Riley, R. T. (2003). Organizational issues = change. *International Journal of Medical Informatics, 69*(2/3), 197–203. doi:10.1016/S1386-5056(02)00105-3

Lu, M. T., Liu, C. H., Jing, J., & Huang, L. J. (2005). Internet banking: Strategic responses to the accession of WTO by Chinese banks. *Industrial Management & Data Systems, 105*(3/4), 429–443. doi:10.1108/02635570510592343

Ma'ai, D., & Leeming, D. (2008). People first network and the potential of ICT in rural development. *Pacific Economic Bulletin, 23*(3), 141–154.

Malcolm, B., Carlson, R. A., Tucker, C. L., et al. (2000). *Veterans Affairs: Eliminating medication errors through point-of-care devices.* Paper presented at Healthcare Information and Management Systems Society Conference, Dallas, TX.

Malhotra, A., Gosain, S., & Sawy, O. A. (2005). Absorptive capacity configurations in supply chains: Gearing for partner-enabled market knowledge creation. *Management Information Systems Quarterly, 29*(1), 145–187.

Maltz, A. (2003). Codes of ethics in the wake of Sarbanes-Oxley. *Trustee, 56*(10), 31.

Marketing, P. (2008). *Fewer sales reps lead to higher costs.* Retrieved June 5, 2010, from http://pharmamkting.blogspot.com/2008/11/less-sales-reps-lead-to-higher-costs.html

Marr, A. (2006). The limitations of group-based microfinance and ways to overcome them. *Small Enterprise Development, 17*(3), 28–40. doi:10.3362/0957-1329.2006.029

Marshall, P., & McKay, J. (2002). Evaluating the benefits of electronic commerce in small and medium enterprises. *The Australian Journal of Information Systems, 9*(2).

Martin, A. G. (2004). Making SOX work for your company is just good business. *Fort Worth Business Press, 17*(24), 34.

Maruyama, Y., & Akahori, K. (2008). Effect of bichrome prints on memory in the middle-school educational environment. In K. McFerrin, et al. (Eds.), *Proceedings of Society for Information Technology and Teacher Education International Conference,* (pp. 4650-4655). Chesapeake, VA: AACE.

Masanet, E., Price, L., Can, S. R., Brown, R., & Worrell, E. (2005). *Optimization of product life cycles to reduce greenhouse gas emissions in California.* California Energy Commission, Public Interest Energy Research Program.

Matlay, H., & Addis, M. (2003). Adoption of ICT and e-commerce in small businesses: An HEI-based consultancy perspective. *Journal of Small Business and Enterprise Perspective., 10*(3), 321–335. doi:10.1108/14626000310489790

Mattelart, A. (1996). *The invention of communication.* London, Minneapolis: University Minnesota Press.

Matthäus-Maier, I., & von Pischke, J. D. (Eds.). (2009). *New partnerships for innovation in microfinance.* Springer Publishers.

Matthews, H. S., Williams, E., Tagami, T., & Hendrickson, C. T. (2002). Energy implications of online book retailing in the United States and Japan. *Environmental Impact Assessment Review, 22*(5), 493–507. doi:10.1016/S0195-9255(02)00024-0

Matthews, H. S., Hendrickson, C. T., & Lave, L. (2001). How much did Harry Potter cost? E-commerce is growing and environmentalists should be happy. Or should they? Retrieved from http://www.oecdobserver.org/news/fullstory.php/aid/415/How_much_did_Harry_Potter_cost_.html

Maxwell, J. C. (2002). [What every leader needs to know. Nashville, TN: Thomas Nelson Publishing.]. *Leadership*, 101.

Mayer, R. C., Davis, J. H., & Schoorman, F. D. (1995). An integrative model of organizational trust. *Academy of Management Review*, *20*(3), 709–734. doi:10.2307/258792

Mayer, R. E. (2001). *Multimedia learning*. New York: Cambridge University Press.

Mayer, R. E., & Gallini, J. K. (1990). When is an illustration worth ten thousand words? *Journal of Educational Psychology*, *82*, 715–726. doi:10.1037/0022-0663.82.4.715

McAllister, D. J. (1995). Affect- and cognition-based trust as foundations for interpersonal cooperation in organizations. *Academy of Management Journal*, *38*(1), 24–59. doi:10.2307/256727

McCracken, D. D., & Wolfe, R. J. (2004). *User-centered website development: A human-computer interaction approach*. Upper Saddle River, NJ: Pearson Prentice Hall Inc.

McCulloch, E. B., Pingali, P. L., & Stamoulis, K. G. (2008). Small farms and the transformation of food systems: An overview. In McCulloch, E. B., Pingali, P. L., & Stamoulis, K. G. (Eds.), *The transformation of agri-food systems. Globalisation, supply chains and smallholder farms* (pp. 3–47).

McKay, J., & Marshall, P. (2004). *Strategic management of e-business*. Queensland, Australia: John Wiley & Sons.

McKnight, D. H., Cummings, L. L., & Chervany, N. L. (1998). Initial trust formation in new organizational relationships. *Academy of Management Review*, *23*(3), 473–490. doi:10.2307/259290

McMillan, R. (2007). *Researchers: Web 2.0 security seriously flawed*. Retrieved June 10th, 2010, from http://www.pcworld.com/article/131215/researchers_web_20_security_seriously_flawed.html

Mead, N., & Stehney, T. (2005). Security Quality Requirements Engineering (SQUARE) methodology. *SESS '05: Proceedings of the 2005 workshop on Software engineering for secure systems-building trustworthy applications,* (pp. 1-7). New York: ACM Press.

Mehrabian, A., & Russell, J. A. (1974). *An approach to environmental psychology*. Cambridge, MA: MIT Press.

Mell, P., & Grance, T. (2009). *Effectively and securely using the cloud computing paradigm*. Retrieved March 02, 2010, from http://csrc.nist.gov/groups/SNS/cloud-computing/

Melville, N., Kraemer, K., & Gurbaxani, V. (2004). Review: Information Technology and organizational performance: An integrative model of IT business value. *Management Information Systems Quarterly*, *28*(2), 283–322.

Melville, N., Kraemer, K., & Gurbaxani, V. (2004). Information Technology and organisational performance: An integrative model of it business value. *Management Information Systems Quarterly*, *28*(2).

Memon, N., & Wong, P. (1998). Protecting digital media content. *Communications of the ACM*, *41*(7), 35–43. doi:10.1145/278476.278485

Mercado, S., Welford, R., & Prescot, K. (2001). *European Business* (4th ed.). Prentice Hall, Financial Times.

Meyer, B. J. F. (1975). *The organization of prose and its effect on memory*. Amsterdam: North Holland Publishing Company.

Meyer, B. J. F., Brandt, D. M., & Bluth, G. J. (1980). Use of top-level structure in text: Key for reading comprehension of ninth-grade students. *Reading Research Quarterly*, *16*, 72–101. doi:10.2307/747349

Miles, M. B., & Huberman, A. M. (1994). *Qualitative data analysis: An expanded sourcebook*. Sage Publications.

Milkovich, G., & Newman, J. (2004). *Compensation*. New York: The McGraw-Hill Companies.

Miller, P., & Wilsdon, J. (2001). Digital futures-an agenda for a sustainable digital economy. *Corporate Environmental Strategy*, *8*(3), 275–280. doi:10.1016/S1066-7938(01)00116-6

Mills, M. P. (1999). *The Internet begins with coal*. Greening Earth Society.

Mindszenthy, B., & Roberts, G. (2000/2001). *Team leaders and the communication loop*. Strategic Communication Management.

Miranda, M., & Vedenov, D. V. (2001). Innovations in agriculture and natural disaster insurance. *American Journal of Agricultural Economics*, *83*(3), 650–655. doi:10.1111/0002-9092.00185

Mirani, R., & Lederer, A. L. (1993). Making promises: The key benefits of proposed IS systems. *Journal of Systems Management*, *44*(10), 10–15.

Mittal, S., Gandhi, S., & Tripathi, G. (2010). *Socio-economic impact of mobile phones on Indian agriculture*. Working Paper 246, Indian Council for Research on International Economic Relations.

Molla, A., & Licker, P. S. (2001). E-commerce systems success: An attempt to extend and respecify the Delone and Maclean model of IS success. *Journal of Electronic Commerce Research*, *2*(4), 131–141.

Möller, K., & Halinen, A. (2000). Relationship marketing theory: Its roots and direction. *Journal of Marketing Management*, *16*(1), 29–54. doi:10.1362/026725700785100460

Montgomery, R. (1996). Disciplining or protecting the poor? Avoiding the social costs of group pressure in micro-credit schemes. *Journal of International Development*, *8*, 289–305. doi:10.1002/(SICI)1099-1328(199603)8:2<289::AID-JID368>3.0.CO;2-2

Moore, D. M., & Dwyer, F. M. (1997). Effect of colour-coding on locus of control. *International Journal of Instructional Media*, *24*, 145–151.

Moore, A., Cappelli, D., & Trzeciak, R. (2008). The big picture of insider IT sabotage across U.S. critical infrastructures. In S. Stolfo, et al. (Eds.), *Insider attack and cyber security: Beyond the hacker.* New York: Springer Science + Business Media, LLC. Also published as an SEI technical report, CMU/SEI-2008-TR-009. Retrieved from http://www.cert.org/archive/pdf/08tr009.pdf.

Moores, T. (2005). Do consumers understand the role of privacy seals in e-commerce? *Communications of the ACM*, *48*(3), 86–91. doi:10.1145/1047671.1047674

Moorman, C., Zaltman, G., & Deshpande, R. (1992). Relationships between providers and users of market research: The dynamics of trust within and between organizations. *JMR, Journal of Marketing Research*, *29*, 314–328. doi:10.2307/3172742

Moran, L., & Myringer, B. (1999). Flexible learning and university change. In Harry, K. (Ed.), *Higher education through open and distance learning* (pp. 57–72). Londres: Routledge.

Morduch, J. (2000). The microfinance schism. *World Development*, *28*, 617–629. doi:10.1016/S0305-750X(99)00151-5

More, E., & McGrath, M. (2002). An Australian case in e-health communication and change. *Journal of Management Development*, *21*(7/8), 612–632.

Morgan, R., & Hunt, S. (1994). The commitment-trust theory of relationship marketing. *Journal of Marketing*, 58.

Morgan, J. M., & Liker, J. K. (2006). *The Toyota product development system: Integrating people, process and technology*. New York: Productivity Press.

Mosse, E., & Sahay, S. (2003). Counter networks, communication and health Information Systems: A case study from Mozambique. In Korpela, M., Montealegre, R., & Poulymenakou, A. (Eds.), *Organizational Information Systems in the context of globalization* (pp. 271–290). Boston: Kluwer Academic Publishers.

Mouzas, S. (2006). Efficiency versus effectiveness in business networks. *Journal of Business Research*, *59*, 1124–1132. doi:10.1016/j.jbusres.2006.09.018

Murphy, K. E., & Simon, S. J. (2002). Intangible benefits valuation in ERP Projects. *Information Systems Journal*, *12*, 301–320. doi:10.1046/j.1365-2575.2002.00131.x

Murray, K., & Schlacter, J. (1990). The impact of services versus goods on consumers' assessment of perceived risk and variability. *Journal of the Academy of Marketing Science*, *18*(1), 51–65. doi:10.1007/BF02729762

My HealtheVet website. (2010). *About my HealtheVet.*

Myers, K. (1996). *Total contingency planning for disasters: Managing risk, minimizing loss, ensuring business continuity*. New York: John Wiley & Sons.

Nachtigal, S. (2007). eBPSM—a new security paradigm for e-business organisations (e-business process security model). *ICEC '07: Proceedings of the Ninth International Conference on Electronic Commerce,* (pp. 101-106). Minneapolis: ACM.

Nachtigal, S. (2007). E-business security design using process security requirements septet. *ICETE - International Joint Conference on E-Business and Telecommunications, SECRYPT.* (pp. 408-413). Barcelona, Spain: INSTICC Press. Co-sponsored by IEEE. In cooperation with ACM SIGMIS.

Nachtigal, S., & Mitchell, C. (2006). Modelling e-business security using business process. *ICETE- SECRYPT 2006, International Conference on Security and Cryptography,* (pp. 459-464). Setubal, Portugal: INSTICC Press, in cooperation with IEEE Systems, Man and Cybernetics (SMC) Society.

Nagarajan, M., et al. (2006). Semantic interoperability of Web Services–challenges and experiences. *Proceedings of the Fourth IEEE International Conference on Web Services* (ICWS 2006), Chicago, IL.

Nappinai, N. S. (2010). Cyber crime law in India: Has law kept pace with emerging trends? An empirical study. *Journal of International Commercial Law and Technology, 5*(1), 22–28.

National Committee of Inquiry into Higher Education. (2010). *Chair information.* Retrieved on May 10, 2010, from https://bei.leeds.ac.uk/Partners/NCIHE/

Navajas, S., Schreiner, M., Meyer, R. L., Gonzalez-Vega, C., & Rodriguez-Meza, J. (2000). Microcredit and the poorest of the poor: Theory and evidence from Bolivia. *World Development, 28,* 333–346. doi:10.1016/S0305-750X(99)00121-7

Nazi, K. M., & Woods, S. S. (2008). *MyHealtheVet PHR: A description of users and patient portal use.* AMIA Annual Symposium.

New, S. (2004). Supply chains: Construction and legitimation. In Westbrook, S. N. (Ed.), *Understanding supply chains.* Oxford University Press.

Ngai, E., & Wat, F. (2002). A literature review and classification of electronic commerce research. *Information & Management, 39,* 415–429. doi:10.1016/S0378-7206(01)00107-0

Niemiec & Otte. (2009). An administrator's guide to the whys and hows of blended learning. *Journal of Asynchronous Learning Networks, 13*(1).

NIST EDI Standards. (1996). Electronic data interchange (EDI). *Federal Information Processing Standards Publications Series of the National Institute of Standards and Technology.* Retrieved on April 30, 2006, from http://www.itl.nist.gov/fipspubs/fip161-2.htm

Noble, D. (2001). *Digital diploma mills: The automation of higher education.* New York: Monthly Review Press.

Nolan, R. L. (1979). Managing the crises in data processing. *Harvard Business Review, 57*(2), 115–126.

Norris, G. D. (1996). Post-investments appraisal. In Willcocks, L. (Ed.), *Investing in Information Systems: Evaluation and management* (pp. 193–223). UK: Chapman & Hall.

Northen, J. R. (2001). Using farm assurance schemes to signal food safety to multiple food retailers in the UK. *International Food and Agribusiness Management Review, 4,* 37–50. doi:10.1016/S1096-7508(01)00069-6

O'Keefe, M. (1998). Establishing supply chain partnerships: Lessons from Australian agribusiness. *Supply Chain Management, 3*(1), 5–9. doi:10.1108/13598549810200799

O'Neill, M. (2003). *Web Services security.* New York: McGraw Hill Publications.

OECD. (2000). *Schooling for tomorrow: Learning to bridge the digital divide. Centre for Educational Research and Innovation. The National centre on Adult Literacy.* NCAL.

Okunbor W. & Okunbor D. (2005). Internet and e-business cultural dynamics.

Okunoye, A., & Karsten, H. (2003). Global access to knowledge: Findings from academic research organisations in Sub-Saharan Africa. *Information Technology & People, 16*(3), 353–373. doi:10.1108/09593840310489421

Oliver, M., & Conole, G. (2000). Assessing and enhancing quality using toolkits. *Quality Assurance in Education, 8*(1), 32–37. doi:10.1108/09684880010312677

Oliver, M., & Dempster, J. A. (2003). Embedding e-learning practices. In Blackwell, R., & Blackmore, P. (Eds.), *Towards strategic staff development in Higher Education.* SRHE/OUP.

Olsen, F. (2000). E-commerce may help colleges cut costs and paperwork. *The Chronicle of Higher Education*, *46*(33), A45–A46.

O'Malley, L., & Tynan, C. (2000). Relationship marketing in consumer markets. Rhetoric or reality? *European Journal of Marketing*, *34*(7), 797–815. doi:10.1108/03090560010331225

O'Reilly, T. (2007). What is Web 2.0: Design patterns and business models for the next generation of software. *Communications & Strategies, First Quarter, 17*.

Orton-Johnson, K. (2009). I've stuck to the path, I'm afraid: Exploring student non-use of blended learning. *British Journal of Educational Technology*, *40*(5), 837–847. doi:10.1111/j.1467-8535.2008.00860.x

Osborne, M. (2006). *How to cheat at managing Information Security*. Syngress Publishing, Inc.

Osheroff, J. A. (2009). *Improving medication use and outcomes with clinical decision support: Step-by-step guide*. Chicago: HIMSS Press.

Osterwalder, A., & Pigneur, Y. (2002). An e-business model ontology for modeling e-business. *15th Bled Electronic Commerce Conference e-Reality: Constructing the e-Economy*, (pp. 73-86).

Ozok, A. A., & Salvendy, G. (2003). The effect of language inconsistency on performance and satisfaction in using the Web: Results from three experiments. *Behaviour & Information Technology*, *22*(3), 155–163. doi:10.1080/0144929031000107081

Parry, E. (2003). *Gartner releases top 11 list of security issues. SearchCIO*. Retrieved February 26, 2010, from http://searchcio.techtarget.com/news/article/0,289142,sid182_gci891984,00.html

Pastoor, S. (1990). Legibility and subjective preference for color combinations in text. *Human Factors*, *32*, 157–171.

Pateli, A., & Giaglis, G. (2003). A framework for undestanding and analysing e-business models. *16th Bled Electronic Commerce Conference - eTransformation (CD-ROM Proceedings)*.

Patton, M. Q. (1990). *Qualitative evaluation and research methods* (2nd ed.). Newbury Park, CA: Sage.

Pavon, F., & Casanova, J. (2007). Experiencias docentes apoyadas en aulas virtuales. *Revista Iberoamericana de Educación a Distancia*, *10*(2), 149–163.

Pearce, J., & Robinson, R. (2004). *Strategic management* (9th ed.). New York: The McGraw-Hill Companies.

Pelet, J.-É. (2008). *Effets de la couleur des sites marchands sur la mémorisation et sur l'intention de l'internaute*. Thèse de doctorat en Sciences de Gestion, Université de Nantes

Pérez, A. (2002). Nuevas estrategias didácticas en entornos digitales para la enseñanza superior. In: J. Salinas & A. Batista (Eds.), *Didáctica y tecnología educativa para una universidad en un mundo digital*. Universidad de Panamá: Imprenta Universitaria.

Perez, S. (2009). *Top 8 Web 2.0 security threats*. Retrieved June 10th, 2010, from http://www.readwriteweb.com/enterprise/2009/02/top-8-web-20-security-threats.php

Periasamy, K. P., Ho, W. M., & Bok, C. W. (2002). Organization structure for electronic commerce: A qualitative investigation. *Proceedings of 10th European Conference on Information Systems, Gdańsk, Poland*, (pp. 596-610).

Peters, M., & May, T. (2004). Universities, regional policy and the knowledge economy. *Policy Futures in Education*, *2*(4), 263–277. doi:10.2304/pfie.2004.2.2.4

Petersen, M., & Rajan, R. (1994). The benefits of lending relationships: Evidence from small business data. *The Journal of Finance*, *49*(1). doi:10.2307/2329133

Petit, J., & Mason, R. (2007). Virtual space, real learning: An introduction to VLEs. In Fry, H., Ketteridge, S., & Marshall, S. (Eds.), *A handbook for teaching and learning in higher education* (2nd ed., pp. 148–161). Oxford: Routledge-Falmer.

Petrova, K., & Sinclair, R. (2008). *E-learning value and student experiences: A case study, advances in E-learning: Experiences and methodologies*. Hershey, PA: IGI Global.

Pew Internet & American Life Project. (2010). *Internet, broadband, and cell phone statistics*.

Phillips, P. A., & Wright, C. (2009). E-business's impact on organizational flexibility. *Journal of Business Research*, (62): 1071–1080. doi:10.1016/j.jbusres.2008.09.014

Pikkarainen, T., Pikkarainen, K., Karjaluoto, H., & Pahnila, S. (2004). Consumer acceptance of online banking: An extension of the technology acceptance model. *Internet Research*, *14*(3). doi:10.1108/10662240410542652

Point-Carbon. (2006). [*-towards a truly global market.*]. *Carbon*, 2006.

Poirier, C., & Bauer, M. (2001). *E-supply chain*. Berrett-Koehler Publishers, Inc.

Pons, A., & Aljifri, H. (2005). Data protection using watermarking in e-business. *Journal of Database Management*, *14*(4).

Pons, A. & Aljifri, H. (2002). An active watermarking system. *IACIS, Issues in Information Systems*.

Porter, M. (1985). *Competitive advantage: Creating and sustaining superior performance*. New York: The Free Press.

Porter, M. E. (2001). Strategy and the Internet. *Harvard Business Review*, *79*(3), 63–78.

Porter, M. (1998). *Competitive advantage: Creating and sustaining superior performance*. Free Press.

Porter, M. & Millar, V. (1985). How information gives you a competitive advantage. (63), 149-161.

Porter, M. (2001). Strategy and the Internet. *Harvard Business Review*.

Pota, H., Eaton, R., Katupitiya, J., & Pathirana, S. D. (2007). Agricultural robotics: A streamlined approach to realization of autonomous farming. In. *Proceedings of the International Conference on Industrial and Information Systems, ICIIS, 2007*, 85–90. doi:10.1109/ICIINFS.2007.4579153

Power, M. (2008). The emergence of a blended online learning environment. *Journal of Online Learning and Teaching*, *4*(4), 503–514.

Power, D. (2005). Supply chain management integration and implementation: A literature review. *Supply Chain Management-An International Journal*, *10*, 252–263. doi:10.1108/13598540510612721

Prègent, R. (1990). *La preparation d'un cours*. Montreal: Editions de l'Ecole Polytechnique de Montréal.

Price Waterhouse. (1997). *Technology forecast*. Menlo Park, CA: Price Waterhouse World Technology Centre.

PriceWaterhouseCooper. (2003). *Benefits realization online survey*. PriceWaterhouseCooper. Retrieved from http://www.pwcglobal.com

PriceWaterHouseCooper. (2006). *Information Technology enhances consistency, transparancy efficiency in EU-ETS*.

PRNewswire. (2010). *Security legislation will help protect U.S. economy, Says VirnetX researcher*. Retrieved March 1st, 2010, from http://markets.chron.com/chron/?GUID=11788469&Page=MediaViewer&Ticker=VHC

Prosser, M., & Trigwell, K. (1999). *Understanding learning and teaching: The experience in Higher Education*. Maidenhead, UK: The Open University Press.

Pruitt, D. G. (1981). *Negotiation Behavior*. New York: Academic Press.

Puckett, J., Byster, L., Westervelt, S., Gutierrez, R., Davis, S., & Hussain, A. (2002). *Exporting harm-the high-tech trashing of Asia. The Basel Action Network (BAN), Silicon Valley Toxics Coalition*. SVTC.

Pyun, C., Scruggs, L., & Nam, K. (2002). *Internet banking in the US, Japan and Europe*. Multinational Business Review.

Quaddus & Archjari. (2005). A model for electronic commerce success. *Telecommunications Policy*, *20*(2-3), 127–152.

Race, P. (2007). *The lecturer's toolkit-a practical guide to assessment, learning and teaching* (3rd ed.). Oxford: Routledge.

Raisinghani, M. S., & Young, E. (2008). Personal health records: Key adoption issues and implications for management. *International Journal of Electronic Healthcare*, *4*(1), 67–77. doi:10.1504/IJEH.2008.018921

Ralston, G., Burke, K., Nakayama, D., & Tolani, A. (2000). Electronic mail. In *Encyclopedia of Computer Science* (4th ed., pp. 637–642). London: Nature Publishing Group.

Ramburuth, P., & Mladenovic, R. (2004). Exploring the relationship between students' orientations to learning. *Accounting Education*, *13*, 507–527. doi:10.1080/0963928042000306774

Rammell, W. (2009). *MP, Minister of State (lifelong learning, further and higher education).* Retrieved on July 23, 2009, from http://www.theabs.org.uk/files/PILLARS-AR

Rao, S. (2004). Building inter-firm relationships online: Perspectives from the Australian Internet banking industry. *International Journal of Internet Marketing and Advertising, 1*(4), 371–387. doi:10.1504/IJIMA.2004.006330

Rao, S. S., Metts, G., & Monge, C. A. M. (2003). Electronic commerce development in small and medium sized enterprises: A stage model and its implications. *Process Management Journal, 9*(1), 11–32. doi:10.1108/14637150310461378

Ratnasingam, P., & Pavlou, P. (2002). *Technology trust: The next value creator in B2B electronic commerce*. Paper presented at the International Resource Management Association Conference: Seattle, Washington, May 21-23.

Reardon, J. L., & Davidson, E. (2007). An organizational learning perspective on the assimilation of electronic medical records among small physician practices. *European Journal of Information Systems, 16*, 681–694. doi:10.1057/palgrave.ejis.3000714

Rees, J., Bandyopadhyay, S., & Spafford, E. (2003). PFIRES: A policy framework for information security. *Communications of the ACM, 46*(7), 101–106. doi:10.1145/792704.792706

Reichheld, F. (1993). Loyalty-based management. *Harvard Business Review, 71*(2), 64–73.

Reisman, S. (2003). *Electronic learning communities*. Connecticut: Information Age Publishing.

Remenyi, D., Sherwood-Smith, M., & White, T. (1997). *Achieving maximum value from Information Systems: A process approach*. Chichester, UK: John Wiley & Sons.

Rhee, C., Moon, J., & Choe, Y. (2005). Web interface consistency in e-learning. *Online Information Review, 30*(1), 53–69. doi:10.1108/14684520610650309

Richard, M. O. (2005). Modeling the impact of internet atmospherics on surfer behavior. *Journal of Business Research, 58*, 1632–1642. doi:10.1016/j.jbusres.2004.07.009

Ricketts, C., & Rawlins, O. (2001). *Introduction to agribusiness*. Albany, NY: Delmar.

Robert, P. C. (2002). Precision agriculture: A challenge for crop nutrition management. *Plant and Soil, 247*(1), 143–149. doi:10.1023/A:1021171514148

Roberts, T., Romm, C. & Jones, D. (2000). *Current practice in web-based delivery of IT courses*. APWeb2000.

Robinson, M. (2001). *The microfinance revolution: Sustainable banking for the poor*. Washington, D.C.: The World Bank.

Rombel, A. (2006). Banks build on twin pillars of security and service. *Global Finance, 20*(8), 39–43.

Romm, J. (2002). The Internet and the new energy economy. *Resources, Conservation and Recycling, 36*(3), 197–210. doi:10.1016/S0921-3449(02)00084-8

Romm, J., Rosenfeld, A., & Herrmann, S. (1999). *The Internet economy and global warming: A scenario of the impact of e-commerce on energy and the environment. Center for Energy and Climate Solutions (CECS), Global Environment & Technology Foundation*. GETF.

Rotter, J. (1967). A new scale for the measurement of interpersonal trust. *Journal of Personality*, 651–665. doi:10.1111/j.1467-6494.1967.tb01454.x

Rouse, A., & Dick, M. (1994). The use of computerized tools in qualitative Information Systems studies. *Proceedings of the 5th Australasian Conference on Information Systems (ACIS 1994),* Monash University, Melbourne, Victoria, (pp. 209-220).

Rowley, T. (1997). Moving beyond dyadic ties: A network theory of stakeholder influences. *Academy of Management Review, 22*(4), 887–911. doi:10.2307/259248

Rubens, P., & Krull, R. (1985). Application of research on document design to online displays. *Technical Communication, 32*(4), 29–34.

Rudestam, K., & Schoenholtz, J. (Eds.). (2002). *Handbook of online learning*. London: Sage Publications.

Salinas, J. (2002). Modelos flexibles como respuesta de las universidades a la sociedad de la información. *Acción Pedagógica, 11*(1), 4–13.

Sallas, B., Lane, S., Mathews, R., Watkins, T., & Wiley-Patton, S. (2007). An iterative assessment approach to improve technology adoption and implementation decisions by healthcare managers. *Information Systems Management*, *24*, 43–57. doi:10.1080/10580530601036802

San Diego, J., Laurillard, D., Boyle, T., Bradley, C., Ljubojevic, D., & Neumann, T. (2008). Towards a user-oriented analytical approach to learning design. *ALT-J: Research in Learning Technology*, *16*(1), 15–29.

Sandelowski, M. (1995). Sample size in qualitative research. *Research in Nursing & Health*, *18*, 179–183. doi:10.1002/nur.4770180211

Sarathy, R., & Robertson, C. J. (2003). Strategic and ethical considerations in managing digital privacy. *Journal of Business Ethics*, *46*(2), 111–126. doi:10.1023/A:1025001627419

Sathye, M. (1999). Adoption of Internet banking by Australian consumers: An empirical investigation. *International Journal of Bank Marketing*, *17*(7). doi:10.1108/02652329910305689

Sayed, E. H., & Westrup, C. (2003). Egypt and ICTs bring national initiatives. Global actors and local companies together. *Information Technology & People*, *16*(1), 93–110. doi:10.1108/09593840310463041

Schiefer, G. (2003). New technologies and their impact on agriculture, environment and the food industry. In *EFITA 2003 Conference Proceedings*. (pp. 3-11). Debrecen, Hungary, Schiefer, G. (2006). Information management in agri-food chains. In C.J.M. Ondersteijn, J.H.M. Wijnands, R.B.M. Huirne & O. Van Kooten (Eds.), *Quantifying the agri-food supply chain*. (pp. 137-146). The Netherlands: Springer.

Schmidt, K. E., & Liu, Y. (2005). *Design of consumer product Web pages: Experimental investigations of aesthetic and performance factors* (pp. 1743–1746). Human Factors and Ergonomics Society.

Schneider, G. P. (2003). *Electronic commerce* (4th ed.). Boston: Thomson Learning Inc.

Schneier, B. (2004). *Secrets and lies*. Wiley.

Schuh, R. G., & Leviton, L. C. (2006). A framework to assess the development and capacity of non-profit agencies. *Evaluation and Program Planning*, *29*(2), 171–179. doi:10.1016/j.evalprogplan.2005.12.001

Schumacher, S. (2007). Probabilistic versus deterministic data matching: making an accurate decision. *Information Management*.

Scornavacca, E., & Hoehle, H. (2007). Mobile banking in Germany: A strategic perspective. *International Journal of Electronic Finance*, *1*(3), 304–320. doi:10.1504/IJEF.2007.011501

Scott, B., Shurville, S., Maclean, P., & Cong, C. (2007). Cybernetic principles for learning design. *Kybernetes*, *36*(9/10), 1497–1514. doi:10.1108/03684920710827445

Seddon, P., Graeser, V., & Willcocks, L. (2002). Measuring organizational IS effectiveness: An overview and update of senior management perspectives. *The Data Base for Advances in Information Systems*, *33*(2), 11–28.

Sen, A. (1985). *Commodities and capabilities*. Amsterdam: North-Holland.

Sen, A. (1987). *The standard of living*. Cambridge, UK: Cambridge University Press. doi:10.1017/CBO9780511570742

Sen, A. (1992). *Inequality re-examined*. Oxford: Oxford University Press.

Sen, A. (1999). *Development as freedom*. New York: Knopf.

Servon, L. J., & Kaestner, R. (2008). Consumer financial literacy and the ompact of online banking on the financial behavior of lower-income bank customers. *The Journal of Consumer Affairs*, *42*(2), 271–305. doi:10.1111/j.1745-6606.2008.00108.x

Setboonsarng, S., Sakai, J., & Vancura, L. (2009). *Food safety and ICT traceability systems: Lessons from Japan for developing countries*. ADBI Working Paper 139. Tokyo: Asian Development Bank Institute. Retrieved March 14, 2010, from http://www.adbi.org/working-paper/2009/05/28/3012.ict.food.traceability.system/

Shafer, S., Smith, J., & Linder, J. (2005). The power of business models. *Business Horizons*, *48*(3), 199–207. doi:10.1016/j.bushor.2004.10.014

Sharpe, R., Benfield, G., Roberts, G., & Francis, R. (2006). *The undergraduate experience of blended e-learning: A review of UK literature and practice*. A report to the Higher Education Academy. Retrieved November 20, 2009, from http://www.heacademy.ac.uk/assets/York/documents/ourwork/teachingandresearch/Sharpe_Benfield_Roberts_Francis.pdf

Sharples, M. (2005). *Learning as conversation: Transforming education in the mobile age*. Paper presented at the 'Seeing, Understanding, Learning in the Mobile Age' Conference, Budapest, Hungary.

Shaw, J., Brain, K., Bridger, K., Foreman, J., & Reid, I. (2007). *Embedding widening participation and promoting student diversity*. Retrieved on May 10, 2010, from http://www.heacademy.ac.uk/assets/York/documents/resources/publications/exchange/embedding_wp_business_case_approach_july07.pdf

Shelly, G.B., Cashman, T.J. & Vermaat (2001). *Discovering computer concepts for a digital world.* Boston: Course Technology, Thompson Learning.

Sherwood, J., Clark, A., & Lynas, D. (2005). *Enterprise security architecture: A business-driven approach*. CMP Books.

Shieh, K. K., Chen, M. T., & Chuang, J. H. (1997). Effects of color combination and typography on identification of characters briefly presented on VDTs. *International Journal of Human-Computer Interaction*, *9*(2), 169–181. doi:10.1207/s15327590ijhc0902_5

Shin, K. H. (2008). Is e-banking a competitive weapon? A causal analysis. *International Journal of Electronic Finance*, *2*(2), 180–196. doi:10.1504/IJEF.2008.017539

Silva, L., & Figueroa, E. B. (2002). Institutional intervention and the expansion of ICTs in Latin America: The case of Chile. *Information Technology & People*, *15*(1), 8–25. doi:10.1108/09593840210421499

Silverman, D. (2001). *Interpreting qualitative data* (2nd ed.). London: Sage.

Silverstein, L. D. (1987). Human factor for color CRT display systems: Concepts, methods and research. In Durrett, H. J. (Ed.), *Color and the computer* (pp. 27–61). Orlando: Academic.

Sinclaire, J. (2005). Current research in information security and privacy. *Proceedings of the 2005 Southern Association of Information Systems Conference.*

Sinfield, S., Burns, T., & Holley, D. (2004). Outsiders looking in or insiders looking out? Widening participation in a post-1992 university. In Satterthwaite, J., Atkinson, A., & Martin, W. (Eds.), *The disciplining of education: New languages of power and resistance* (pp. 137–152). Trentham Books.

Singh, S. N., Dalal, N., & Spears, N. (2005). Understanding Web home page perception. *European Journal of Information Systems*, *14*, 288–302. doi:10.1057/palgrave.ejis.3000525

Siponen, M., & Willison, R. (2009). Information Security management standards: Problems and solutions. *Information & Management*, *46*(5), 267–270. doi:10.1016/j.im.2008.12.007

Siqueira, S. W. M., Braz, M. H. L. B., & Melo, R. N. (2007). Modeling e-learning content. *International Journal of Web Information Systems*, *3*(1/2), 140–152. doi:10.1108/17440080710829261

Sitting, D. F., Campbell, E., Guappone, K., Dystra, R., & Ash, J. S. (2007). Recommendations for monitoring and evaluation of in-patient computer-based provider order entry systems: Results of a Delphi survey. In *proceedings of the AMIA Annual Symposium*. (pp. 671–675).

Slade, R. (2006). *Dictionary of information security*. Syngress.

Sluiter, J. (2006). Services Oriented Architecture security and deperimeterisation. *Information Security Bulletin*, *11*(1), 65–72.

Smith, A. D. (2008). Online payment service providers and customer relationship management. *International Journal of Electronic Finance*, *2*(3), 257–283. doi:10.1504/IJEF.2008.020597

Smith, H., & Oliver, M. (2002). University teachers' attitudes to the impact of innovations in ICT on their practice." In C. Rust (Ed), *Proceedings of the 9th International Improving Student Learning Symposium*, (pp. 237-246). Oxford: Oxford Centre for Staff and Learning Development.

Smith, J. N. (2005). Specialized logistics for a longer perishable supply chain. *World Trade Magazine, November 2005*. Retrieved on March 12, 2010, from http://www.worldtrademag.com/Articles/Feature_Article/52de170abaaf7010VgnVCM100000f932a8c0

Sobel, P. J., & Reding, K. F. (2004). Aligning corporate governance with enterprise risk management. *Management Accounting Quarterly*, *5*(2), 29.

Sohal, A. S., & Ng, L. (1998). The role and impact of Information Technology in Australian business. *Journal of Information Technology*, *13*, 201–217. doi:10.1080/026839698344846

Soliman, K. S., & Janz, B. D. (2004). An exploratory study to identify the critical success factors affecting the decision to establish Internet-based interorganizational Information Systems. *Information & Management*, *41*(6), 697–706. doi:10.1016/j.im.2003.06.001

Soulier, J. (1988). *The design and development of computer based instruction*. Toronto, ON: Allyn and Bacon.

Spitzner, L. (2003). *Honeypots: Tracking hackers*. Reading, MA: Addison-Wesley Professional.

Srinivasan, R., Rangaswamy, A. & Lilien, G.L. (2005). Turning adversity into advantage: Does proactive marketing during a recession pay off? *International Journal of Research in Marketing*.

Srinivasan, S. (2003). On piracy and privacy. *Computer - IEEE, (36)*7, 36-38.

Stamoulis, D. S. (2000). How banks fit in an Internet commerce business activities model. *Journal of Internet Banking & Commerce*, *5*(1).

Stamoulis, D., Kanellis, P., & Martakos, D. (2002). An approach and model for assessing the business value of e-banking distribution channels: Evaluation as communication. *International Journal of Information Management*, *22*, 247–261. doi:10.1016/S0268-4012(02)00011-7

Standing, C., & Lin, C. (2007). Organizational evaluation of the benefits, constraints, and satisfaction with business-to-business electronic commerce. *International Journal of Electronic Commerce*, *11*(3), 107–153. doi:10.2753/JEC1086-4415110304

Standing, S., Standing, C., & Love, P. (2010). A review of research on e-marketplaces 1997–2008. *Decision Support Systems*, *49*(1), 41–51. doi:10.1016/j.dss.2009.12.008

Steinberg, E. R. (1991). *Computer-assisted instruction: A synthesis of theory. Practice, and technology*. Hillsdale, NJ: Lawrence Erlbaum.

Stern, N. (2006). *The economics of climate change: The Stern Review*. Retrieved from http://www.hm-treasury.gov.uk/independent_reviews/stern_review_economics_climate_change/stern_review_report.cfm

Stewart, K. J. (2003). Trust transfer on the World Wide Web. *Organization Science*, *14*(1), 5–17. doi:10.1287/orsc.14.1.5.12810

Stiglitz, J. (1990). Peer monitoring and credit markets. *The World Bank Economic Review*, *4*(3), 351–366. doi:10.1093/wber/4.3.351

Stockdale, R., & Standing, C. (2006). An interpretive approach to evaluating information systems: A content, context, process framework. *European Journal of Operational Research*, *173*(3), 701–1188. doi:10.1016/j.ejor.2005.07.006

Stone, M., & Woodcock, N. (1997). *Winning new customers in financial services: Using relationship marketing and Information Technology in consumer financial services*. London: Pitman Publishing.

Strassmann, P. A. (2003). Does IT matter? An HBR debate: Web exclusive. *Harvard Business Review*, 7–9.

Stratopoulos, T., & Dehning, B. (2000). Does successful investments in Information Technology solve the productivity paradox? *Information & Management*, *38*, 103–117. doi:10.1016/S0378-7206(00)00058-6

Straub, D. W., Hoffman, D. L., Weber, B. W., & Steinfield, C. (2002). Measuring e-commerce in Net-enabled organizations: An introduction to the Special Issue. *Information Systems Research*, *13*(2), 115–124. doi:10.1287/isre.13.2.115.86

Subramani, M. (2004). How do suppliers benefit from Information Technology use in supply chain relationships. *Management Information Systems Quarterly*, *28*(1), 45–73.

Subramaniam, C., & Shaw, M., J. (2002). A study of the value and impact of B2B e-commerce: The case of Web based procurement. *International Journal of Electronic Commerce*, *6*(4), 19–40.

Sugumaran, V., & Arogyaswamy, B. (2004). Measuring IT performance: Contingency variables and value modes. *Journal of Computer Information Systems*, *44*(2), 79–86.

Sustainable-Development-Commision. (2006). Carbon trading for individuals? Retrieved from http://www.sd-commission.org.uk/pages/carbontrading.html

Swann, J. (2008). Catering to consumer expectations online. *Community Banker.*, *17*(2), 55–56.

Sweeney, A., & Morrison, M. (2004). Clicks vs. bricks: Internet-facilitated relationships in financial services. *International Journal of Internet Marketing and Advertising*, *1*(4), 350–370. doi:10.1504/IJIMA.2004.006329

Sweller, J. (1999). *Instructional design in technical areas*. Melbourne, Australia: Acer Press.

Sweller, J. (2002). *Visualisation and instructional design.* International Workshop on Dynamic Visualizations and Learning, July, 18.-19., Knowledge Media Research Center (KMRC), Tübingen, Germany.

Szabo, M., & Poohkay, B. (1994). So what if it's in color and moves? A critique of multimedia. *Proceedings of the World Conference on Ed-Media/EdTelecomm 94*. Vancouver, Canada. Retrieved on June 29, 2006, from http://www.quasar.ualberta.ca/IT/research/Szabo/poohkay.html

Tanenbaum, A. S. (2003). *Computer networks* (4th ed.). New Jersey: Pearson Education, Inc.

Tapscott, D. (1998). *Growing up digital: The rise of the Net generation*. New York: McGraw-Hill.

Tapscott, D., & Ticoll, D. (2003). *The naked corporation: How the age of transparency will revolutionise business*. Dow Jones & Company, Inc.

Tax, S., Brown, S., & Chandrashekaran, N. (1998). Customer evaluations of service complaint experiences: Implications for relationship marketing. *Journal of Marketing*, *62*(2), 60–76. doi:10.2307/1252161

Technology Forecast. (2003). *Enterprise applications: Enterprise suites*. Menlo Park, CA: PriceWaterHouseCoopers Technology Center.

Tettero, O. (2000). *Intrinsic information security. Embedding security issues in the design process of telematics systems*. Telematics Institute Fundamental Research Series.

Thatcher, M. D., & Pingry, D. E. (2004). Understanding the business value of Information Technology investments: Theoretical evidence from alternative market and cost structure. *Journal of Management Information Systems*, *21*(2), 61–85.

Thompson, J., Millstone, E., Scoones, I., Ely, A., Marshall, F., Shah, E., et al. (2007). *Agri-food system dynamics: Pathways to sustainability in an era of uncertainty*. STEPS Working Paper 4. Brighton, UK: STEPS Centre.

Thuraisingham, B. (2005). Directions for security and privacy for semantic e-business applications. *Communications of the ACM*, *48*(12), 71–73. doi:10.1145/1101779.1101812

Timmers, P. (1999). *Electronic commerce*. Chichester, UK: John Wiley.

Timmons, H. (2000). Online banks can't go it alone. *Business Week Online*.

Torkzadeh, G., & Dhillon, G. (2002). Measuring factors that influence success of Internet commerce. *Information Systems Research*, *13*(2), 187–204. doi:10.1287/isre.13.2.187.87

Traynor, I. (2007, May 17). Russia accused of unleashing cyber war to disable Estonia. *The Guardian*.

Tricot, A. & Plégat-Soutjis, F. (2003). Pour une approche ergonomique de la conception d'un dispositif de formation à distance utilisant les TIC. *Sciences et Technologies de l'Information et de la Communication pour l'Éducation et la Formation, 10*.

Trout, B. (2007). *Cyber law: A legal arsenal for online business*. World Audience, Inc.

Trout, P. (2000). Flunking the test: The dismal record of student evaluations. *Academe: Bulletin of the AAUP*, *86*(4). Retrieved on November 2, 20015 from http://www.aaup.org/AAUP/pubsres/academe/2000/JA/Feat/trou.htm

Truax, J. (1997). *Investing with benefits in mind: Curing investments myopia*. DMR White Paper. (pp. 1-6).

Tsao, H., Lin, K., & Lin, C. (2004). An investigation of critical success factors in the adoption of B2B e-commerce by Taiwanese companies. *The Journal of American Academy of Business*, *5*(1/2), 198–202.

Tullberg, J. N., Yule, D. F., & McGarry, D. (2007). Controlled traffic farming-from research to adoption in Australia. *Soil & Tillage Research*, *97*, 272–281. doi:10.1016/j.still.2007.09.007

Turban, E., Aronson, J. E., Liang, T., & Sharda, R. (2007). *Decision support and business intelligence systems* (5th ed.). Upper Saddle River, NJ: Pearson Education, Inc.

Twigg, C. A. (2001). *Innovations in online learning: Moving beyond no significant difference. National Center for Academic Transformation (NCAT)*. Monograph.

United Nations Population Fund. (2009) World population to exceed 50 billion by 2050. Retrieved on February 16, 2010, from http://www.un.org/esa/population/publications/wpp2008/pressrelease.pdf

United States Department of Defense. (2008). *DoD Launches Personal Health Record Pilot*. Military health systems press release.

United States Government Accountability Office. (2005). *Federal agencies face challenges in implementing initiatives to improve public health infrastructure*. Retrieved October 13, 2009, from http://www.gao.gov/new.items/d05308.pdf

Universities and Colleges Information Systems Association Survey. (2008). Retrieved from http://www.ucisa.ac.uk/publications/~/media/groups/tlig/vle_surveys/TEL%20survey%202008%20pdf.ashx

University of Greenwich. (2007). *Business School Strategic Plan*.

University of Pittsburgh Medical Center. (2009). *Biosecurity*. Retrieved from www.upmc-biosecurity.org

Unwin, A. (2007). The professionalism of the higher education teacher: What's ICT got to do with it? *Teaching in Higher Education*, *12*(3), 295–308. doi:10.1080/13562510701278641

USDA. (2002). Traceability for food marketing & food safety: What's the next step? *Economic Research Service/ USDA Agricultural Outlook/January-February 2002*.

Van Akkeren & Rowlands, B. (2007). An epidemic of pain in an Australian radiology practice. *European Journal of Information Systems*, *16*(6), 695–711. doi:10.1057/palgrave.ejis.3000715

van Hofwegen, G., Becx, G., & van den Broek, J. (2005). *Drivers for competitiveness in agri-food chains: A comparative analysis of 10 EU food product chains*. Report No: WP4 EUMercoPol. Retrieved March 14 2010 from: http://www.eumercopol.org/Files/Report%20EU%20Agri-food%20Chains.pdf

van Slyke, C., & Belanger, F. (2003). *E-business technologies supporting the net-enhanced organization*. Danvers, MA: John Wiley & Sons, Inc.

Vandenbosch, B., & Lyytinen, K. (2004). Much ado about IT: A response to the corrosion of IT advantage by Nicholas G. Carr. *The Journal of Business Strategy*, *25*(6), 10–12. doi:10.1108/02756660410569157

Vickery, A.W. & Lake, F.R. (2005). Teaching on the run: Giving feedback. *eMJA*, 183(5), 267-268.

Victor, C. R., & Wilburn, R. L. (2006). Electronic data interchange and electronic funds transfer. Retrieved on April 29, 2006, from http://www.referenceforbusiness.com/management/De-Ele/Electronic-Data-Interchange-and-Electronic-Funds-Transfer.html

Violino, B. (2002). Building B2B trust. *Computerworld*, *36*(25), 32.

von Pischke, J. (1996). Measuring the trade-off between outreach and sustainability of microenterprise lenders. *Journal of International Development*, *8*, 225–239. doi:10.1002/(SICI)1099-1328(199603)8:2<225::AID-JID370>3.0.CO;2-6

W.K. Kellogg Foundation. (2004). *Logic model development guide*. Battle Creek, MI: WK Kellogg Foundation.

Wakeford, R. (2007). Principles of student assessment. In Fry, H., Ketteridge, S., & Marshall, S. (Eds.), *A handbook for teaching and learning in higher education* (2nd ed.). Oxford: Routledge-Falmer.

Walker, S., & Ryan, M. (2008). A post conference reflection and report. In Walker, S., Ryan, M., & Teed, R. (Eds.), *Designing for learning – the e-learning@greenwich/ conference*. University of Greenwich.

Walsham, G., Robey, D., & Sahay, S. (2007). Forward: Special issue on Information Systems in developing countries. *Management Information Systems Quarterly, 31*(2), 317–326.

Walsham, G., & Sahay, S. (2006). Research on Information Systems in developing countries: Current landscape and future prospects. *Information Technology for Development, 12*(1), 7–24. doi:10.1002/itdj.20020

Walter, Z., & Scott, G. (2006). Management issues of Internet/Web systems. *Communications of the ACM, 49*(3), 87–91. doi:10.1145/1118178.1118181

Wang, N., Zhang, N., & Wang, M. (2006). Wireless sensors in agriculture and food industry-recent development and future perspective. *Computers and Electronics in Agriculture, 50*, 1–14. doi:10.1016/j.compag.2005.09.003

Wang, W., Bailey, A., Hidvegi, Z., & Whinston, A. (2001). *A framework for proactive, automated and continuous e-commerce control and assurance*. Goizueta Business School Papers Series.

Wang, W., Hidvegi, Z., Bailey, A., & Whinston, A. (2000). E-process design and assurance using model checking. *Computer, 33*(10), 48–53. doi:10.1109/2.876292

Wang, W., Yuan, Y., & Archer, N. (2006). A contextual framework for combating identity theft. *IEEE Security and Privacy, 4*(2), 30–38. doi:10.1109/MSP.2006.31

Wang, W., Hidvegi, Z., Bailey, A. & Whinston, A. (2001). Model checking-a rigorous and efficient tool for e-commerce internal control and assurance. *Knowledge Emory* (DIA-2001-007).

Wang, X., & Chen, D. (2005). Research on e-business management thought. *Proceedings of the ICEC'05*. Xi'an, China: ACM Press.

Ward, J., & Daniel, E. (2006). *Benefits management: Delivering value from IS & IT investments*. Chichester, UK: John Wiley and Sons Ltd.

Ward, J., & Griffiths, P. (1996). *Strategic planning for Information Systems*. Chichester, UK: John Wiley & Sons Ltd.

Ward, J., Taylor, P., & Bond, P. (1996). Evaluation and realization of IS/IT benefits: An empirical study of current practice. *European Journal of Information Systems, 4*, 214–225. doi:10.1057/ejis.1996.3

Wareham, J., Zheng, J., & Straub, D. (2005). Critical themes in electronic commerce research: A meta-analysis. *Journal of Information Technology, 20*, 1–19. doi:10.1057/palgrave.jit.2000034

Wasserman, G., & Su, Z. (2007). Sound and precise analysis of Web applications for injection vulnerabilities. *Proceedings of the PLDI'07*. ACM.

Watson, E. (2005). *Tesco admits failures*. Retrieved May 28, 2010, from http://www.foodmanufacture.co.uk/Business-News/Tesco-admits-failures

Web, M. D. (2008). *Migranes and headaches*. Retrieved January 24, 2010, from http://www.webmd.com/migraines-headaches/triptans-serotonin-receptor-agonists-for-migraine-headaches

Webometrics. (2009). *Ranking of world universities*. Retrieved from http://www.webometrics.info/top6000.asp

Websense. (2009). *State of Internet security*. Q1-Q2 2009. Retrieved June 10th, 2010, from http://www.websense.com/assets/reports/report-wsl-state-of-internet-security-q1-q2-2009.pdf

Weinreich, H., Obendort, H., Herder, E., & Mayer, M. (2006). Off the beaten tracks: Exploring three aspects of Web navigation. *Proceedings of the 15th international conference on World Wide Web table of contents*, Edinburgh, Scotland, (pp. 133-142).

Wescott, C. G. (In press). E-commerce and the application of technology to government services. In Rondinelli, D. A., & Heffron, J. M. (Eds.), *Globalization in transition: Forces of adjustment in the Asia Pacific region*.

White, S., & Davis, H. (2002). Harnessing Information Technology for learning. In Ketteridge, S. M., & Fry, H. (Eds.), *The effective academic: A handbook for enhanced acdemic practice*. London: Kogan Page.

Wieczorek-Zeul, H. (2009). The importance of new partnership. In Matthäus-Maier, I., & von Pischke, J. D. (Eds.), *New partnerships for innovation in microfinance*. Springer Publishers.

Wilder, C., Caldwell, B., & Dalton, G. (1997). More than electronic commerce. *Information Week*, *661*, 30–34.

Williams, E. (2004). Energy intensity of computer manufacturing: Hybrid analysis combining process and economic input-output methods. *Environmental Science & Technology*, *28*(22), 6166–6174. doi:10.1021/es035152j

Williams, E. D., Ayres, R. U., & Heller, M. (2002). The 1.7 kilogram microchip and material use in the production of semiconductor devices. *Environmental Science & Technology*, *36*(24), 5504–5510. doi:10.1021/es025643o

Williams, A. (2010). *Cloud computing security: What the big guns have in store*. Retrieved March 2nd, 2010, from http://www.readwriteweb.com/cloud/2010/03/cloud-computing-security-what.php

Williamson, O. E. (1975). *Markets and hierarchies, analysis and anti-trust implications*. New York: Free Press.

Wisdom, J. (2006). Developing higher education teachers to teach creatively. In Jackson, N., Oliver, M., Shaw, M., & Wisdom, M. (Eds.), *Developing creativity in higher education* (pp. 183–196). London: Routledge.

Wolter, R., & Haselden, K. (2008). *The what, why, and how of master Data Management*. Microsoft Corporation. Retrieved February 23, 2010, from http://msdn2.microsoft.com/en-us/library/bb190163.aspx

Woon, I., & Kankanhalli, A. (2007). Investigation of IS professionals intention to practise secure development of applications. *International Journal of Human-Computer Studies*, *65*(1), 29–41. doi:10.1016/j.ijhcs.2006.08.003

World Privacy Forum. (2008). *A legal and policy analysis of Personal Health Records: Why many PHRs threaten privacy.*

World Report, I. B. I. S. (2009). *Bread and cake retailing in Australia*. ANZSIC Code: G5124, (p. 31).

Wright, P., Lickorish, A., & Milory, R. (2000). Route choices, anticipated forgetting, and interface design for online reference documents. *Journal of Experimental Psychology. Applied*, *6*(2), 158–167. doi:10.1037/1076-898X.6.2.158

Wu, C.-S., Cheng, F.-F., & Yen, D. C. (2008). The atmospheric factors of online storefront environment design: An empirical experiment in Taiwan. *Information & Management*, *45*, 493–498. doi:10.1016/j.im.2008.07.004

Yee, G. (2006). Measuring privacy protection in Web Services. *Proceedings of the ICWS'06, International Conference on Web Services,* Chicago, IL, (pp 647–654).

Yi, L., & Thomas, H. R. (2006). E-business and sustainable development. *International Journal of Environment and Sustainable Development*, *5*(3), 262–274. doi:10.1504/IJESD.2006.010896

Yi, L., & Thomas, H. R. (2007). A review of research on the environmental impact of e-business and ICT. *Journal of Environment International*, *33*(6), 841–849. doi:10.1016/j.envint.2007.03.015

Yin, L. (2001). An empirical study about development and monitoring issues in online banking development in China. *Finance Research*, *1*(1), 1–7.

Yin, R. K. (2002). *Case study research, design and methods* (3rd ed.). Newbury Park, CA: Sage Publications.

Young, L., & Wilkinson, I. (1989). The role of trust and cooperation in marketing channels: A preliminary study. *European Journal of Marketing*, *23*, 109–122. doi:10.1108/EUM0000000000550

Zaheer, A., McEvily, B., & Perrone, V. (1998). Does trust matter? Exploring the effects of interorganizational and interpersonal trust on performance. *Organization Science*, *9*, 141–159. doi:10.1287/orsc.9.2.141

Zaltman, G., Duncan, R., & Holbeck, J. (1973). *Innovations and organisations*. New York: Wiley and Sons.

Zawacki-Richter, O., Baecker, E., & Vogt, S. (2009). Review of distance education research: Analysis of research areas, methods, and authorship patterns. *The International Review of Research in Open and Distance Learning*, Retrieved on January 22, 2009, from http://www.irrodl.org/index.php/irrodl/article/view/741/1433

Zhao, H. (2002). Rapid Internet development in China: A discussion of opportunities and constraints on future growth. *Thunderbird International Business Review*, *44*(1), 119–138. doi:10.1002/tie.1042

Zheng, Y. (2009). Different spaces for e-development: What can we learn from the capability approach? *Information Technology for Development, 15*(2), 66–82. doi:10.1002/itdj.20115

Zhu, K. (2004). Information transparency of business-to-business electronic markets: A game-theoretic analysis. *Management Science, 50*(5), 670–685. doi:10.1287/mnsc.1040.0226

Zhu, K., & Kraemer, K. (2005). Post-adoption variations in usage and value of e-business by organizations: Cross-country evidence from the retail industry. *Information Systems Research, 16*(1), 61–84. doi:10.1287/isre.1050.0045

Zirpins, C., Weinreich, H., Bartelt, A., & Lamersdorf, W. (2001). Advanced concepts for next generation portals. *Proceedings of the 12th International Conference on Database and Expert Systems Applications*, (pp. 501-506). Munich, Germany.

Zucker, L. G. (1986). Production of trust: Institutional sources of economic structure. In Staw, B. M., & Cummings, L. L. (Eds.), *Research in organizational behavior* (pp. 53–111). Thousand Oaks, CA: Sage Publications.

About the Contributors

Ozlem Bak currently holds a Senior Lecturer position at University of Brighton Business School. She has worked previously in consultancy, and mainly in multinational automotive corporations. She has received her PhD from University of East Anglia, Norwich Business School for which she has been awarded a highly Commended Award (Emerald/EFMD Outstanding Doctoral Research Award). She has several publications in the area of e-supply chains and automotive industry.

Nola Stair is a Principal Lecturer in International Collaborations and Learning Technology for the University of Greenwich Business School in London, England. Previously, she held the position of Senior Information and Instructional Technology Specialist at Johns Hopkins University in the Center for Educational Resources in Baltimore, Maryland. Nola has over 20 years of experience in both administrative/academic positions and assisted with the development and implementation of one of the first Web-based RN to BSN degree programs in the United States for the University of Maryland Baltimore, School of Nursing. Her educational background includes a BSc in Education/Educational Computing and an MBA focusing on the management of Information Technology. She has published articles/book chapters in the area of e-learning and presented at numerous international conferences. Nola's research interests involve the integration and assessment of technology in administrative, teaching, and learning processes.

***

David Anyiwo is the Chair of the Department of Management Information Systems at Bowie State University in Bowie, Maryland. He earned his PhD in Systems Engineering from the University of Virginia and completed a comprehensive program of Advanced Study in Intelligent Decision Systems at the National Defense University. Dr. Anyiwo is an internationally recognized scholar who has led innovative research efforts at several U.S. and international universities, public agencies and corporations. He provided technical leadership in the building of the electronic commerce domain of the DARPA Agent Mark-up Language (DAML). His work spans the fields of integrated systems and software engineering, robust/sustainable system designs and architectures, secure/intelligent network centric systems, agent-based computing, and Semantic Web development. He has published several technical papers in leading professional journals and conference proceedings, and has served on the program and advisory committees of numerous national and international conferences.

Virginia Barba-Sánchez is an Associate Professor of Business Organization at the Department of Business Administration of the University of Castilla-La Mancha (Spain), where she received her PhD with a European mention and doctorate award. She's currently the head of the research group ENSITMA (Environmental Strategy Interdisciplinary Team & Management), and has authored a large number of refereed articles and conference papers in the field of business organization with regard to the environmental management, ITCs applications in management, and entrepreneurship.

Kim Bryceson is an Associate Professor in the School of Integrative Systems at the University of Queensland, Australia. Kim's current research focuses on value chain analysis and ICT applications in agri-food supply and value chains along with their impact on traceability systems including compliance issues associated with the retail sector and automation, the role they have in developing the sustainability of food industry chains in relation to 'carbon footprint' and lean (waste) management, and their role as platforms for information and knowledge dissemination through the development of innovative training environments. Additionally she is very interested in the modeling of agri-food supply and value networks to investigate emergent behaviours and the development of risk and/or performance management strategies.

Wilhelmina Djoleto is the founder, President, and CEO of the CulRitzWil Co., a management and computer Information Systems consultancy and a culinary services company. Wilhelmina holds a Bachelor of Science degree in Computer Science, a Masters degree in Applied Computer Science and a Doctor of Philosophy degree in Organizational Leadership from the University of Maryland Eastern Shore. Dr. Djoleto also held the positions of Database Manager/Administrator and Lecturer at the University of Maryland Eastern Shore and as Adjunct Professor at Bowie State University, taught both graduate and undergraduate students. She has authored and co-authored several research publications spanning symplectic integration for power systems and the impact of e-business on institutions and organisations. Dr. Djoleto serves as a reviewer for numerous journals.

Brian Gugerty DNS, MS, RN has been in the clinical informatics field for 22+ years holding academic, management and leadership roles in healthcare delivery organizations (Director on Nursing Informatics), healthcare information technology (HIT) companies (Product Manager; Sr. Research Analyst) and academia (Assistant Professor). Dr. Gugerty is presently in independent consultancy in clinical informatics where he assists healthcare delivery organizations, HIT companies, educational institutions and governments with electronic health record implementation, change management, evaluation of HIT, clinical terminology/ health information exchange initiatives, design of clinical Information Systems and informatics business opportunities in the US and internationally. He is active in the American Medical Informatics Association as well as other professional societies and initiatives. Dr. Gugerty has published widely and presented on informatics topics including clinical documentation, evaluation of healthcare information systems and HIT project management.

Debbie Holley is a Principal Lecturer in Teaching and Learning at London Metropolitan University Business School. Since joining the University from industry, she has become very interested in the use of technology to facilitate learning and encourage students to engage with their studies both on and off campus. As part of the Reusable Learning Object Centre of Excellence for Teaching and Learning (www.rlo-cetl.ac.uk) team, she is working with students to develop interactive learning materials. These resources are peer reviewed and freely available to the educational sector. Her research focuses on the student experience of blended learning and developing the digital curriculum. Debbie was recently awarded a 3 year University Teaching Fellowship for excellence in teaching.

Petros Ieromonachou is a Principal Lecturer based at the University of Greenwich Business School in London. Before joining Greenwich in 2007, Petros worked at the Open University, where he is currently acting as Visiting Research Fellow. His doctorate was in transport policy implementation and his varied research interests include sustainable transport and urban development, tourism transport, and intelligent transport systems. As well as transport research, Petros also undertakes studies in pedagogy, including e-learning, technology and management of higher education. He has recently studied technology aspects of e-learning and their relevance with planning and management. Petros has nearly ten years of working experience in higher education institutions; from conducting research and contributing to the writing of CPD courses at the Open University, to teaching and supervising at undergraduate and postgraduate levels at the University of Greenwich and contributing policy papers and guidelines relating to the Business School's teaching and learning strategy.

Alicia Izquierdo-Yusta is Associate Professor at the Department of Marketing of the University of Castilla-La Mancha (Spain). She has participated in different conferences and seminars worldwide and has written several articles in different high standing international journals in different international journals (Innovar, European Journal of Marketing etc.). Her main research lines are e-commerce, technologies adaptation, marketing communications, sales promotions, and product and services innovation. Dr. Izquierdo-Yusta is a member of the following associations: European Marketing Academy (EMAC) and The European Association for Education and Research in Commercial Distribution (EAERCD).

Ana Isabel Jimenez-Zarco is Associate professor at the Economic and Business Studies Department of the Open University of Catalonia (Spain). She has participated in different conferences and seminars worldwide and has written several articles in different international journals (e.g. Academy of Marketing Science Review, Journal of Marketing Channels, The Marketing Review, European Journal of Innovation Management, Innovar, European Journal of Marketing etc.). Her main research lines are brand identity, image management, product innovation, and ICT applications in management and marketing. She is a member of the following associations: Product Development Management Association (PDMA); European Marketing Academy (EMAC), reviewer of some journal and international congress and associate editor of journals as Innovar, Journal of Marketing Trends or Revista da Micro e Pequena Empresa). She is a member of the research group i2TIC (UOC), focused on the interdisciplinary analysis of the interaction between ICT use and information, communication and knowledge flow on individuals and organizations.

Raechel Johns is an Assistant Professor in Marketing at The University of Canberra in Australia. Author of numerous publications, including two books, Raechel is a passionate researcher. Her main areas of interest include relationship marketing and service dominant logic. More broadly, she researches business-to-business marketing, Internet marketing and services marketing, however she has an interest in most aspects of marketing. Raechel has taught marketing in Australia, Singapore, China, and Hong Kong and has consulted to Australian and international organizations. Prior to her academic career, Raechel worked in marketing for the retail, telecommunications and IT industries.

Georg Juelke is a senior consultant with Capgemini in the Netherlands, one of the leading technology and business consultancies, where he designs, builds and implements data warehouse solutions for large domestic and international clients. Prior to his work as a consultant, he held the post of Vice President of Business Intelligence at Adecco, the world's largest staffing and recruitment company. Working out of the UK and Spain, he headed the design and development of Adecco's global client and candidate data warehouse systems, enabling the company to grow its international business. By building worldwide business intelligence systems, Adecco introduced international customer and candidate analytics to the industry and was able to rapidly expand its international servicing capabilities. He holds an MA in Literature and Psychology from the Ludwig Maximilian University in Munich, Germany, and sat on the board of editors for the SAIS (School for Advanced International Studies) Review at the John Hopkins University in Washington D.C.

Chad Lin is a Research Fellow at the Centre for Behavioural Research in Cancer Control, Curtin University, Australia. Dr. Lin has conducted extensive research in the areas of: e-commerce, e-health, health communication, health informatics, IS/IT investment evaluation and benefits realization, IS/IT outsourcing, IT adoption and diffusion, RFID, social marketing, strategic alliance in healthcare, and virtual teams. He has authored more than 100 internationally refereed journal articles (e.g. Decision Support Systems, European Journal of Information Systems, Information and Management, International Journal of Electronic Commerce, European Journal of Marketing, Technovation, Medical Journal of Australia, and ANZ Journal of Public Health), book chapters, and conference papers in the last five years. He has served as an associate editor or a member of editorial review board for 7 international journals and as a reviewer for 11 other international journals. He is currently a member of the Research & Development Committee in the Faculty of Health Sciences at Curtin University.

Michael Maranda has over 25 years of experience in applied research. He holds a Doctorate in Sociology and a Master's of Psychology from NYU, and a graduate certificate in UN Studies from Long Island University. His experience includes conducting applied research and policy analysis. Topics he has studied have been related to health, healthcare, substance abuse, education, criminal justice, poverty, and ethnicity. He has worked for the University of Maryland's School of Nursing, Battelle Centers for Public Health Research and Evaluation, and NYS Office of Alcoholism and Substance Abuse Services/NDRI. Also, he has been a consultant to the UNHCR/the Ministry of Health of the Republic of Cyprus, the EU's Public Health Executive Agency, the European Anti-smoking Network, and a visiting scientist at the National Institute for Alcohol and Drug Research in Oslo, Norway. Recently, he has been collaborating with Gugerty Consulting on evaluating the implementation of a healthcare informatics projects.

Ana Marr BSc (PUCP, Peru), MSc (LSE, UK), PhD (SOAS, UK) is Reader in International Development Economics and Director of the International Microfinance Research Group at the Business School and Senior Research Fellow at the Natural Resources Institute, University of Greenwich. Ana has over 15 years working experience in microfinance and financial market development, conducting research, consultancy and lecturing in various UK institutions. Prior to joining the University of Greenwich, she worked at the London School of Economics, the Institute of Development Studies (University of Sussex), and the Overseas Development Institute. Her extensive overseas experience includes long- and short-term assignments in Africa (Tanzania, South Africa and Zimbabwe), Asia (Bangladesh, India,

Thailand), and Latin America (Bolivia, Peru, Colombia). She is currently principal investigator and project director of a major Leverhulme-funded research that includes looking at the role of technology in microfinance outreach to poor communities.

Sharon Nachtigal obtained her Bachelor in statistics, her Masters in Information Systems at Tel-Aviv University, and her PhD degree at Royal Holloway, University of London, UK. Her PhD thesis was concerned with introducing a new security paradigm for e-business security and developing a model and methodology for e-business organisation information security. Sharon works as a staff member at Engineering College, Tel-Aviv, at Royal Holloway (DL program in E-commerce Security), University of London, UK, and is also involved in business information security policies and management projects. Her research work is focused on information security management; in particular e-business security, and her professional and academic interests include information security compliance, user privacy, risk management, cyber terror, and all aspects of organisational security.

Martin Oliver is a Reader in Information and Communication Technologies (ICT) in Education at the London Knowledge Lab, which is part of the Institute of Education at the University of London and is seconded part-time to the UK's Higher Education Academy. He is an editor of the journal Learning, Media and Technology, and serves on the editorial boards for Innovations in Education and Teaching International; Research and Practice in Technology Enhanced Learning; and the Association for Learning Technology's journal, ALT-J: Research in Learning Technology. Martin's research focuses on the staff and student uses of technology in Higher Education, and the ways in which this affects their roles and experiences.

Panagiota Papadopoulou holds a BSc (Hons) in Informatics from the National and Kapodistrian University of Athens, an MSc (Distinction) in Distributed and Multimedia Information Systems from Heriot-Watt University and a PhD in Information Systems from the National and Kapodistrian University of Athens. She is a research fellow in the Department of Informatics and Telecommunications at the National and Kapodistrian University of Athens. She has extensive university-level teaching experience, as an adjunct faculty member at the University of Athens, the University of Pireaus, the University of Peloponnese, the University of Central Greece and other educational institutions in Greece. Dr. Papadopoulou has also actively participated in a number of European community and national research projects. She has published more than 30 papers in international journals and conferences, with her current research interests focusing on online trust, e-commerce, interface design, Web-based Information Systems, and social computing.

Jean-Eric Pelet obtained a doctorate in marketing with distinction (Nantes University, France) and a MBA in Information Systems with distinction (Laval University, Quebec, Canada). He works as an assistant professor at SupAgro Montpellier (France) on problems dedicated to the interface and to the consumer behaviour facing a website or any Information System (e-learning, knowledge management, e-commerce platforms). Its main interest lies on the variables that enhance the navigation in order to help people to be more efficient on these. He works as a visiting professor in several places in France thanks to its Knowledge Management and Content Management System platform (kmcms.net) in Design School (Nantes), Business Schools (Paris, Reims), and Universities (Paris Dauphine – Nantes), on lectures focused on e-marketing, ergonomics, usability, and consumer behaviour. Dr. Pelet has also actively participated in a number of European community and national research projects. His current research interests focus on Web-based Information Systems, social networks, interface design and usability.

María Pilar Martínez-Ruiz is Associate Professor at the Department of Marketing of the University of Castilla-La Mancha (Spain). She has participated in different conferences and seminars worldwide and has written several articles in different high standing international journals (e.g., Journal of the Operational Research Society, European Journal of Marketing, The International Journal of Market Research, European Journal of Operational Research, etc.). Her main research lines are retailing, marketing communications, sales promotions, and product and services innovation. She is a member of the following associations: Academy of Marketing Science (AMS), European Marketing Academy (EMAC) and The European Association for Education and Research in Commercial Distribution (EAERCD).

Gary A. Stair is currently the National Sales Director for Arista Marketing (a Publicis Selling Solutions Company). His newly created role was added to help design, create and implement a platform focused on providing "Virtual Sales Representatives" to the pharmaceutical, medical device, and women's health marketplaces. Gary was previously the Director of Sales, for LifeWatch, Inc (a Medical Telemetry Monitoring company). His team was responsible for the management, delivery, and service coordination of newly developed technology in the cardiac telemetry and sleep apnea marketplaces. Gary has been involved in a variety of executive sales leadership positions and marketing functions with large pharmaceutical companies, bio-tech start-ups, and medical device firms. In addition, he has led several successful worldwide product launches and developed domestic/international new product business endeavors. He is also an executive board member for ManCave, a classic/luxury automotive and condominium startup company.

Sharma Sumana is a Lecturer in the Department of Management Information Systems at Bowie State University. She teaches undergraduate and graduate courses in Information Systems. Sumana graduated with a PhD in Information Systems from Virginia Commonwealth University in December 2008. Her research interests pertain to data mining and information security and has published in journals such as Expert Systems and Applications and Knowledge Engineering Review. She has also published conference proceedings in various Information Systems Conferences such as HICSS, DSI, IACIS, Annual Security Conference and IBIMA. Her dissertation research was nominated for the 2009 ACM-SIGMIS Best doctoral dissertation award and the 2009 INFORMS ISS Design Science Research Award. She has been inducted into several Honor Societies including the Phi Kappa Phi, Beta Gamma Sigma, and Golden Key International Honor Society. She is a member of the Decision Science Institute and the Association of Information Systems.

Lin Yan BSc (Tianjin, China), MPhil (Cambridge, UK), PhD (Cambridge, UK) Lin is Senior Lecturer in Entrepreneurship and Innovation at the Business School, University of Greenwich. Lin has taught extensively on courses on small business development, entrepreneurship and international entrepreneurship. Her research focuses on the social aspect of international entrepreneurship, particularly the intra- and inter-organisational collaboration of international SMEs facilitated by Information Technology.

Jiaqin Yang is a Professor of Management at the Georgia College & State University, Milledgeville, Georgia. He received his PhD in Operations Management from Georgia State University. Dr. Yang's current research interests include: operations strategy, e-business, and quality management. Dr. Yang has served as the guest editor for a special issue of International Journal of Electronic Finance on the recent research issues of e-banking, and served as one of the guest editors for a special issue of International Journal of Management in Education on the research issues of e-learning. Dr. Yang has published in the International Journal of Production Economics, European Journal of Operations Research, Decision Sciences, Journal of Computers & Industrial Engineering, Mathematical and Computer Modeling, Integrated Manufacturing Systems, Environment International, International Journal of Business, Industrial Management & Data Systems, Facilities, Electronic Government, International Journal of Management and Enterprise Development, as well as numerous professional conference proceedings.

Index

C

D

E

O

P

R

S

V

W

Y

Breinigsville, PA USA
09 March 2011
257180BV00006BA/1/P